Charles Augustus Young

Lessons in astronomy including uranography

A brief intoductory course without mathematics

Charles Augustus Young

Lessons in astronomy including uranography
A brief intoductory course without mathematics

ISBN/EAN: 9783744737364

Printed in Europe, USA, Canada, Australia, Japan

Cover: Foto ©berggeist007 / pixelio.de

More available books at **www.hansebooks.com**

The Great Telescope of the Lick Observatory. Aperture, 36 inches; Length, 57 feet.

LESSONS IN ASTRONOMY

INCLUDING URANOGRAPHY

A BRIEF INTRODUCTORY COURSE

WITHOUT MATHEMATICS

BY

CHARLES A. YOUNG, Ph.D., LL.D.

PROFESSOR OF ASTRONOMY IN THE COLLEGE OF NEW JERSEY (PRINCETON)
AUTHOR OF A "GENERAL ASTRONOMY FOR COLLEGES AND SCIENTIFIC
SCHOOLS," AND OF "ELEMENTS OF ASTRONOMY."

BOSTON, U.S.A., AND LONDON
GINN AND COMPANY, PUBLISHERS

PREFACE.

THIS volume has been prepared to meet the want of certain classes of schools which find the author's "Elements of Astronomy" rather too extended and mathematical to suit their course and pupils. It is based upon the Elements, but with many condensations, simplifications, and changes of arrangement : everything has been carefully worked over and rewritten, in order to adapt it to those whose mathematical attainments are not sufficient to enable them to use the larger work to advantage.

Of course, such pupils cannot gain the same insight into the mechanism of the heavens as those who take up the subject at a more advanced stage in their education. They must often be contented with the bare statement of a fact without any explanation of the manner in which its truth is established, and thus will necessarily miss much that is most valuable in the discipline to be derived from the study of Astronomy.

But enough remains — surely there is no other science which, apart from all questions of How or Why, supplies so much to widen the student's range of thought, and to make him comprehend his place in the infinite universe.

The most important change in the arrangement of the book has been in bringing the Uranography or " constellation-tracing," into the body of the text, and placing it near the beginning; a change in harmony with the accepted principle that those whose minds are not mature succeed best in the study of a new subject by beginning with what is concrete, and appeals to the senses, rather than with the abstract principles.

vi PREFACE.

It has been thought well also to add brief notes on the legendary mythology of the constellations for the benefit of such pupils as are not likely to become familiar with it in the study of classical literature.

In the preparation of the book great pains have been taken not to sacrifice accuracy and truth to compactness; and no less to bring everything thoroughly down to date. * * * * *

The Appendix contains in its first chapter descriptions of the most used astronomical instruments, and where time permits, might profitably be brought into the course. The second chapter of the Appendix is designed only for the use of teachers and the more advanced pupils. Arts. 431–434, however, explaining how the sun's distance may be found in the simplest way, might well be read by all.

My warmest thanks are due to my friend and assistant, Mr. Taylor Reed, who has gone over all the proofs of the book, and has given me many valuable suggestions.

PREFACE TO THE EDITION OF 1895.

Since the first publication of this work the progress of astronomy has been so rapid that in order to keep abreast of the times it has become necessary to give the book a thorough revision. This has been done: numerous minor changes and corrections have been made, some articles have been rewritten, and others added. At the same time the alterations have been so managed that they will cause no serious inconvenience in using the older editions in connection with the new one in class-room work.

It is hoped that, so far as its scope permits, the book now presents a satisfactory summary of the existing state of the science.

AUGUST, 1895.

CONTENTS.

PAGES

CHAPTER I. — INTRODUCTION ; Fundamental Notions and Definitions. — The Celestial Sphere and its Circles. — Altitude and Azimuth. — Right Ascension and Declination. — Celestial Latitude and Longitude 1–16

CHAPTER II. — URANOGRAPHY : Globes and Star-maps. — Star Magnitudes. — Names and Designations of Stars. — The Constellations in Detail 17–54

CHAPTER III. — FUNDAMENTAL PROBLEMS : Latitude and the Aspect of the Celestial Sphere. — Time, Longitude, and the Place of a Heavenly Body 55–67

CHAPTER IV. — THE EARTH : Its Form and Dimensions ; its Rotation, Mass, and Density ; its Orbital Motion and the Seasons. — Precession. — The Year and the Calendar . . 68–90

CHAPTER V. — THE MOON : Her Orbital Motion and the Month. — Distance, Dimensions, Mass, Density, and Force of Gravity. — Rotation and Librations. — Phases. — Light and Heat. — Physical Condition. — Telescopic Aspect and Surface . . 91–110

CHAPTER VI. — THE SUN : Its Distance, Dimensions, Mass, and Density. — Its Rotation, Surface, and Spots. — The Spectroscope and the Solar Spectrum ; the Chemical Constitution of the Sun. — The Chromosphere and Prominences. — The Corona. — The Sun's Light. — Measurement and Intensity of the Sun's Heat. — Theory of its Maintenance, and Speculations regarding the Age and Duration of the Sun . . . 111–143

vii

PAGES

CHAPTER VII. — Eclipses and the Tides : Form and Dimensions of Shadows. — Eclipses of the Moon. — Solar Eclipses, Total, Annular, and Partial. — Number of Eclipses in a Year. — Recurrence of Eclipses, and the Saros. — Occultations. — The Tides 144–158

CHAPTER VIII. — The Planetary System : The Planets in General. — Their Number, Classification, and Arrangement. — Bode's Law. — Orbits of the Planets. — Kepler's Laws and Gravitation. — The Apparent Motions of the Planets and the Systems of Ptolemy and Copernicus. — Determination of the Planets' Diameters, Masses, etc. — Herschel's Illustration of the System. — Description of Individual Planets : the ' Terrestrial' Planets, Mercury, Venus, and Mars . . . 159–189

CHAPTER IX. — Planets (continued) : The Asteroids. — Intra-Mercurian Planets and the Zodiacal Light. — The Major Planets, Jupiter, Saturn, Uranus, and Neptune. — Ultra-Neptunian Planet 190–212

CHAPTER X. — Comets and Meteors : Comets, their Number, Designation, and Orbits ; their Constituent Parts and Appearance ; their Spectra, Physical Constitution, and Probable Origin ; Remarkable Comets ; Photography of Comets ; . Aerolites, their Fall and Characteristics ; Shooting Stars and Meteoric Showers ; Connection between Meteors and Comets 213–249

CHAPTER XI. — The Stars : Their Nature, Number, and Designation. — Star Catalogues and Charts. — Their Proper Motions, and the Motion of the Sun in Space. — Stellar Parallax. — Star Magnitudes and Photometry. — Variable Stars. — Stellar Spectra 250–274

CHAPTER XII. — The Stars (continued) : Double and Multiple Stars ; Clusters and Nebulæ ; the Milky Way, and Distribution of Stars in Space ; the Stellar Universe. — Cosmogony and the Nebular Hypothesis 275–301

APPENDIX.

PAGES

CHAPTER XIII. — ASTRONOMICAL INSTRUMENTS: The Telescope, Simple Refracting, Achromatic, and Reflecting. — The Equatorial. — The Filar Micrometer. — The Transit Instrument. — The Clock and the Chronograph. — The Meridian Circle. — The Sextant 302–320

CHAPTER XIV. (FOR THE MOST PART SUPPLEMENTARY TO ARTICLES IN THE TEXT). — Hour-angle and Time. — Twilight. — Determination of Latitude. — Place of a Ship at Sea. — Finding the Form of the Earth's Orbit. — The Ellipse. — Illustrations of Kepler's ' Harmonic ' Law. — The Equation of Light, and the Sun's Distance determined by it. — Aberration of Light. — De l'Isle's Method of getting the Sun's Parallax from a Transit of Venus. — The Parabola and the Conic Sections. — Determination of Stellar Parallax 321–339

QUESTIONS FOR REVIEW 340

TABLES OF ASTRONOMICAL DATA:
    I. Astronomical Constants 347
    II. The Principal Elements of the Solar System . . . 348
    III. The Satellites of the Solar System 349
    IV. The Principal Variable Stars 350
    V. The Best Determined Stellar Parallaxes 351
    VI. The Greek Alphabet and Miscellaneous Symbols . . 352

INDEX 353

STAR–MAPS 367

CHAPTER I.

INTRODUCTION. — FUNDAMENTAL NOTIONS AND DEFINI-
TIONS. — THE CELESTIAL SPHERE AND ITS CIRCLES.
— ALTITUDE AND AZIMUTH. — RIGHT ASCENSION AND
DECLINATION. — CELESTIAL LATITUDE AND LONGITUDE.

1. ASTRONOMY[1] is the science which deals with the heavenly
bodies.

As it is the oldest of the sciences, so also it is one of the
most perfect, and in certain aspects the noblest, as being the
most "unselfish" of them all. And yet, although not bearing
so directly upon the material interests of life as the more
modern sciences of Physics and Chemistry, it is of high utility.
By means of Astronomy the latitudes and longitudes of places
upon the earth's surface are determined, and by such determinations alone is it possible to conduct vessels upon the sea.
Moreover, all the operations of surveying upon a large scale,
such as the determination of the boundaries of countries, depend more or less upon astronomical observations. The same
is true of operations which, like the railway service, require
an accurate knowledge and observance of time; for the fundamental timekeeper is the diurnal revolution of the heavens, as
determined by the astronomer's transit-instrument.

In ancient times the science was supposed to have a still higher
utility. It was believed that human affairs of every kind, the welfare
of nations, and the life history of individuals alike, were controlled,

---

[1] The term is derived from two Greek words: *astron*, a star, and
nomos, a law.

1

or at least prefigured, by the motions of the stars and planets ; so
that from the study of the heavens it ought to be possible to predict
futurity. This belief is embodied in the pseudo-science of " Astrol-
ogy," long since shown to be a baseless delusion though it still
retains a hold upon the credulous.

2. The heavenly bodies include, *first*, the solar system, —
that is, the sun and the planets which revolve around it, with
their attendant satellites ; *second*, the comets and the meteors,
which also revolve around the sun, but are bodies of a very
different nature from the planets, and move in different kinds
of orbits ; and, *thirdly*, the stars and nebulæ. The earth on
which we live is one of the planets, and the moon is the earth's
satellite. The stars which we see are bodies of the same kind
as the sun, shining like him with fiery heat, while the planets
and the satellites are dark and cool like the earth, and visible
to us only by the sunlight they reflect. As for the comets and
nebulæ, they appear to be mere *clouds*, composed of heated gas
or swarms of little particles of more solid substances, perhaps
not very hot, but luminous from some cause or other. The
telescope reveals millions of stars invisible to the naked eye,
and there are others, possibly thousands of them, that do not
shine, but manifest their existence by affecting the motions
of their neighbors.

3. As we look off from the earth at night, the stars appear
to be all around us, like glittering points fastened to the inside
of a huge hollow globe. Really they are at very different dis-
stances, all enormous as compared with any distances with
which geography makes us familiar. Even the moon is eighty
times as far away as New York from Liverpool, and the sun
is nearly four hundred times as distant as the moon, and the
nearest of the stars is more than two hundred thousand times
as distant as the sun ; as to the remoter stars, some of them
are certainly thousands of times as far away as the nearer
ones, — so far that light itself is thousands of years in coming

to us from them. These are facts which are *certain*, not mere guesses or beliefs.

Then, too, as to their motions. Although the heavenly bodies seem to us for the most part to be at rest, except as the earth's rotation makes them appear to rise and set, yet really they are all moving, and with a swiftness of which we can form no conception. A cannon-ball is a snail to the slowest of them. The earth itself in its revolution around the sun is flying eighteen and a half miles in a second, which is more than thirty times as fast as the swiftest rifle bullet. We fail to perceive the motion simply because it is so smooth and so unresisted. The space outside our air contains nothing that can sensibly obstruct either sight or motion.

4. But this knowledge as to the real distance and motions of the heavenly bodies was gained only after long centuries of study. If we go out to look at the stars some moonless night we find them apparently sprinkled over the dome of the sky in groups or constellations, which are still substantially the same as in the days of the earliest astronomers. At first these constellations were figures of animals and other objects, and many celestial globes and maps still bear grotesque pictures representing them. At present, however, a constellation is only a certain region of the sky, limited by imaginary lines which divide it from the neighboring constellations, just as countries are divided in geography. As to the exact boundaries of these constellations, and even their number, there is no precise agreement among astronomers. Forty-eight of them have come down to us from the time of Ptolemy,[1] and even in his day many of them were already ancient.

About twenty more, which have been proposed by more recent astronomers, are now recognized, besides a considerable number which have been abandoned.

---

[1] Ptolemy, the greatest astronomer of antiquity, flourished at Alexandria about 130 A.D.

5. Uranography, or Description of the Visible Heavens. — The study of the constellations, or the apparent arrangement of the stars in the sky, is called Uranography.[1] It is not an *essential* part of Astronomy, but it is an easy and pleasant study; and in becoming familiar with the constellations and their principal stars, the pupil will learn more readily and thoroughly than in any other way the most important facts in relation to the apparent motions of the heavenly bodies, and the principal points and circles of the celestial sphere. For this reason the teacher is urged to take the earliest opportunity to have his pupils trace such of the constellations as happen to be visible in the evening sky when they begin the study of Astronomy.

6. The Celestial Sphere.[2] — The sky appears like a hollow vault, to which the stars seem to be attached, like specks of gilding upon the inner surface of a dome. We cannot judge of the distance of this surface from the eye, further than to perceive that it must be very far away. It is therefore natural and extremely convenient to regard the distance of the sky as everywhere the same and unlimited. The '*celestial sphere*,' as it is called, is conceived of as so enormous that the whole world of stars and planets lies in its centre like a few grains of sand in the middle of the dome of the Capitol. Its diameter is assumed to be immeasurably greater than any actual distance known, and greater than any quantity assignable. In technical language it is taken as *mathematically infinite.*

Since the celestial sphere is thus infinite, any two parallel lines drawn from distant points on the surface of the earth, or even from points as distant as the earth and the sun, will seem *to meet at one point* on the surface of the sphere. If the two

---

[1] From the Greek, *ouranos* (heavens), and *graphe* (description).

[2] The study of the celestial sphere and its circles is greatly facilitated by the use of a globe, or armillary sphere. Without some such apparatus it is not easy for a young person to get clear ideas upon the subject.

lines were anywhere a million miles apart, for instance, they will, of course, still be a million miles apart when they reach the surface of the sphere; but at an infinite distance even a million miles is a mere nothing, so that the two lines make apparently but a single point[1] where they pierce the sphere.

7. The Apparent Place of a Heavenly Body. — This is simply the point where a line drawn from the observer through the body in question, continued outward, pierces the celestial sphere. It depends solely upon the *direction* of the body, and is in no way affected by its distance from us. Thus, in Fig. 1, *A, B, C*, etc., are the apparent places of *a, b, c*, etc., the observer being at *O*. Objects that are nearly in line with each other, as *h, i, k*, will appear close together. The moon, for instance,

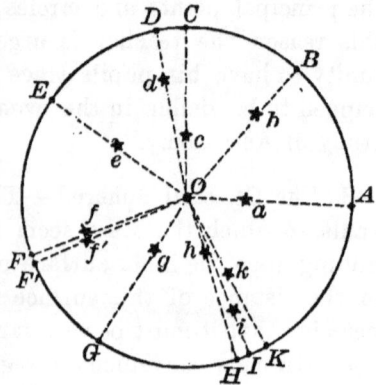

FIG. 1.

often looks to us very near a star, which is really of course at an enormous distance beyond her.

8. Angular Measurement. — It is clear that we cannot properly describe the apparent distance of two points upon the celestial sphere from each other by feet or inches. To say that two stars are about five *feet* apart, for instance, — and it is not very uncommon to hear such an expression, — means nothing unless we know how far from the eye the five-foot measure is to be held. The proper units for expressing apparent distance in the sky are those of *angle*, viz.: degrees (°), minutes ('), and seconds ("); the circumference of a circle being divided into 360 degrees, each degree into 60 minutes, and each minute into 60 seconds. Thus, the Great Bear's tail,

[1] This is the same as the 'vanishing-point' of perspective.

or Dipper-handle, is about 16° long, and the long side of the Dipper-bowl is about 10°; the moon and the sun are each about half a degree, or 30', in diameter.

It is very important that the student in Astronomy should become accustomed as soon as possible to estimate celestial measures in this way. A little practice soon makes it easy, though at first one is apt to be embarrassed by the fact that the sky looks to the eye not like a true hemisphere but like a flattened vault, so that the estimates of distances for all objects near the horizon are apt to be too large. The moon, when rising or setting, looks to most persons much larger than when overhead; and the Dipper-bowl, when underneath the pole, seems to cover a much larger area than when above it.

9. Circles and Principal Points of the Celestial Sphere. — Just as the surface of the earth in Geography is covered with a net-work of imaginary lines, — meridians and parallels of latitude, — so the sky is supposed to be marked off in a somewhat similar way. Two such sets of points and reference circles are in common use to describe the apparent places of the stars, and a third was used by the ancients and is still employed for some purposes. The first system depends upon the direction of the force of gravity shown by a plumb-line at the point where the observer stands; the second upon the direction of the axis of the earth, which points very near the so-called Pole-star; and the third depends upon the position of the orbit in which the earth travels around the sun.

10. The Gravitational or Up-and-Down System. — (*a*) The Zenith and Nadir. The point in the sky directly above the observer is called the *zenith;* the opposite point, under the earth and of course invisible, the *nadir.*[1]

(*b*) The Horizon (pronounced ho-rī'-zon, not hor'-ĭ-zon).

---

[1] These are Arabic terms. About 1100 A.D. the Arabs were the world's chief astronomers, and have left their mark upon the science in numerous names of stars and astronomical terms.

This is a 'great circle'[1] around the sky, half-way between the zenith and the nadir, and therefore everywhere 90° from the zenith. The word is derived from a Greek word which means a 'boundary'; i.e., the line where the earth or sea limits the sky. The actual line of division, which on the land is always more or less irregular, is called the *visible* horizon, to distinguish it from the true horizon defined above. We may also define the horizon as the great circle where a plane which passes through the observer's eye perpendicular to the plumbline cuts the celestial sphere.

11. Vertical Circles and the Meridian; Altitude, and Azimuth. — Circles drawn from the zenith to the nadir cut the horizon at right angles, and are known as *vertical circles*. Each star has at any moment its own vertical circle.

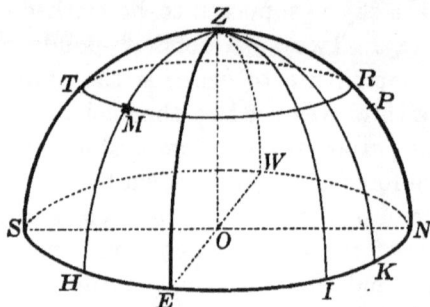

FIG. 2. — The Horizon and Vertical Circles.

O, the place of the Observer.
OZ, the Observer's Vertical.
Z, the Zenith; *P*, the Pole.
SWNE, the Horizon.
SZPN, the Meridian.
EZW, the Prime Vertical.

M, some Star.
ZMH, arc of the Star's Vertical Circle.
TMR, the Star's Almucantar.
Angle *TZM*, or arc *SH*, Star's *Azimuth*.
Arc *HM*, Star's *Altitude*.
Arc *ZM*, Star's *Zenith Distance*.

That particular vertical circle which passes north and south is known as the *celestial* MERIDIAN; while the vertical circle at right angles to this is called the *prime vertical*. Small circles

[1] '*Great* Circles' are those which divide the sphere into two equal parts.

drawn parallel to the horizon are known as *parallels of altitude,* or *almucantars.* Fig. 2 illustrates these circles.

By their help we can easily define the apparent position of a heavenly body.

Its *Altitude* is its apparent elevation above the horizon; that is, the number of degrees between it and the horizon, measured on a vertical circle. Thus, in Fig. 2, the vertical circle *ZMH* passes through the point *M.* The arc *MH,* measured in degrees, is the altitude of *M,* and the arc *ZM* is called its zenith distance.

The *Azimuth* of a heavenly body is the same as its 'bearing' in Surveying, but measured from the true meridian and not from the magnetic.[1] It is the arc of the horizon, measured in degrees, intercepted between the south point and the foot of the vertical circle which passes through the object.

There are various ways of reckoning azimuth. Many writers express it in the same way as the 'bearing' in Surveying, *i.e.,* so many degrees east or west of north or south. In the figure, the azimuth of *M* thus expressed is about *S, 50° E.* The more usual way at present, however, is to reckon clear around from the south, through the west, to the point of beginning. Expressed in this way the azimuth of *M* would be about 310°, — *i.e.,* the arc *S WNEH.*

Altitude and azimuth, however, are inconvenient for many purposes, because they continually change for a celestial object as it moves across the sky.

12. The Apparent Diurnal Rotation of the Heavens. — If we go out on some clear evening in the early autumn, say about the 22d of September, and face the north, we shall find the appearance of that part of the heavens directly before us substantially as shown in Fig. 3. In the north is the constellation of

---

[1] The reader is reminded that the magnetic needle does not point exactly north. Its direction varies widely at different parts of the earth, and, moreover, is continually changing to some extent.

the Great Bear (Ursa Major), characterized by the conspicuous group of seven stars known as the "Great Dipper." It now lies with its handle sloping upward to the west. The two easternmost stars of the four which form its bowl are called

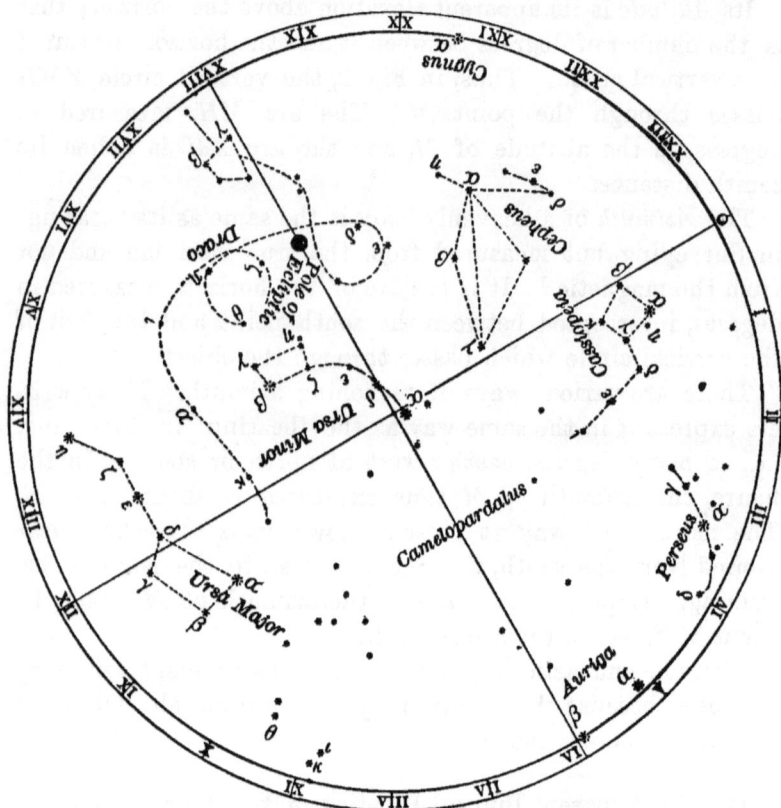

Fig. 3. — The Northern Circumpolar Constellations.

the "Pointers," because they point to the Pole-star, which is a solitary star not quite half-way from the horizon to the zenith (in the latitude of New York), and about as bright as the brighter of the two Pointers.

High up on the opposite side of the Pole-star from the Great Dipper, and at nearly the same distance, is an irregular zigzag of five stars, each about as bright as the Pole-star itself. This is the constellation of Cassiopeïa.

If now we watch these stars for only a few hours, we shall find that while all the forms remain unaltered, their places in the sky are slowly changing. The Great Dipper slides downward towards the north, so that by eleven o'clock (on Sept. 22) the Pointers are directly under the Pole-star. Cassiopeïa still keeps opposite, however, rising towards the zenith; and if we continue the watch through the whole night, we shall find that all the stars appear to be moving in circles around a point near the Pole-star, revolving in the opposite direction to the hands of a watch (as we look towards the north) with a steady motion which takes them completely around once a day, or, to be more exact, once in 23^h 56^m 4.1^s of ordinary time. They behave just as if they were attached to the inner surface of a huge revolving sphere.

To indicate the position of the stars as it will be at midnight of Sept. 22, the figure must be held so that XII in the margin is at the bottom; at 4 A.M. the stars will have come to the position indicated by bringing XVI to the bottom, and so on. But at eight o'clock on the next night we shall find things in their original position very nearly.

If instead of looking toward the north we now look southward, we shall find that in that part of the sky also the stars appear to move in the same kind of way. All that are not too near the Pole-star rise somewhere in the eastern horizon, ascend obliquely to the meridian, and descend to their setting at points on the western horizon. The next day they rise and set again at precisely the same points, and the motion is always in an arc of a circle, called the star's diurnal circle, the size of which depends upon its distance from the pole. Moreover, all of these arcs are strictly concentric.

The ancients accounted for these fundamental and obvious facts by supposing that the stars are really fastened to the celestial sphere, and that this sphere really turns daily in the manner indicated. According to this view there must really be upon the sphere two opposite points which remain at rest, and these are the poles.

13. Definition of the Poles. — The Poles, therefore, may be defined as *those two points in the sky where a star would have no diurnal motion.* The exact position of either pole may be determined with proper instruments, by finding the centre of the small diurnal circle described by some star near it, as, for instance, by the Pole-star.

The student must be careful not to confound the *Pole* with the *Pole-star.* The pole is an imaginary point; the Pole-star is only that one of the conspicuous stars which happens *now*[1] to be nearest to that point. The Pole-star at present is about $1\frac{1}{4}°$ distant from it. If we draw an imaginary line from the Pole-star to the star Mizar (the one at the bend of the Dipper-handle), it will pass almost exactly through the pole itself; the distance of the pole from the Pole-star being very nearly one-quarter of the distance between the two "Pointers."

This definition of the pole is that which would be given by one familiar with the sky, but ignorant of the earth's rotation, and it is still perfectly correct; but knowing, as we now do, that this apparent revolution of the celestial sphere is due to the real spinning of the earth on its axis, we may also define the poles as *the two points where the earth's axis of rotation, produced indefinitely, would pierce the celestial sphere.*

Since the two poles are diametrically opposite in the sky, only one of them is usually visible from any given place. Observers north of the earth's equator see only the north pole, and *vice versa* for observers in the southern hemisphere.

14. The Celestial Equator, or Equinoctial; Declination. — The Equator is a great circle of the celestial sphere drawn half-way between the poles, everywhere 90° from each of them,

---

[1] See Article 126.

and is the great circle in which the plane of the earth's equator
cuts the celestial sphere. It is often called the *Equinoctial.*
Fig. 4 shows how the plane of the earth's equator produced
far enough would mark out such a circle in the heavens.

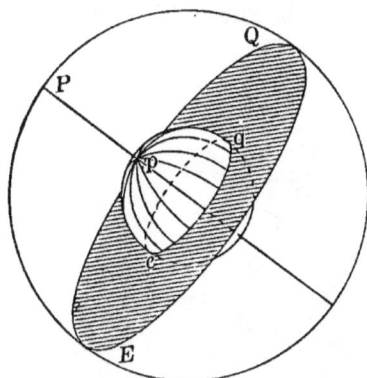

Small circles drawn parallel
to the equinoctial, like the parallels of latitude on the earth, are
known as '*Parallels of Declination,*' the *Declination* of a star
being its distance in degrees
north or south of the celestial
equator, + if north, — if south.
It corresponds precisely with
the latitude of a place on the
earth's surface; but it cannot be

Fig. 4.—The Plane of the Earth's Equa- called celestial *latitude*, because
tor produced to cut the Celestial Sphere. that term has been preoccupied
for an entirely different quantity (Art. 20). A star's parallel
of declination is identical with its diurnal circle.

15. Hour-Circles. — The great circles of the celestial sphere
which pass through the poles like the meridians on the earth,
and are therefore perpendicular to the celestial equator, are
called *Hour-Circles.* Some writers call them celestial meridians, but the term is objectionable since it is sometimes used
to indicate an entirely different set of circles. That particular hour-circle which at any moment passes through the zenith
of course coincides with the celestial *meridian* already defined
in Art. 11.

16. The Celestial Meridian and the Cardinal Points. — The
best definition of the celestial meridian is, however, *the great
circle which passes through the zenith and the poles.* The points
where this meridian cuts the horizon (the circle of level), are
the north and south points, and the east and west points of

the horizon lie half-way between them, the four being known as the "Cardinal Points." The student is especially cautioned against confounding the north *point* with the north *pole.* The north *point* is on the *horizon;* the north pole is high up in the sky.

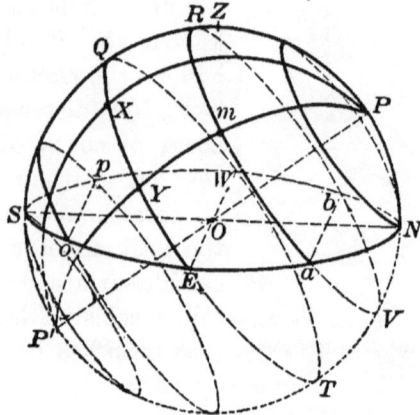

FIG. 5.— Equator, Hour-Circles, etc.

O, place of the Observer; *Z*, his Zenith.
SENW, the Horizon.
POP', line parallel to the axis of the Earth.
P and *P'*, the two Poles of the Heavens.
EQWT, the Celestial Equator, or Equinoctial.
X, the Vernal Equinox, or "First of Aries."
PXP', the Equinoctial Colure, or Zero Hour-Circle.

m, some Star.
Ym, the Star's *Declination; Pm*, its *North-Polar Distance.*
Angle *mPR* = arc *QY*, the Star's (eastern) *Hour-Angle;* = 24ʰ *minus* Star's western Hour-Angle.
Angle *XPm* = arc *XY*, Star's *Right Ascension.*
Sidereal time at the moment = 24ʰ *minus XPQ.*

In Fig. 5, *P* is the north celestial pole, *Z* is the zenith, and *SQZPN* is the celestial meridian. *P* and *P'* are the poles, *PmP'* is the hour-circle of *m*, and *amRbV* is its parallel of declination, or diurnal circle. *N* and *S* are the north and south points respectively. In the figure, *mY* is the *declination* of *m*, and *mP* is called its *polar distance.*

17. The Vernal Equinox, or First of Aries. — In order to use this system of circles as a means of designating the places of stars in the sky, it is necessary to fix upon some one hour-circle, to be reckoned from in the same way that the meridian of Greenwich is used on the earth's surface. The "Greenwich of the sky" which has thus been fixed upon, is the point where the sun crosses the celestial equator in the spring. The sun and moon and the planets do not behave as if they, like the stars, were firmly fixed upon the celestial sphere, but rather as if they were glow-worms crawling slowly about upon its surface while it carries them in its diurnal rotation. As every one knows, the sun in winter is far to the south of the equator, and in the summer far to the north, apparently completing a yearly circuit of the heavens on a path known as the ecliptic. It crosses the equator, therefore, twice a year, passing from the south side of it to the north about March 20th, and *always at the same point* (neglecting for the present the effect of what is known as 'precession'). This point is called the '*Vernal Equinox*,' and is made the starting-point. Unfortunately it is not marked by any conspicuous star; but a line drawn from the Pole-star through Beta Cassiopeïæ (the westernmost or "preceding" star in the zigzag) (see Map I.) and continued 90° from the pole, strikes very near it. In Fig. 5, *X* represents this point. It is often called the "*First of Aries.*"

18. Right Ascension. — The *right ascension* of a star is *the arc of the celestial equator intercepted between the vernal equinox and the point where the star's hour-circle cuts the equator*, and is reckoned always *eastward* from the equinox and completely around the circle. It may be expressed either in degrees or in hours.[1] A star one degree *west* of the equinox has a right ascension of 359°, or of 23^h 56^m. Evidently the diurnal motion does not affect the right ascension of a star, but this, like the declination, remains practically unchanged for years. In Fig. 5, if *X* be the vernal equinox the right ascension of *m* is the arc *XY* measured from *X* eastward.

---

[1] Twenty-four hours = 360°; one hour = 15°.

19. Thus we can define the position of a star either by its altitude and azimuth, which tell how high it is in the sky, and how it "bears," as a sailor would say; or we may use its right ascension and declination, which do not change from day to day (not perceptibly at least), and so are better adapted to mapping purposes, corresponding as they do precisely to latitude and longitude upon the surface of the earth.

Perhaps the easiest way to think of these celestial circles is the following: Imagine a tall pole standing straight up from the observer, having attached to it at the top (the zenith) two half circles coming down to the level of the observer's eye, one of them running north and south (the meridian), and the other east and west (the prime vertical). The bottoms of these two semicircles are connected by a complete circle, the horizon, at the level of the eye. This framework, immense but fortunately only imaginary and so not burdensome, the observer takes with him wherever he goes, keeping always at its centre, while over it turns the celestial sphere ; more strictly, he and the earth and his framework turn together *under* the celestial sphere.

The circles of the other set are drawn upon the celestial sphere itself (the equator and the hour-circles) and are not affected at all by the observer's journeys, but are as fixed as the poles and meridians upon the earth; the stars also, to all ordinary observation, are fixed upon the sphere just as cities are upon the earth. They really move, of course, and swiftly, as has been said before, but they are so far away that it takes centuries, as a rule, to produce the slightest apparent change of place.

20. Celestial Latitude and Longitude. — A different way of designating the positions of the heavenly bodies in the sky has come down to us from very ancient times. Instead of the equator it makes use of another circle of reference in the sky, known as the '*Ecliptic.*' This is simply the apparent path described by the sun in its annual motion among the stars; for the sun appears to creep around the celestial sphere in a circle once every year, and the Ecliptic may be defined as the intersection of the plane of the earth's *orbit* with the celestial sphere, just as the celestial equator is the intersection of the earth's *equator;* the vernal equinox is one of the points where the two circles cross. Before the days of clocks, the Ecliptic was in many respects a more convenient circle of reference than the equator and was almost universally used as such by the old astronomers. Celestial *longitude* and *latitude* are measured with reference to the Ecliptic, in the same way that right ascension and declination are measured with respect to the equator. Too much care can hardly be taken to avoid confusion between *terrestrial* latitude and longitude and the celestial quantities that bear the same name.

CHAPTER II.

URANOGRAPHY.

GLOBES AND STAR-MAPS. — STAR MAGNITUDES. — DESIG-NATION OF THE STARS. — THE CONSTELLATIONS.

NOTE. — It is hardly necessary to say that this chapter is to be treated by the teacher differently from the rest of the book. It is to be dealt with, not as recitation matter, but as *field-work:* to be taken up at different times during the course as the constellations make their appearance in the evening sky.

For convenience of reference we add the following alphabetical list of the constellations described or mentioned in the chapter: —

	ARTICLE		ARTICLE
Andrómĕda	35	Cepheus	29
Anser, see Vulpécŭla	69	Cetus	39
Antínoüs, see Aquila	71	Coma Berĕnīces	57
Antlia	62	Columba	45
Aquarius	78	Corŏna Borealis	60
Aquïla	71	Corvus	55
Argo Navis	51	Crater	55
Aries	38	Cygnus	68
Aurīga	41	Delphīnus	74
Boötes	59	Draco	30
Camelopárdus	31	Equúleus	75
Cancer	52	Erídănus	44
Canes Venatici	58	Gemini	47
Canis Major	49	Grus	79
Canis Minor	48	Hércŭles	66
Capricornus	73	Hydra	55
Cassiŏpéia	28	Lacerta	76
Centaurus	62	Leo	53

	ARTICLE		ARTICLE
Leo Minor	54	(Pleiades) . . .	42
Lepus	45	Sagitta	70
Libra	61	Sagittarius	72
Lupus	62	Scorpio	63
Lynx	46	Sculptor	39
Lyra	67	Serpens	65
Monóceros	50	Serpentarius, *see* Ophiuchus .	65
Norma	64	Sextans	54
Ophiūchus	65	Taurus	42
Oríon	43	Taurus Poniatovii . . .	65
Pégăsus	77	Triángŭlum	37
Perseus . . .	40	Ursa Major	26
Phœnix . . .	39	Ursa Minor	27
Pisces . . .	36	Virgo	56
Piscis Australis . . .	79	Vulpécŭla	69

21. Globes and Star-Maps. — In order to study the constellations conveniently, it is necessary to have either a celestial globe or a star-map, by which to identify the stars. The globe is better and more accurate, if of sufficient size; but is costly and rather inconvenient. (For a figure and description of the globe, see Appendix, Art. 400.) For most purposes a star-map will answer just as well as the globe, but it can never represent any considerable portion of the sky correctly without more or less distortion of all the lines and figures near the margin of the map. Such maps are made on various systems, each presenting its own advantages. In all of them the heavens are represented as seen from the inside, and not as on the globe, which represents the sky as seen from the outside.

22. Star-Maps of this Book. — We present a series of four small maps, which, though hardly on a large enough scale to answer every purpose of a complete celestial atlas, are quite sufficient to enable the student to trace out the constellations, and to identify the principal stars. In the map of the north circumpolar regions, Map I., the pole is in the centre, and at the circumference are numbered the twenty-four right ascension hours. The parallels of declination are represented by equidistant concentric circles. On the three other rectangular maps, which show the equatorial belt of the heavens lying between 50° north and 50° south of the equator, the parallels of declination are horizontal lines, while the hour-circles are represented by vertical lines, also equidistant, but spaced at a distance which is correct, not at the equator but for declination 35°. This keeps the distortion within reasonable bounds, even near the margin of the map, and makes it very easy to lay off the places of any object for which the right ascension and declination are given. The ecliptic is the curved line which extends across the middle of the map. The top of the map is north; and the east, instead of being at the right hand, as in a map of the earth's surface, is to the *left*, so that if the observer faces the south, and holds the map up before and above him, the constellations which are near the meridian will be pretty truly represented.

The hours of right ascension are indicated on the central horizontal line, which is the celestial equator, and at the top of the map are given the names of the months. The word "September," for instance, means that the stars which are directly under it on the map will be near the meridian about 9 o'clock in the evening during that month.

23. Star Magnitudes. — To the eye the principal difference in the appearance of the different stars is in their brightness, or their so-called 'magnitude.' Hipparchus (B.C. 125) and Ptolemy divided the visible stars into six classes, the brightest fifteen or twenty being called first-magnitude stars, and the faintest which can be seen by the naked eye being called sixth.

It has since been found that the light of the average first-magnitude star is just about 100 times as great as that of the sixth; and at this rate, the light of a first-magnitude star is just a trifle more than equal to two and a half second-magnitude stars, and a second-magnitude star to two and a half third-magnitude stars, etc.

Our maps show all the stars down to about 4½ magnitude, about a thousand in number, and all which can be seen in a moonlight night. A few smaller stars are also inserted where they mark some particular configuration or point out some interesting telescopic object. Such double stars as can be observed by a three or four inch telescope are marked on the map by underscoring : two underscoring lines denote a triple star, and three a multiple. A variable star is denoted by a circle enclosing the star symbol. A few clusters and nebulæ are also indicated. The letter M. against one of these stands for 'Messier,' who made the first catalogue of 103 such objects in 1784; e.g., 97 M. designates No. 97 on Messier's list.

For reference purposes and for study of the heavens in detail, the more elaborate star-atlases of Proctor, Heis, or Klein are recommended, especially the latter, which contains a great amount of useful information in addition to the maps, and is very cheap compared with the others. The student or teacher who possesses a telescope will also find an invaluable accessory to it in Webb's "Celestial Objects for Common Telescopes." (*Published by Longmans, Green and Co., N. Y.*)

24. Designation of the Stars. — A few of the brighter stars are designated by names of their own, and upon the map those names which are in most common use are indicated. Generally, however, the designation of visible stars is by the letters of the Greek alphabet, on a plan proposed in 1603 by Bayer, and ever since followed. The letters are ordinarily applied nearly in the order of brightness, Alpha being the brightest star in the constellation and Beta the next brightest; but they are sometimes applied to the stars in their order of position rather than in that of brightness. When the stars of a constellation are so numerous as to exhaust the letters of the Greek alphabet, the Roman letters are next used, — and then, if necessary, we employ the numbers which Flamsteed assigned a century later. At present every star visible to the naked eye can be referred to and identified by its number or letter in the constellation to which it belongs. For the Greek Alphabet, see page 344 (Appendix).

25. We begin our study of Uranography with the constellations which are circumpolar (*i.e.*, within 40° of the north pole), because these are always visible in the United States and so can be depended on to furnish land (or rather *sky*) marks to aid in tracing out the others. Since in the latitude of New York the elevation of the pole is about 41°, it follows that there (and this is nearly enough true of the rest of the United States) all the constellations which are within 41° of the north pole will move around it once in twenty-four hours without setting. For this reason they are called circumpolar. Map I. contains them all.

26. Ursa Major, the Great Bear (Map I.). — Of these circumpolar constellations none is more easily recognized than Ursa Major. Assuming the time of observation as about 8 o'clock in the evening on Sept. 22d, it will be found below the pole and to the west. Hold the map so that VIII. is at the bottom and it will be rightly placed for the time assumed.

The familiar Dipper is sloping downward in the northwest, composed of seven stars, all of about the second magnitude, excepting Delta (at the junction of the handle to the bowl), which is of the third magnitude. The stars Alpha (*Dubhe*), and Beta (*Merak*), are known as the "Pointers," because a line drawn from Beta through Alpha and produced about 30° passes very near the Pole-star. The dimensions of the Dipper furnish a convenient scale of angular measure. From Alpha to Beta is 5°; from Alpha to Delta is 10°; and from Alpha to Eta, at the extremity of the Dipper-handle, (which is also the Bear's tail,) is 26°. The Dipper (known also in England as the "Plough" and as the "Wain," or wagon) comprises but a small part of the whole constellation. The head of the Bear, indicated by a small group of scattered stars, is nearly on the line from Delta through Alpha, carried on about 15°; at the time assumed (Sept. 22d, 8 o'clock) it is almost exactly under the pole.

Three of the four paws of the creature are marked each by a pair of third or fourth magnitude stars 1½° or 2° apart. The three pairs are nearly equidistant, about 20° apart, and almost on a straight line parallel to the diagonal of the Dipper-bowl from Alpha to Gamma, but some 20° south of it. At the time assumed they are all three very near the horizon for an observer in latitude 40°, but during the spring or summer, when the constellation is high in the sky, they can be easily made out.

The star Zeta (or *Mizar*), at the bend in the handle, is easily recognized by the little star *Alcor* near it. Mizar itself is a double star, easily seen as double with a small telescope, and one of the most interesting recent astronomical results is the discovery that it is really *triple*, the larger of the two stars being itself double, invisibly so to the telescope, but revealing its double character by means of the lines in its spectrum (see Art. 373). The star Xi, the southern one of the pair, which marks the left-hand paw, is also double and binary, *i.e.*, the two stars which compose it revolve about their common centre of gravity in about sixty-one years. (For diagram of the orbit, see Fig. 77, Art. 369.) It was the first binary whose orbit was computed.

According to the ancient legends, Ursa Major is Callisto, the daughter of Lycaon, king of Arcadia. The jealousy of Juno [1] changed her into a bear, and afterwards Jupiter placed her among the constellations with Arcas her son, who became Ursa Minor. One of the quaint

---

[1] We have followed throughout the Roman nomenclature of the gods and heroes, as used by Virgil and Ovid ; but the reader should be reminded that, in many important respects, these Roman personages differ from the Greek divinities who were identified with them. It should be said, also, that in many cases the old legends are greatly confused and often contradictory, as, for instance, in the case of Hercules.

old authors explains the very un-bearlike length of the creatures' tails, by saying that they stretched as Jupiter lifted them to the sky.

27. Ursa Minor, the Lesser Bear (Map I.). — The line of the "Pointers" unmistakably marks out the Pole-star (*Polaris*), a star of the second magnitude, standing quite alone. It is at the end of the tail of Ursa Minor, or at the extremity of the handle of the "Little Dipper"; for in Ursa Minor, also, the seven principal stars form a dipper, though with the handle bent in a different way from that of the other dipper. Beginning at *Polaris*, a curved line (concave towards Ursa Major) drawn through Delta and Epsilon brings us to Zeta, where the handle joins the bowl. Two bright stars (second and third magnitude), Beta and Gamma, correspond to the Pointers in the large Dipper, and are known as the "Guardians of the Pole"; Beta is named *Kochab*. The pole now lies about $1\frac{1}{4}°$ from the Pole-star, on the line joining it to Mizar (at the bend in the handle of the large Dipper).

It has not always been so. Some 4000 years ago the star *Thuban* (Alpha Draconis) was the Pole-star, and 2000 years ago the present Pole-star was very much farther from the pole than now. At present the pole is coming nearer to the star, and towards the close of the next century it will be within half a degree of it. Twelve thousand years hence the bright star Alpha Lyræ will be the Pole-star, — and this not because the stars change their positions, but because the axis of the earth slowly changes its direction, owing to '*precession*' (see Art. 125).

The Greek name of the Pole-star was *Cynosura*, which means the 'tail of the Dog,' indicating that at one time the constellation was understood to represent a *Dog* instead of a Bear.

As already said (Art. 26) this constellation is by many writers identified with Arcas, Callisto's son. But more generally Arcas is identified with Boötes.

The Pole-star is double, having a small companion barely visible with a telescope of two or three inches diameter.

28. Cassiopeia (Map I.).—This constellation lies on the opposite side of the pole from the Dipper, and at about the same distance from it as the "Pointers." It is easily recognized by the zigzag, "rail-fence" configuration of the five or six bright stars that mark it. With the help of the rather inconspicuous star Kappa, one can make out of them a pretty good chair with the feet turned away from the pole. But this is wrong. In the recognized figures of the constellation the lady sits with feet towards the pole, and the bright star Alpha is in her bosom, while Zeta and the other faint stars south of Alpha are in her head and uplifted arms; Iota, on the line from Delta to Epsilon produced, is in the foot. The order of the principal stars is easily remembered by the word 'Bagdei,' *i.e.*, Beta, Alpha, Gamma, Delta, Epsilon, Iota.

Alpha, which is slightly variable in brightness, is known as *Schedir ;* Beta is called *Caph.* The little star Eta, which is about half-way between Alpha and Gamma, a little off the line, is a very pretty double star,—the larger star orange, the smaller one purple. It is binary (*i.e.*, the two stars revolve around each other), with a period of about 206 years.

In the year 1572 a famous temporary star made its appearance in this constellation, at a point on the line drawn from Gamma through Kappa, and extended about half its length. It was carefully observed and described by Tycho Brahe, and at one time was bright enough to be seen easily in broad daylight. There has been an entirely unfounded notion that this was identical with the Star of Bethlehem, and there has been an equally unfounded impression that its reappearance may be expected about the present time.

Cassiopeia was the wife of Cepheus, king of Libya, and the mother of Andromeda, who was rescued from the sea-monster, Cetus, by Perseus, who came flying through the air, and used the head of Medusa, (which he still holds in his hand,) to turn his adversaries to stone. Cassiopeia had indulged in too great boasting of her daughter's beauty, and thus excited the jealousy of the Nereids, at whose instigation the sea-monster was sent by Neptune to ravage the kingdom.

29. Cepheus (Map I.). — This constellation, though large, contains very few bright stars. At the assumed time (8 o'clock, Sept. 22d) it is above Cassiopeia and to the west, not having quite reached the meridian above the pole. A line carried from Alpha Cassiopeia through Beta, and produced 20°, will pass very near to Alpha Cephei, a star of the third magnitude in the king's right shoulder. Beta Cephei is about 8° north of Alpha, and Gamma about 12° from Beta, both also of the third magnitude. Gamma is so placed that it is at the obtuse angle of a rather flat isosceles triangle of which Beta Cephei and the Pole-star form the other two corners. Cepheus is represented as sitting behind Cassiopeia (his wife) with his feet upon the tail of the Little Bear, Gamma being in his left knee. His head is marked by a little triangle of fourth-magnitude stars, of which Delta is a remarkable variable with a period of $5\frac{1}{8}$ days. It is also a spectroscopic binary (Art. 373). There are also several other telescopic variables in the same neighborhood. Beta is a very pretty and easy double-star.

30. Draco, the Dragon (Map I.). — The constellation of Draco is characterized by a long, winding line of stars, mostly small, extending half-way around the pole and separating the two Bears. A line from Delta Cassiopeia drawn through Beta Cephei and extended about as far again will fall upon the head of Draco, marked by an irregular quadrilateral of stars, two of which are of the $2\frac{1}{2}$ and 3 magnitude. These two bright stars about 4° apart are Beta and Gamma. The latter (named *Etanin*), in its daily revolution, passes almost exactly through the zenith of Greenwich, and it was by observations upon it that the "aberration of light" was discovered (see Art. 435). The nose of Draco is marked by a smaller star, Mu, some 5° beyond Beta, nearly on the line drawn through it from

Gamma. From Gamma we trace the neck of Draco, eastward
and downward[1] toward the Pole-star, until we come to Delta
and Epsilon and some smaller stars near them.

There the direction of the line is reversed, as shown upon the
map, so that the body of the monster lies between its own head
and the bowl of the Little Dipper, and winds around this bowl
until the tip of the tail is reached, at the middle of the line
between the Pointers and the Pole-star. The constellation
covers more than 12 hours of right ascension.

One star deserves special notice, the star Alpha or *Thuban*,
a star of 3½ magnitude, which lies half-way between Zeta Ursæ
Majoris (Mizar) and Gamma Ursa Minoris. Four thousand
seven hundred years ago it was the Pole-star, and then within
a quarter of a degree of the pole, much nearer than Polaris is
at present or ever will be. It is probable also that its brightness has considerably fallen off within the last 200 years, since
among the ancient astronomers it was always reckoned as of
the second magnitude and is not now much above the fourth.
The so-called 'Pole of the Ecliptic' is in this constellation,
i.e., the point which is 90° distant from every point in the
Ecliptic, the circle annually described by the sun. This point
(see map) is the centre around which precession causes the
pole to move nearly in a circle (see Art. 126) once in 25,800
years.

The mythology of this constellation is doubtful. According to
some it is the dragon which Cadmus slew, afterwards sowing its teeth,
from which sprung up the harvest of armed men who fought and slew
each other, leaving only the five survivors who were the founders of
Thebes. Others say that it was the dragon who watched the golden
apples of the Hesperides, and was killed by Hercules when he captured that prize. This accords best with the fact that in the heavens
Hercules has his foot on the dragon's head.

---

[1] The description applies strictly only at the time assumed, 8 o'clock,
Sept. 22d.

31. Camelopardus. — This is the only remaining one of the strictly circumpolar constellations, — a modern one containing no stars above fourth magnitude, and established by Hevelius (1611–1687) simply to cover the great empty space between Cassiopeia and Perseus on one side, and Ursa Major and Draco on the other. The animal stands on the head and shoulders of Auriga, and his head is between the Pole-star and the tip of the tail of Draco.

The two constellations of Perseus (which at the time assumed is some 20° below Cassiopeia), and of Auriga, are partly circumpolar, but on the whole can be more conveniently treated in connection with the equatorial maps. Capella, the brightest star of Auriga, and next to Vega and Arcturus the brightest star in the northern hemisphere, is at the time assumed (Sept. 22d, 8 o'clock) a few degrees above the horizon in the N.E. Between it and the nose of Ursa Major lies part of the constellation of the *Lynx*, a modern one, made, like Camelopardus, by Hevelius, merely to fill a gap, and without any large stars.

32. The Milky Way in the Circumpolar Region. — The only circumpolar constellations traversed by the Milky Way are Cassiopeia and Cepheus. It enters the circumpolar region from the constellation of Cygnus, which at this time is just in the zenith, sweeps down across the head and shoulders of Cepheus, and on through Cassiopeia and Perseus to the northeastern horizon in Auriga. There is one very bright patch a few degrees north of Beta Cassiopeiæ, and half way between Delta Cassiopeiæ and Gamma Persei there is another bright cloud in which is the famous double cluster of the "Sword-handle of Perseus," — a beautiful object for even the smallest telescope.

33. For the most part the constellations shown upon the circumpolar map (I.) will be visible every night in the northern part of the United States. At places farther south the constellations near the rim of the map will stay below the horizon for a short time every twenty-four hours, since the height of the pole always equals the latitude of the observer, and therefore only those stars which have a polar distance less than the latitude will remain constantly visible. In other words, if, with the pole as a centre, we draw a circle with a radius equal to the height of the pole above the horizon, all the stars within this circle will remain continually above the horizon. This is called the circle of 'Perpetual Apparition.' (Art. 85.) At New Orleans, in latitude 30°, its radius, therefore, is only 30°, and only those stars which are within 30° of the pole will make a complete circle without setting. At stations in the northern part of the United States, as Tacoma, it is nearly as large as the whole map.

34. Before proceeding to consider the other constellations, the student should be reminded that he will have to select those that are conveniently visible at the time of the year when he happens to be studying the subject, and that, if he wishes to cover the whole sky, he will have to take up the subject more than once, and at various seasons of the year. The constellations near the southern limits of the map especially can be seen only a few weeks in each year.

He will also be likely to be occasionally perplexed by finding in the heavens certain conspicuous stars not given on the maps, — stars much brighter than any that are given. These are the planets Venus, Jupiter, Mars, and Saturn, called *planets*, *i.e.*, 'wandering stars,' just because they continually change their place, and so cannot be mapped. The student will find it interesting and instructive, however, to dot down upon the star-map every clear night the places of any planets he may notice, and thus to follow their motion for a month or two.

Remember also that on these maps east always lies on the *left hand*, so that the map should be held between the eye and the sky in order to represent things correctly. We begin with Andromeda at the N.W. corner of Map II.

35. Andromeda (Maps II. and IV.). Nov.— Andromeda will be found exactly overhead in our latitudes about 9 o'clock in the middle of November. Its characteristic configuration is the line of three second-magnitude stars, Alpha, Beta, and Gamma, extending east and north from Alpha, (*Alpheratz*) which itself forms the N.E. corner of the so-called "Great Square of Pegasus," and is sometimes lettered as Delta Pegasi. This star may readily be found by extending an imaginary line from Polaris through Beta Cassiopeiæ, and producing it about as far again : Alpha is in the head of Andromeda, Beta (*Mirach*) in her waist, and Gamma (*Almaach*) in her left foot. A line drawn northwesterly from Beta, nearly at right angles to the line Beta Gamma, will pass through Mu at a distance of about 5°, and produced another 5° will strike the "great nebula," which is visible to the naked eye like a little cloud of light, and forms a small obtuse-angled triangle with Nu and a little sixth-magnitude star. Andromeda has her mother, Cassiopeia, close by on the north, with her father, Cepheus, not far away, while at her feet is Perseus, her deliverer. Her head rests upon the shoulder of Pegasus. In the south, beyond the constellations of Aries and Pisces, Cetus, the sea-monster, who was to have devoured her, stretches his ungainly bulk.

We have already mentioned the nebula. Another very pretty object is Gamma, which in a small instrument is a double star, the larger one orange, the smaller a greenish blue. The small star is itself double, making the system really triple, but as such is beyond the reach of any but very large instruments.

When Neptune sent the leviathan, Cetus, to ravage Libya, the oracle of Ammon announced that the kingdom could be delivered only if Cepheus would give up his daughter. He assented and chained the poor girl to a rock to await her destruction. But Perseus, returning through the air from the slaying of the Gorgon, Medusa, saw her, rescued her, won her love, and made her his wife.

36. Pisces, the Fishes (Maps II. and IV.). Nov.—Immediately south of Andromeda lies Pisces, the first of the constellations

of the Zodiac,[1] which is a belt 16° wide (8° on each side of
the ecliptic) encircling the heavens, and including the space
within the limits of which the sun, the moon, and all the principal planets perform their apparent motions. At present, in
consequence of precession, it occupies the *sign* of Aries (see
Art. 126). It has not a single conspicuous star, and is notable
only as now containing the *Vernal Equinox*, or "*First of
Aries*," which lies near the southern boundary of the constellation in a peculiarly starless region. A line from Alpha Andromedæ through Gamma Pegasi, continued as far again, strikes
about 2° east of the point. The body of one of the two fishes
lies about 15° south of the middle of the southern side of the
"Great Square of Pegasus," and is marked by an irregular
polygon of small stars, 5° or 6° in diameter. A long, crooked
"ribbon" of little stars runs eastward for more than 30°,
terminating in Alpha Piscium, (called *El Rischa*, or 'the
knot,') a star of the fourth magnitude 20° south of the head
of Aries. From there another line of stars leads up northwest in the direction of Delta Andromedæ to the northern
fish, which lies in the vacant space south of Beta Andromedæ.

Alpha is a very pretty double star, the two components being about
2″ apart.

The mythology of this constellation is not very well settled. One
story is that the fishes are Venus and her son Cupid, who once were
thus transformed when endeavoring to escape from the giant Typhon.
The northern fish is Cupid, the southern his mother.

37. Triangulum or **Deltōton, the Triangle** (Map II.). December. — This little constellation, insignificant as it is, is one
of Ptolemy's ancient forty-eight. It lies half-way between
Gamma Andromedæ and the head of Aries, and is character-

---

[1] The word is derived from the Greek word *zoön*, a living creature,
and indicates that all the constellations in it (Libra alone excepted) are
animals. The zodiacal constellations are for the most part of remote
antiquity, antedating by many centuries even the Greek mythology.

ized by three stars of the third and fourth magnitude, easily
made out by the help of the map.

It may be regarded as a canonization of "Divine Geometry," but
has no special mythological legend connected with it.

38. Aries, the Ram (Map II). December. — This is the sec-
ond of the zodiacal constellations, now occupying the *sign* of
Taurus.   It lies just south of Triangulum and Perseus.   Its
characteristic star-group is that composed of Alpha (*Hamal*),
Beta, and Gamma (see map), about 20° due south of Gamma
Andromedæ.   Alpha, a star of 2½ magnitude, is fairly conspic-
uous, forming a large isosceles triangle with Beta and Gamma
Andromedæ.

Gamma Arietis is a very pretty double star with the components
about 9″ apart.   It is probably the first double star discovered, having been noticed by Hooke in 1664.

The star 41 Arietis (3½ magnitude), which forms a nearly equilateral triangle with Alpha Arietis and Gamma Trianguli, constitutes,
with two or three other stars near it, the constellation of *Musca*
(Borealis), a constellation, however, not now generally recognized.

According to the Greeks, Aries is the ram which bore the golden
fleece and dropped Helle into the Hellespont, when she and her brother,
Phrixus, were flying on its back to Colchis.   Long afterwards the
Argonautic Expedition, with Jason as its head and Hercules as one
of its members, sailed from Greece to Colchis to recover the fleece,
and finally succeeded after long endeavors.

39. Cetus, the Sea-monster (Maps II. and IV.).   November-
December. — South of Aries and Pisces lies the constellation
of Cetus, the sea-monster, which backs up into the sky from
the southeastern horizon.   The head lies some 20° southeast
of Alpha Arietis, and is marked by an irregular five-sided figure of stars, each side being some 5° or 6° long.   The southern
edge of this pentagon is formed by the stars Alpha or *Menkar*
(2½ magnitude) and Gamma (3½ magnitude) ; Delta lies southwest of Gamma.   Beta (*Deneb Ceti*), the brightest star of the

constellation (2 magnitude), stands by itself nearly 40° west
and south of Alpha. Gamma is a very pretty double star, but
rather close for a small telescope, the components being only
2.5″ apart, yellow and blue.

Cetus is the leviathan that was sent by Neptune to ravage Libya
and devour Andromeda. Perseus turned him into stone by showing
him the head of the Gorgon, Medusa. On the globes he is usually
represented as a nondescript sort of beast, with a face like a puppy's,
and a tightly curled tail; as if the Gorgon's head had frightened out
all his savageness.

South of Cetus lies the modern constellation of *Sculptoris Apparatus* (usually known simply as Sculptor), which, however, contains
nothing that requires notice here. South of · Sculptor, and close to
the horizon, even when on the meridian, is *Phœnix*. It has some
bright stars, but none easily observable in the United States.

40. Perseus (Maps I. and II.). January. — Returning now
to the northern limit of the map, we come to the constellation of Perseus. Its principal star is Alpha (*Algenib*), rather
brighter than the standard second magnitude, and situated
very nearly on the prolongation of the line of the three chief
stars of Andromeda. A very characteristic configuration is
the so-called "segment of Perseus" (Map I.), a curved line
formed by Delta, Alpha, Gamma, and Eta, with some smaller
stars, concave towards the northeast, and running along the
line of the Milky Way towards Cassiopeia. The remarkable
variable star, Beta, or *Algol*, is situated about 9° south and a
little west of Alpha, at the right angle of a right-angled triangle
which it forms with Alpha Persei and Gamma Andromedæ.
Algol and a few small stars near it form "Medusa's Head,"
which Perseus carries in his hand. For further particulars
and recent discoveries regarding this star, see Arts. 358 and
360.

Epsilon is a very pretty double star with the components about 8″
apart; but the most beautiful telescopic object in the constellation,

perhaps the finest, indeed, in the whole heavens for a small telescope, is the pair of clusters about half-way between Gamma Persei and Delta Cassiopeiæ, visible to the naked eye as a bright knot in the Milky Way, and already referred to in Art. 32.

Perseus was the son of Danaë by Jupiter, who won her in a shower of gold. He was sent by his enemies on the desperate venture of capturing the head of Medusa, the only mortal one of the three Gorgons, which were frightful female monsters with wings, tremendous claws, and brazen teeth, and serpents for hair; of such aspect that the sight turned all who looked at them to stone. The gods helped Perseus by various gifts which enabled him to approach his victim, invisible and unsuspected, and to deal the fatal blow without looking at the sight himself. From the blood of Medusa, where her body fell, sprang Pegasus, the winged horse, and where the drops fell on the sands of Libya, as Perseus was flying across the desert, thousands of venomous serpents swarmed. On his way, returning home, he saw and rescued Andromeda, as already mentioned (Arts. 28 and 35). Hercules was one of their descendants.

41. Auriga, the Charioteer (Maps I. and II.). January. — Proceeding east from Perseus we come to Auriga, who is represented as holding in his arms a goat and her kids. The constellation is instantly recognized by the bright yellow star, Capella (the Goat), and her attendant 'Hœdi' (the Kids). Alpha Aurigæ (*Capella*) is, according to Pickering, precisely of the same brightness as Vega, both of them being about $\frac{1}{5}$ of a magnitude fainter than Arcturus, but distinctly brighter than any other stars visible in our latitudes except Sirius itself. The spectroscope shows that Capella is very similar in character to our own sun, though probably vastly larger. About 10° east of Capella is Beta Aurigæ (*Menkalinan*) of the second magnitude; Epsilon, Zeta, and Eta, which form a long triangle 4° or 5° south of Alpha, are the Kids.

There seems to be no well-settled mythological history for this constellation, though some say that he is the charioteer of Œnomaus, king of Elis; while others connect him with the story of Phaëton, the son of Apollo, who borrowed the horses of his father and was overthrown in mid-heaven. The goat is supposed to be Amalthea, the goat which suckled Jupiter in his infancy. Capella and the Kids were always regarded by astrologers as of kindly influence, especially towards sailors.

42. Taurus, the Bull (Map II.). January. — This, the third of the zodiacal constellations, lies directly south of Perseus and Auriga, and north of Orion. It is unmistakably characterized by the Pleiades, and by the V-shaped group of the Hyades which forms the face of the bull, with the red *Aldebaran* (Alpha Tauri), a standard first-magnitude star, blazing in the creature's eye, as he charges down upon Orion. His long horns reach out towards Gemini and Auriga, and are tipped with the second and third magnitude stars, Beta and Zeta. As in the case of Pegasus, only the head and shoulders appear in the constellation. Six of the Pleiades are easily visible, and on a dark night a fairly good eye will count nine of them. With a three-inch telescope about 100 stars are visible in the cluster, which is more fully described with a figure in Art. 376. The brightest of the Pleiades is called '*Alcyone*,' and was assigned to the dignity of the 'Central Sun' by Maedler (Art. 386).

About 1° west and a little north of Zeta is a nebula (Messier 1), which has many times been discovered by tyros with a small telescope as a new comet: it is an excellent imitation of the real thing.

According to the Greek legends, Taurus is the milk-white bull into which Jupiter changed himself when he carried away Europa from Phœnicia to the island of Crete, where she became the mother of Minos and the grandmother of Deucalion, the Noah of Greek mythology. But Taurus, like most of the other zodiacal constellations, is really far older than the Greek mythology, and appears in the most ancient zodiacs of Egypt, where it was probably connected with the worship of the bull, Apis; so also in the ancient Astronomy of Chaldea and India.

The Pleiades were daughters of the giant Atlas. Of the seven sisters, one, who married a mortal, lost her brightness, according

to the legend, so that only six remain visible. Some say that Merope
was the one who thus gave up her immortality for love, but her star is
still visible, while Celæno and Asterope are both faint. The now recognized names of the stars in the group (see map, Art. 376) include
Atlas and Pleione, the parents of the family, as well as the seven sisters. As for the Hyades, who were half-sisters of the Pleiades, there
is less legendary interest in their case. They are always called by
the poets " the rainy Hyades."

43. Orīon (not *O'rion*) (Map II.). February. — This is the
most splendid constellation in the heavens. As the giant
stands facing the bull, his shoulders are marked by the two
bright stars, Alpha (*Betelgeuze*) and Gamma (*Bellatrix*), the
former of which in color closely matches Aldebaran, though
its brightness is somewhat variable. In his left hand he holds
up the lion skin, indicated by the curved line of little stars
between Gamma and the Hyades. The top of the club, which
he brandishes in his right hand, lies between Zeta Tauri and
Mu and Eta Geminorum. His head is marked by a little triangle of stars of which Lambda is the chief. His belt, through
the northern end of which passes the celestial equator, consists
of three stars of the second magnitude, pointing obliquely
southeast toward Sirius. It is very nearly 3° in length, and
is called the " Ell and Yard " or " Jacob's Staff." From the belt
hangs the sword, composed of three smaller stars lying more
nearly north and south: the middle one of them is the multiple, Theta, in the great nebula, which even in a small telescope is a beautiful object, the finest nebula in the sky. Beta
Orionis, or *Rigel*, a magnificent white star, is in the left foot,
and Kappa is in the right knee. Orion has no right foot, or if
he has, it is hidden behind Lepus. The quadrilateral Alpha,
Gamma, Beta, Kappa, with the diagonal belt, Delta, Eta, Zeta,
once learned can never be mistaken for anything else in the
heavens.

Rigel is a very pretty double star, the larger star having a very
small companion about 10″ distant. The two stars at the extremities
of the belt are also double.

Orion was a giant and mighty hunter, son of Neptune, and beloved by both Aurora and Diana. The legends of his life and exploits are numerous, and often contradictory. He conquered every wild beast except the Scorpion, which stung and killed him. As a winter constellation his influence was counted stormy, and he was greatly dreaded by sailors.

44. Eridanus, the River Po (Map II.). January. — This constellation lies south of Taurus, in the space between Cetus and Orion, and extends far below the southern horizon. The portion near the south pole has a pair of bright stars, which, of course, are never visible at the United States. Starting with Beta (*Cursa*, as it is called), of the third magnitude, about 3° north and a little west of Rigel, one can follow a sinuous line of stars westward to the paws of Cetus, where the stream turns at right angles, and runs southward and southwest to the horizon. One can trace it, however, only by the help of a map on a larger scale than the one we present.

45. Lepus and Columba (Map II.). February. — The constellation of Lepus, the *Hare*, one of Orion's victims, is one of the ancient forty-eight, and lies just south of the giant, occupying a space of some 15° square. Its characteristic configuration is a quadrilateral of third and fourth magnitude stars, with sides from 3° to 5° long, about 10° south of Kappa Orionis, and 15° west of Sirius.

Columba, the Dove, lies next south of Lepus, too far south to be well seen in the Northern States. Its principal star, Alpha (*Phact*) is of 2½ magnitude, and is readily found by drawing a line from Procyon to Sirius and prolonging it about the same distance. In passing, we may note that a similar line drawn from Alpha Orionis through Sirius, and produced, will strike near Zeta Argus, or *Naos*, a star about as bright as Phact, — the two lines which intersect at Sirius making the so-called " Egyptian X."

Columba is a modern constellation, commemorating Noah's dove returning to the ark with the olive branch.

46. Lynx (Maps I., II., and III.). February. — Returning now to the northern limit of the map, we find the modern constellation of the Lynx lying just east of Auriga, and enveloping it on the north and in the circumpolar region, as shown on the map. It contains no stars above the fourth magnitude, and is of no importance except as occupying an otherwise vacant space.

47. Gemini, the Twins (Map II.). February and March. — This is the fourth of the zodiacal constellations, now lying mostly in the *sign* of Cancer. It contains the summer solstitial point — the point where the sun turns from its northern motion to its southern in the summer. At present it is about 2° west and a little north of the star Eta. Gemini lies northeast of Orion and southeast of Auriga, and is sufficiently characterized by the two stars Alpha and Beta (about 4½° apart), which mark the heads of the twins. The southern one, Beta, or *Pollux*, is now the brighter; but Alpha, *Castor*, is much more interesting, as being double (easily seen with a small telescope). The feet are marked by the third-magnitude stars Gamma and Mu, some 10° east of Zeta Tauri.

Castor and Pollux were the sons of Jupiter by Leda, and ancient mythology, especially that of Rome, is full of legends relating to them. Many of our readers will remember Macaulay's ballad of "The Battle of Lake Regillus," when they won the fight for Rome. They were regarded as the special patrons of the sailor, who relied much on their protection against the evil powers of Orion and the Hyades.

48. Canis Minor, the Little Dog (Map III.). March. — This constellation, about 20° south of Castor and Pollux, is marked by the bright star *Procyon*, which means "before the dog," because it rises about half an hour before the Dog Star, Sirius. Alpha, Beta, and Gamma form together a configuration closely resembling that formed by Alpha, Beta, and Gamma Arietis. Procyon, Alpha Orionis, and Sirius form nearly an equilateral triangle, with sides of about 25°.

The animal is supposed to have been one of Orion's dogs, though some say the dog of Icarus, whom they identify with Boötes.

49. Canis Major, the Great Dog (Map II.). February. — This glorious constellation hardly needs description. Its Alpha is the Dog Star, *Sirius*, beyond all comparison the brightest star in the heavens, and one of our nearer neighbors, — so distant, however, that it requires more than eight years for light to come to us from it. It is nearly pointed at by a line drawn through the three stars of Orion's belt. Beta, at the extremity of the uplifted paw, is of the second magnitude, and so are several of the stars farther south in the rump and tail of the animal, who sits up watching his master Orion, but with an eye out for Lepus.

50. Monoceros, the Unicorn (Map II.). March. — This is one of the modern constellations organized by Hevelius to fill the gap between Gemini and Canis Minor on the north, and Argo Navis and Canis Major on the south. It lies just east of Orion, and has no conspicuous stars, but is traversed by a brilliant portion of the Milky Way. The Alpha of the constellation (fourth magnitude) lies about half-way between Alpha Orionis and Sirius, a little west of the line that joins them. 11 Monocerōtis, a fine triple star (see Fig. 76, Art. 366), fourth magnitude, is very nearly pointed at by a line drawn from Zeta Canis Majoris northward through Beta, and continued as far again.

51. Argo Navis, the Ship Argo (genitive *Argûs*) (Maps II. and III.). March. — This is one of the largest, oldest, and most important of the constellations, lying south and east of Canis Major. Its brightest star, Alpha Argûs, *Canopus*, ranks next to Sirius, and is visible in the Southern States, but not in the Northern. The constellation, huge as it is, is only a half one, like Pegasus and Taurus, — only the stern of a vessel, with mast, sail, and oars; the stem being wanting. In the part of the constellation covered by our maps there are no very conspicuous stars, though there are some of third and fourth magnitude which lie east and southeast of the rump and tail of Canis Major. We have already mentioned Zeta, or *Naos*, at the southeast extremity of the "Egyptian X."

According to the Greek legends, this is the miraculous ship in which Jason and his fifty companions sailed from Greece to Colchis to recover the Golden Fleece. It had in its bow a piece of oak from the sacred grove of Dodona, which enabled the ship to talk with its commander and give him advice.

Some see in the constellation the ark of Noah.

52. Cancer, the Crab (Maps II. and III.) March. — This is the fifth of the zodiacal constellations, lying just east of Canis Minor. It does not contain a single conspicuous star, but is easily recognizable from its position, and in a dark night by the nebulous cloud known as *Præsepe*, or the "Manger," with the two stars Gamma and Delta near it, — the so-called *Aselli*, or "Donkeys." Præsepe, sometimes also called the "Beehive," is really a coarse cluster of seventh and eighth magnitude stars, resolvable by an opera-glass. The line from Castor through Pollux, produced about 12°, passes near enough to it to serve as a pointer.

The star Zeta is a very pretty triple star, though with a small telescope it can be seen only as double. It is easily found by a line from Castor through Pollux, produced 2½ times as far.

By the Greeks this was identified as the Crab who attacked Hercules when he was fighting the Lernæan Hydra. In the old Egyptian zodiacs the Crab is replaced by the Scarabæus, or Beetle; and in some of the more recent zodiacs by a pair of asses, still recognized in the name, Aselli, given to the two stars Gamma and Delta.

53. Leo, the Lion (Map III.). April. — East of Cancer lies the noble constellation of Leo, which adorns the evening sky in March and April; it is the sixth of the zodiacal constellations, now occupying the sign of Virgo. Its leading star, *Regulus*, or "Cor Leonis," is of the first magnitude, and two others, Beta (*Denebola*) and Gamma, are of the second magnitude. Alpha, Gamma, Delta, and Beta form a conspicuous irregular quadrilateral (see map), the line from Regulus to Denebola being about 26° long. Another characteristic configuration is "the Sickle," of which Regulus is the handle, and the curved line Eta, Gamma, Zeta, Mu, and Epsilon, is the blade, the cutting edge being turned towards Cancer.

The "radiant" of the November meteors lies between Zeta and Epsilon. Gamma, in the Sickle, and at the southeast corner of the quadrilateral, is a very pretty double star, binary, with a period of about 400 years.

According to classic writers, this is the Nemæan Lion which was killed by Hercules, as the first of his Twelve Labors; but, like Aries and Taurus, the constellation is far older than the Greeks, and stands in its present form on all the ancient zodiacs.

54. Leo Minor and Sextans (Map II.). April. — Leo Minor, the *Smaller Lion*, is an insignificant modern constellation composed of a few small stars north of Leo, between it and the feet of Ursa Major. It contains nothing deserving special notice. The same remark holds good as to Sextans, the *Sextant*, and even more emphatically.

55. Hydra (Map III.). March to June. — This constellation, with its riders, *Crater* (the Cup) and *Corvus* (the Raven), is a large and important one, though not very brilliant. The head is marked by a group of five or six fourth and fifth magnitude stars just 15° south of Præsepe. A curving line of small stars leads down southeast to Alpha, *Cor Hydræ*, or *Alphard*, a 2½ magnitude star standing very much alone. From there, as the map shows, an irregular line of fourth-magnitude stars running far south and then east, almost to the boundary of Scorpio, marks the creature's body and tail, the whole covering almost six hours of right ascension, and very nearly 90° of the sky. About the middle of the length of Hydra, and just below the hind feet of Leo (30° due south from Denebola), we find the little constellation of *Crater;* and just east of it the still smaller but much more conspicuous one of *Corvus,*

with two second-magnitude stars in it, and four of the third
and fourth magnitudes. It is well marked by a characteristic
quadrilateral (see map), with Delta and Eta together at its
northeast corner. The order of the letters in Corvus differs
widely from that of brightness, suggesting that changes may
have occurred since the letters were applied.

Epsilon Hydræ and Delta Corvi are pretty double stars, the latter
easily seen with a small telescope; colors, yellow and purple.

Hydra, according to the Greeks, is the immense-hundred-headed
monster which inhabited the Lernæan Marsh, and was killed by Hercules as his second labor. But the Hydra of the heavens has only one
head, and is probably much older than the legends of Hercules.

An old legend says that Corvus is Coronis, a nymph who was transformed into a raven to escape the pursuit of Neptune. Another story
is that she was changed into a crow for telling tales of some impru-
dent actions that came under her notice.

56. Virgo (Map III.). May. — East and south of Leo lies
Virgo, the seventh zodiacal constellation, mostly in the sign
of Libra. Its Alpha, *Spica Virginis*, is of the $1\frac{1}{2}$ magnitude,
and, standing rather alone, 10° south of the celestial equator,
is easily recognized as the southern apex of a nearly equilateral triangle which it forms with Denebola (Beta Leonis) to
the northwest, and Arcturus northeast of it. Beta Virginis, of
the third magnitude, is 14° south of Denebola. A line drawn
eastward and a little south from Beta (third magnitude) and
then carried on, curving northward, passes successively (see
map) through Eta, Gamma, Delta, and Epsilon, of the third
magnitude. (Notice the word Begde, like Bagdei in Cassiopeia, Art. 28.)

Gamma is a remarkable binary star, at present easily visible as
double in a small telescope. Its period is one hundred and eighty-five
years, and it has completed pretty nearly a full revolution since its
first discovery. For a diagram of its orbit, see Fig. 77, Art. 369. A
few degrees north of Gamma lies the remarkable nebulous region of

Virgo, containing hundreds of these curious objects ; but for the most part they are very faint, and observable only with large telescopes.

The classic poets recognize Virgo as Astræa, the goddess of justice, who, last of all the old divinities, left the earth at the close of the Golden Age. She holds the Scales of Justice (Libra) in one hand, and in the other a sheaf of wheat.

Some identify her with Erigone, the daughter of Icarus or Boötes. Others recognize in her the Egyptian Isis.

57. Coma Berenices, Berenice's Hair (Map III.). May. — This little constellation, composed of a great number of fifth and sixth magnitude stars, lies 30° north of Gamma and Eta Virginis, and about 15° northeast of Denebola. It contains a number of interesting double stars, but they are not easily found without the help of a telescope equatorially mounted.

The constellation was established by the Alexandrian astronomer Conon, in honor of the queen of Ptolemy Soter. She dedicated her splendid hair to the gods, to secure her husband's safety in war.

58. Canes Venatici, the Hunting-Dogs (Map III.). May. — These are the dogs with which Boötes, the huntsman, is pursuing the Great Bear around the pole : the northern of the two is *Asterion*, the southern *Chara*. Most of the stars are small, but Alpha is of the 2½ magnitude, and is easily found by drawing from Eta Ursæ Majoris (the star in the end of the Dipper-handle) a line to the southwest, perpendicular to the line from Eta to Zeta (Mizar), and about 15° long : in England it is generally known as Cor Caroli (the Heart of Charles), in allusion to Charles I. With Arcturus and Denebola it forms a triangle much like that which they form with Spica.

The remarkable whirlpool nebula of Lord Rosse is situated in this constellation, about 3° west and somewhat south of the star Eta Ursæ Majoris. In a small telescope it is by no means conspicuous, but in a large telescope is a wonderful object.

The constellation is modern, formed by Hevelius.

59. Boötes, the Huntsman (Maps I. and III.). June. — This fine constellation extends more than 60° in declination, from

near the equator quite to Draco, where the uplifted hand holding the leash of the hunting-dogs overlaps the tail of the Bear.
Its principal star, Alpha, *Arcturus* (meaning 'bear-driver'), is
of a ruddy hue, and in brightness is excelled only by Sirius
among the stars visible in our latitudes.   It is at once recognizable by its forming with Spica and Denebola the great triangle already mentioned (Art. 56).   Six degrees west and a
little south of it is Eta, of the third magnitude, which forms
with it, in connection with Upsilon, a configuration like that
in the head of Aries.   Epsilon is about 10° northeast of Arcturus, and in the same direction about 10° farther lies Delta.
The word Boötes means "the shouter," — the huntsman calling to his dogs.

> Epsilon is a fine double star; colors, orange and greenish blue;
> distance, about 3″.
> The legendary history of this constellation is very confused.   One
> legend makes it to be Icarus, the father of Erigone (Virgo).   But the
> one most usually accepted makes it to be Arcas, son of Callisto.   After
> she was changed to a bear (Ursa Major), her son, not recognizing her,
> hunted her with his dogs, and was on the point of killing her, when
> Jupiter interfered and took them both to the stars.

60. Corona Borealis, the Northern Crown (Map III.).   June.
— This beautiful little constellation lies 20° northeast of Arcturus, and is at once recognizable as an almost perfect semicircle composed of half a dozen stars, among which the brightest, Alpha (*Gemma* or *Alphacca*), is of the second magnitude.
The extreme northern one is Theta; next comes Beta, and the
rest follow in the Bagdei order, just as in Cassiopeia.   About a
degree north of Delta, now visible with an opera-glass, is a
small star which in 1866 suddenly blazed out until it became
brighter than Alphacca itself (see Art. 355).

> The little star Eta is a rapid binary with a period of less than forty
> two years.   At times it can be easily divided by a small telescope.
> The constellation is said to be the crown that Bacchus gave to
> Ariadne, before he deserted her on the island of Naxos.

61. Libra, the Balance (Map III.). June. — This is the eighth of the zodiacal constellations, lying east of Virgo, bounded on the south by Centaurus and Lupus, on the east by the upstretched claw of Scorpio, and on the north by Serpens and Virgo. It is inconspicuous, the most characteristic figure being the trapezoid formed by the lines joining the stars Alpha, Iota, Gamma, and Beta. Beta, which is the northern one, is about 30° due east from Spica, while Alpha is about 10° southwest of Beta. The remarkable variable, Delta Libræ, is 4° west and a little north from Beta. Most of the time it is of the $4\frac{1}{2}$ or 5 magnitude, but runs down nearly two magnitudes, to invisibility, once in $2\frac{1}{3}$ days. "Algol" type (Art. 358).

Libra is the Balance of Virgo, the goddess of justice, and was not recognized by the classic writers as a separate constellation until the time of Julius Cæsar; the space now occupied by Libra being then covered by the extended claws of Scorpio.

62. Antlia, Centaurus, and Lupus (Map III.). April to June. — These constellations lie south of Hydra and Libra.

Antlia Pneumatica (the *Air-Pump*) is a modern constellation of no importance and hardly recognizable by the eye, having only a single star as bright as the $4\frac{1}{2}$ magnitude.

Centaurus, on the other hand, is an ancient and extensive asterism, containing in its south (circumpolar) regions, not visible in the United States, two stars of the first magnitude, Alpha and Beta. Alpha Centauri stands next after Sirius and Canopus in brightness, and, as far as present knowledge indicates, *is our nearest neighbor among the stars.* The part of the constellation which becomes visible in our latitudes is not especially brilliant, though it contains several stars of the $2\frac{1}{2}$ and 3 magnitudes in the region lying south of Corvus and Spica Virginis.

Lupus, the *Wolf,* also one of Ptolemy's constellations, lies due east of Centaurus and just south of Libra. It contains a considerable number of third and fourth magnitude stars; but it is too low for any satisfactory study in our latitudes. It is best seen late in June. These constellations contain numerous objects interesting for a southern observer, but not observable by us.

The Centaurs were a fabulous race, half man, half horse, who lived in Thessaly and herded cattle. Chiron was the most distinguished of them, the teacher of almost all the Greek heroes in every manly and noble art, and the friend of Hercules, by whom, however, he was accidentally killed. Jupiter transferred him to the stars. (See Sagittarius, Art. 72.) The wolf is represented as transfixed by the Centaur's spear.

63. Scorpio (or **Scorpius;** genitive *Scorpii*), **the Scorpion** (Map IV.). July. — This, the ninth of the zodiacal constellations and the most brilliant of them all, lies southeast of Libra, which in ancient times used to form its claws (Chelæ). It is recognized at once by the peculiar configuration of the stars, which resembles a boy's kite, with a long streaming tail extending far down to the south and east, and containing several pairs of stars. The principal star of the constellation, *Antares*, is of the first magnitude, and fiery red like the planet Mars. From this it gets its name, which means "the rival of *Ares*" (Mars). Antares is a very pretty double star, with a beautiful little green companion just to the west of it, not very easy to be seen, however, with a small telescope. Beta (second magnitude) is in the arch of the kite bow, about 8° or 9° northwest of Antares, while the star which Bayer lettered as Gamma Scorpii is well within Libra, 20° west of Antares. (There is considerable confusion among uranographers as to the boundary between the two constellations.) The other principal stars of the constellation are easily found on the map.

Many of them are of the second magnitude. One of the finest clusters known, and easily seen with a small telescope, is Messier 80, which lies about half-way between Alpha and Beta.

According to the Greek mythology, this is the scorpion that killed

Orion. It was the sight of this monster of the heavens that frightened the horses of the sun, when poor Phaeton tried to drive them and was thrown out of his chariot. Among astrologers, the influence of Scorpio has always been held as baleful to the last degree.

64. Norma Nilotica, the rule with which the height of the Nile was measured, lies west of Scorpio, while Ara lies due south of Eta and Theta. Both are old Ptolemaic constellations, but are small and of little importance, at least to observers in our latitudes.

65. Ophiūchus and Serpens (Maps III. and IV). July.—Ophiuchus means "serpent-holder," and probably refers to the great physician, Æsculapius. The hero is represented as standing with his feet on Scorpio, and grasping the "serpent." The two constellations, therefore, are best treated together. The head of Serpens is marked by a group of small stars lying just south of Corona and 20° due east of Arcturus. Beta and Gamma are the two brightest stars in the group, their magnitudes 3½ and 4. Delta lies 6° southwest of Beta, and there the Serpent's body bends southeast through Alpha and Epsilon Serpentis (see map) to Delta and Epsilon Ophiuchi in the giant's hand. The line of these five stars carried upwards passes nearly through Epsilon Boötis, and downwards through Zeta Ophiuchi. A line crossing this at right angles, nearly midway between Epsilon Serpentis and Delta Ophiuchi, passes through Mu Serpentis on the southwest and Lambda Ophiuchi to the northeast. The lozenge-shaped figure formed by the lines drawn from Alpha Serpentis and Zeta Ophiuchi to the two stars last mentioned is one of the most characteristic configurations of the summer sky. Alpha Ophiuchi (2¼ magnitude) (*Ras Alaghue*) is easily recognizable in connection with Alpha Herculis, since they stand rather isolated, about 6° apart, on the line drawn from Arcturus through the head of Serpens, and produced as far again. Alpha Ophiuchi is the eastern and the brighter of the two, and forms with Vega and Altair a nearly equilateral triangle. Beta Ophiuchi lies about 9° southeast of Alpha.

Five degrees east and a little south of Beta are five small stars in the Milky Way, forming a V with the point to the south, much like the Hyades of Taurus. They form the head of the now discredited constellation, "Poniatowski's Bull" (*Taurus Poniatovii*), proposed in 1777, and found in many maps. 70 Ophiuchi (the middle star in the eastern leg of the V of Poniatowski's Bull) is a very pretty double star, binary, with a period of ninety-three years. Just at present the star is too close to be resolved by a small instrument, but it will soon open up again.

Ophiuchus is identified with Æsculapius, who was the first great physician, the son of Apollo and the nymph Coronis, educated in the art of medicine by Chiron, the Centaur. The serpent and the cock were sacred to him in his character as a deity. But the constellation is older than the classic legends.

66. Hercules (Maps III. and IV.).   July. — This noble constellation lies next north of Ophiuchus, between it and Draco. The hero is represented as resting on one knee, with his foot on the head of Draco, while his head is close to that of Ophiuchus. The constellation contains no stars of the first or even of the second magnitude, but there are a number of the third. The most characteristic figure is the keystone-shaped quadrilateral formed by the stars Epsilon, Zeta, Eta, with Pi and Rho together at the northeast corner. It lies about midway on the line from Vega to Corona.

On its western boundary, a third of the way from Eta towards Zeta, lies the remarkable cluster, Messier 13, — on the whole the finest of all star clusters, — barely visible to the naked eye on a dark night. Alpha Herculis (*Ras Algethi*), in the head of the giant, is a very beautiful double star, colors orange and blue, distance about 5″. It is slightly variable, and has a remarkable spectrum, characterized by numerous dark bands.

Hercules, the son of Jupiter and Alcmena (a granddaughter of Andromeda), was the Greek incarnation of gigantic strength. His heroic actions and freaks occupy more space in their mythology than those of any personage except Jupiter himself. He was the pupil of Chiron, but by the will of Jupiter, his father, was subjected to the power of Eurystheus, the king of Tiryns, for many years. At his bidding he performed the great enterprises known as the Twelve Labors of Hercules, for which we must refer the reader to the Classical Dictionaries. Among them we have already mentioned the conquest of the Nemæan Lion and of the Lernæan Hydra. Another was to bring from the garden of the Hesperides the golden apples which were guarded by the dragon that he killed, and on which his feet rest in the sky. His last and greatest achievement was to bring to the earth the three-headed dog, Cerberus, the guardian of the infernal regions.

67. Lyra (Map IV.). August. — This constellation is sufficiently marked by the great white or blue star, *Vega,* one of the finest stars in the whole sky, and certainly many times larger than our own sun. It is attended on the east by two fourth-magnitude stars, Epsilon and Zeta, which form with it a little equilateral triangle having sides about 2° long. Epsilon is a double-double or quadruple star. A sharp eye, even unaided by a telescope, divides the star into two, and a large telescope splits each of the components. It is a very pretty object even for a small telescope (Fig. 76). Beta and Gamma, of the third magnitude (Beta is variable), lie about 8° southeast from Vega, 2½° apart. (See Art. 357.)

On the line between Beta and Gamma, one-third of the way from Beta, lies Messier 57, the Annular Nebula, which can be seen as a small hazy ring even by a small telescope, though of course it is much more interesting with a larger one.

According to the legends this constellation is the lyre of Orpheus, with which he charmed the stern gods of the lower world, and persuaded them to restore to him his lost Eurydice.

68. Cygnus (Maps I. and IV.). September. — This constellation lies due east from Lyra, and is easily recognized by the cross that marks it. The bright star Alpha (1½ magnitude) is at the top, and Beta (third magnitude) at the bottom, while Gamma is where the cross-bar from Delta to Epsilon intersects the main piece, which lies along the Milky Way from the northeast to the southwest. Beta (*Albireo*) is a beautiful double star, orange and dark blue, one of the finest of the colored pairs for a small telescope. 61 Cygni, which is memorable as the first star to have its parallax determined (by Bessel in 1838), is easily found by completing the parallelogram of which Alpha, Gamma, and Epsilon are the other three corners. Sigma and Tau form a little triangle with 61, which is the faintest of the three. 61 is a fine double star. Delta is also a fine double, but too difficult for an instrument of less than six inches' aperture.

According to Ovid, Cygnus was a friend of Phaëton's, who mourned his unhappy fate and was changed to a swan. Others see in the constellation the swan in whose form Jupiter visited Leda, the mother of Castor and Pollux and of Helen of Troy.

69. Vulpecula et Anser, the Fox and the Goose (Map IV.). September. — This little constellation is one of those originated by Hevelius, and has obtained more general recognition among astronomers than most of his creations. It lies just south of Cygnus, and is bounded to the south by Delphinus, Sagitta, and Aquila. It has no conspicuous stars, but it contains one very interesting telescopic object, — the " Dumb-bell Nebula" (see map). It may be found on a line from Gamma Lyræ through Beta Cygni, produced as far again.

70. Sagitta (Map IV.). August. — This little constellation, though very inconspicuous, is one of the old 48. It lies south of Vulpecula, and the two stars Alpha and Beta, which mark the feather of the arrow, lie nearly midway between Beta Cygni and Altair, while its point is marked by Gamma, 5° farther east and north. Beta, the middle star of the shaft of the arrow, is a very pretty double star, distance about 8″: the larger star is itself a close double.

71. A'quila (not *A-qui'la*) (Map IV.). August. — This constellation lies on the celestial equator, east of Ophiuchus and north of Sagittarius and Capricornus. Its characteristic configuration is that formed by Alpha, *Altair*, with Gamma to the north and Beta to the south. It lies about 20° south of Beta

Cygni, and forms a fine triangle with Beta and Alpha Ophiuchi. Altair is taken as the *standard* first-magnitude star. Of course, several of those which are called first magnitude, like Sirius and Vega, are very much brighter than this, while others fall considerably below it.

Aquila was the bird of Jupiter, which he kept by the side of his throne and sent to bring Ganymede to him.

The southern part of the region allotted to Aquila on our maps has been assigned to *Antinoüs*, which is recognized on some celestial globes. The constellation existed even in Ptolemy's time, but he declined to adopt it. Hevelius has appropriated the eastern part of Antinoüs for his constellation of *Scutum Sobieski*.

72. Sagittarius, the Archer (Map IV.). August. — This, the tenth of the zodiacal constellations, contains no stars of the first magnitude, but a number of the second and third magnitude, which make it reasonably conspicuous. The most characteristic configuration is the little inverted " milk-dipper," formed by the five stars, Lambda, Phi, Sigma, Tau, and Zeta, of which the last four form the bowl, while Lambda (in the Milky Way) is the handle (see map). Delta, Gamma, and Epsilon, which form a triangle, right-angled at Delta, lie south and a little west of Lambda, the whole eight together forming a very striking group. There is a curious disregard of any apparent principle in the lettering of the stars of this constellation; Alpha and Beta are stars not exceeding in brightness the fourth magnitude, about 4° apart on a north and south line, and lying some 15° south and 5° east of Zeta (see map), while Sigma is now a bright second-magnitude star, strongly suspected of being irregularly variable. (The constellation contains an unusual number of known variables.) The Milky Way in Sagittarius is very bright and complicated in structure, full of knots and streamers and dark pockets, and containing many beautiful and interesting objects.

This constellation is said by many writers to commemorate the Centaur, Chiron, but the same constellation appears on the ancient zodiacs of Egypt and India, and it seems probable, therefore, that, like the Bull and the Lion, it was not representative of any particular individual.

73. Capricornus (Map IV.). September. — This, the eleventh of the zodiacal constellations, follows Sagittarius on the east. It has no bright stars, but the configuration formed by the two Alphas (a_1 and a_2) with each other and with Beta, 3° south, is characteristic, and not easily mistaken for anything else. The two Alphas, a pretty double to the naked eye, lie on the line drawn from Beta Cygni (at the foot of the cross) through Altair, and produced about 25°.

Some say that this constellation represents the god Pan, who was represented by the Greeks as having the legs of a goat and the head of a man. Others find in the goat, Amalthea (the foster-mother of the infant Jupiter), who is also, it will be remembered, represented in the constellation of Auriga.

74. Delphinus, the Dolphin (Map IV.). September. — This constellation, though small, is one of the ancient 48, and is unmistakably characterized by the rhombus of third-magnitude stars known as " Job's Coffin." It lies about 15° east of Altair. There are a few stars visible to the naked eye, in addition to the four that form the rhombus. Epsilon, about 3° to the southwest, is the only conspicuous one.

Gamma, at the northwest angle of the rhombus, is a very pretty double star. Beta is also a very close and rapid binary, beyond the reach of all but large telescopes.

This is the Dolphin that preserved the life of the musician, Arion, who was thrown into the sea by sailors, but carried safely to land upon the back of the compassionate fish, who loved his music.

75. Equuleus, the Little Horse (Map IV.). This little constellation, simply a horse's head, though still smaller than the Dolphin and less conspicuous, is also one of Ptolemy's. It lies about 20° due east of Altair, and 10° southeast of the Dolphin (see map).

76. Lacerta, the Lizard (Maps I. and IV.). This is one of
Hevelius's modern constellations, lying between Cygnus and Andromeda, with no stars above the 4½ magnitude, and of no importance for
our purposes.

77. Pe'gasus (not *Pe-gas'us*) (Map IV.). October. — This
winged horse covers an immense space. Its most notable configuration is the "great square," formed by the second-magnitude stars, Alpha (*Markab*), Beta, and Gamma Pegasi, in
connection with Alpha Andromedæ (sometimes lettered Delta
Pegasi), at its northeast corner. The stars of the square lie in
the body of the horse, which has no hindquarters. A line drawn
from Alpha Andromedæ through Alpha Pegasi, and produced
about an equal distance, passes through Xi and Zeta in the
animal's neck, and reaches Theta in his ear. Epsilon (or
Enif), the bright star 8° northwest of Theta, marks his nose.
The forelegs are in the northwestern part of the constellation
just east of Cygnus, and are marked, one of them by the stars
Eta and Pi, and the other by Iota and Kappa.

This is the winged horse which sprang from the blood of Medusa,
after Perseus had cut off her head. He fixed his residence on Mt.
Helicon, where he was the favorite of the Muses, and after being
tamed by Minerva he was given to Bellerophon to aid him in conquer-
ing the Chimæra. After the destruction of the monster, Bellerophon
attempted to ascend to heaven upon Pegasus, but the horse threw off
his rider, and continued his flight to the stars.

78. Aquarius, the Water-bearer (Map IV.). October. —
This, the twelfth and last of the zodiacal constellations,
extends more than $3\frac{1}{2}^{h}$ in right ascension, covering a considerable region which by rights ought to belong to Capricornus. The most notable configuration is the little Y of
third and fourth magnitude stars which marks the "water-
jar" from which Aquarius pours the stream that meanders
down to the southeast and south for 30°, till it reaches the
Southern Fish. The middle of the Y is about 18° south and

west of Alpha Pegasi, and lies almost exactly on the celestial
equator.

Zeta, the central star of the Y, is a pretty and interesting double
star, distance about 4″. The green nebula, nearly on the line from
Alpha through Beta, produced about its own length, 1¼° west of Nu,
is a planetary nebula, and curious from the vividness of its color (see
map).

There are various opinions respecting the origin of this constellation. According to a Greek legend it represents Deucalion, the hero
of the Greek Deluge; but among the Egyptians it evidently had reference to the rising and falling of the Nile.

79. Piscis Austrinus (or **Australis**), **the Southern Fish** (Map
IV.). October. — This small constellation, lying south of Capricornus and Aquarius in the stream that issues from the
Water-bearer's urn, presents little of interest. It has one
bright star, *Fomalhaut* (pronounced *Fōmalo*), of the 1½ magnitude, which is easily recognized from its being nearly on the
same hour-circle with the western edge of the great square of
Pegasus, 45° to the south of Alpha Pegasi, and solitary, having no star exceeding the fourth magnitude within 15° or 20°.

This constellation is by some said to represent the transformation
of Venus into a fish, when fleeing from Typhon (but see Pisces).

South of the Southern Fish, barely rising above the southern horizon, lie the constellations of *Microscopium* and *Grus*. The former is
of no account. In the southern hemisphere Grus is a conspicuous
constellation, and its two brightest stars, Alpha and Beta, of the second magnitude, rise high enough to be seen in latitudes south of
Washington. They lie about 20° south and west of Fomalhaut.

LIST OF CONSTELLATIONS, SHOWING THEIR POSITION IN THE HEAVENS, AND THE NUMBER OF STARS IN EACH.

(The zodiacal constellations are denoted by italics, non-Ptolemaic constellations by an asterisk.)

R.A. (h.)	+90° to +50°	+50° to +25°	+25° to 0°	0° to -25°	-25° to -50°	-50° to -90°
I, II	Cassiopeia, 46	Andromeda, 18; Triangulum, 5	*Pisces*, 18; *Aries*, 17	Cetus, 32	Phœnix, 32; *App. Sculp. 13	Phœnix, *bis.* 18; Hydrus, 13
III, IV	—	Perseus, 40	*Taurus*, 58	Eridanus, 64	(Eridanus, *bis.*)	*Horologium, 11; *Reticulum, 9
V, VI	*Camelopardus, 36	Auriga, 35	Orion, 37; *Gemini*, 28	Lepus, 18	*Columba, 15	*Dorado, 16; *Pictor, 14; *Mons Mensæ, 12
VII, VIII	—	*Lynx, 28	Canis Minor, 6; *Cancer*, 15	Canis Major, 27; *Monoceros, 12	Argo-Navis, 133	Argo-Navis, *bis.* (Puppis) 9; *Piscis Volans,
IX, X	—	*Leo Minor, 15	*Leo*, 47	Hydra, 49; *Sextans, 3	—	Argo-Navis (Vela)
XI, XII	Ursa Major, 53	—	*Coma Ber. 20	Crater, 9; Corvus, 8	Centaurus, 54	Argo-Navis (Carina); *Chameleon, 13
XIII, XIV	—	*Canes Venat. 15; Boötes, 35	—	*Virgo*, 39	Lupus, 34	Centaurus, *bis.*; *Crux, 13; *Musca, 15
XV, XVI	Ursa Minor, 23	Corona Bor. 19; Hercules, 65	Serpens, 23	*Libra*, 23	Norma, 14	*Circinus, 10
XVII, XVIII	Draco, 80	Lyra, 18	Aquila, 37; Sagitta, 5	*Scorpio*, 34; Ophiuchus, 46	Ara, 15	*Triangul. Aust. 11; *Apus, 15
XIX, XX	—	Cygnus, 67	*Vulpecula, 23; Delphinus, 10	*Sagittarius*, 38	Corona Austr. 7	*Telescopium, 16; Pavo, 37; *Octans, 22
XXI, XXII	Cepheus, 44	*Lacerta, 13	Equuleus, 5	*Capricornus*, 22	Piscis Austr. 10	*Indus, 15; *Octans,
XXIII, XXIV	—	—	Pegasus, 43	*Aquarius*, 43	*Grus, 30	*Toucana, 22; *Octans.

CHAPTER III.

LATITUDE, AND THE ASPECT OF THE CELESTIAL SPHERE. — TIME. — LONGITUDE. — THE PLACE OF A HEAVENLY BODY.

80. Latitude defined. — In geography the latitude of a place is usually defined simply as its distance north or south of the equator, measured in degrees. This is not explicit enough, unless it is stated how the degrees themselves are to be measured. There would be no difficulty if the earth were a perfect sphere; but since the earth is a little flattened at the poles, the degrees (geographical) are of somewhat different lengths at different parts of the earth. The exact definition of the astronomical latitude of a place is the *angle between the direction of the observer's plumb-line and the plane of the earth's equator;* and this is the same as the *altitude of the pole,* as will be clear from Fig. 6. Here the angle *ONQ* is the latitude as defined. If now at *O* we draw *HH'* perpendicular to *OZ*, it will

FIG. 6. — Relation of Latitude to the Elevation of the Pole.

be a level line, and will point to the horizon. From *O* also draw *OP''*, parallel to *CP'*, the earth's axis. Since *OP'* and *CP''* are parallel they will be directed apparently to the same point in the celestial sphere (Art. 6), and this point is the

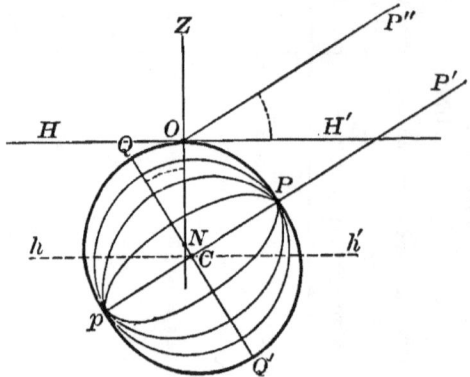

celestial pole. The angle $H'OP'''$ is therefore the altitude of
the pole, as seen at O, and it obviously equals ONQ; and this
is true whether the earth be a sphere, or whatever its form.
This fundamental relation, THAT THE ALTITUDE OF THE POLE
IS IDENTICAL WITH THE OBSERVER'S LATITUDE, cannot be too
strongly impressed on the mind.

81. Method of measuring the Latitude. — The most obvious
method is to observe, with a suitable instrument, the altitude
of some star near the pole (a "circumpolar" star) at the
moment when it is crossing the meridian above the pole, and
again twelve hours later, when it is once more on the meridian
below the pole. In the first position, its elevation is the greatest possible; in the second, the least. The average of these two
altitudes, when corrected for refraction, is the latitude of the
observer. It is exceedingly important that the student understand this simple method of determining the latitude.

The instrument ordinarily used for making observations of this
kind at an observatory is called a meridian circle, and a brief description is given in the Appendix (see Art. 418).

82. Refraction. — When we observe the altitude of a heavenly body with any instrument, we do not find it as it would
be if our atmosphere had no effect upon the rays of light. As
these rays enter the earth's atmosphere they are bent downward by "refraction," excepting only such as come from
exactly overhead. Since the observer sees the object in the
direction in which the rays *enter the eye*, without any reference
to its real position, this bending down of the rays causes every
object seen through the air to look *higher up in the sky* than
it would be if the air were absent; and we must therefore
correct the observed altitude by diminishing it a certain
amount. Under ordinary conditions, refraction elevates a
body at the horizon about $35'$, so that the sun and moon in
rising appear clear of the horizon while they are still wholly

below it. The refraction correction diminishes very rapidly
as the body rises. At an altitude of only 5° the refraction is
but 10′; at 44°, it is about 1′; and at the zenith, zero, of
course.

Its amount at any given time is affected quite sensibly, however, by
the temperature and by the height of the barometer, increasing as the
thermometer falls or as the barometer rises; so that whenever great
accuracy is required in measures of altitude we must have observations of both the barometer and thermometer to go with the reading
of the circle. In works on Practical Astronomy tables are given by
which the refraction can be computed for an object at any altitude
and in any state of the weather. It is hardly necessary to say that
this indispensable correction is very troublesome, and always involves
more or less error.

For other methods of determining the latitude, see Appendix,
Art. 424.

**83. Effect of the Observer's Latitude upon the Aspect of the
Heavens; the Right Sphere.** — If the observer is situated at the
earth's equator, — i.e., in latitude zero, — the celestial poles will
evidently be on the horizon, and the celestial equator will pass
through the zenith and coincide with the prime vertical (Art.
11). At the earth's equator, therefore, all heavenly bodies
will rise and set vertically, and their diurnal circles will be
equally divided by the horizon, so that they will be twelve
hours above it and twelve hours below it, and the length of
the night will always equal that of the day. This aspect of
the heavens is called the *right sphere*.

84. Parallel Sphere. — If the observer is at one of the poles
of the earth, where the latitude equals 90°, then the corresponding celestial pole will be exactly overhead, and the
celestial equator will coincide with the horizon. If he is at
the north pole, all the stars north of the celestial equator will
remain permanently visible, never rising or setting, but sailing
around the sky on parallels of altitude, while the stars south

of the equator will never rise to view. Since the sun and the moon move in such a way that during half the time they are north of the equator and half the time south of it, they will therefore be half the time above the horizon and half the time below it (that is, *approximately*, since refraction has a noticeable effect). The moon will be visible for about a fortnight at a time, and the sun for about six months.

85. The Oblique Sphere. — At any station between the pole and the equator the pole will be elevated above the horizon, and the stars will rise and set in oblique circles, as shown in Fig. 7. Those stars whose distance from the elevated pole is less than *PN*, the latitude of the observer, will never set, the radius of this "*circle of perpetual apparition*" being just equal to the height of the pole, and becoming larger as the latitude increases. On the other hand, stars within the same distance of the depressed pole will lie within the "*circle of perpetual occultation*," and will never rise above the observer's horizon. An object which is exactly on the celestial equator will have its diurnal circle, *EQWQ'*, equally divided by the horizon, and will be above the horizon just as long as below it.

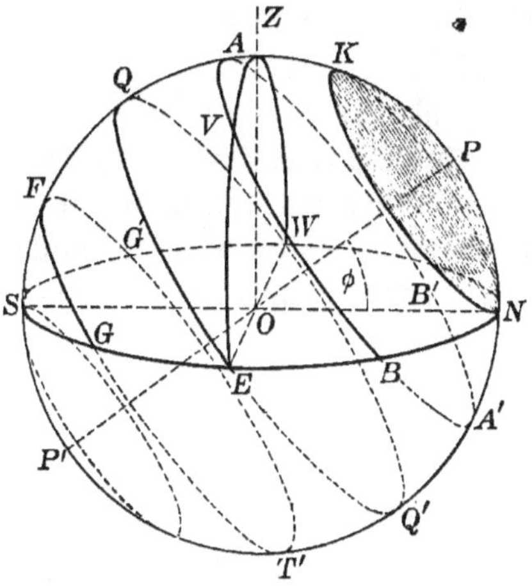

For an observer in the United States, a star north of the equator will have more than half of its diurnal circle above the horizon, and will be visible for more than twelve hours of each day; as, for instance, the star at *A*. Whenever the sun is north of the celestial equator, the day will therefore be longer

Fig. 7. — The Oblique Sphere.

than the night for all stations in northern latitude: how much longer will depend both on the latitude of the place and the sun's distance from the equator (its declination).

86. Moreover, when the sun is north of the equator, it will, in the northern latitudes, rise at a point *north* of east, as at *B* in the figure, and will continue to shine on the north side of every wall that runs east and west until, as it ascends, it crosses the prime vertical, *EZW*, at some point, as *V*. In the latitude of New York the sun in June is south of the prime vertical for only about six hours of the whole fifteen during which it is above the horizon. During nine hours of the day, therefore, it shines into north windows.

If the latitude of the observer is such that *PN*, in the figure, is greater than the sun's polar distance at the time when it is farthest north, the sun at midsummer will make a complete circuit of the heavens without setting, thus producing the phenomenon of the "midnight sun," visible at the North Cape and at all stations within the Arctic Circle.

87. A celestial globe will be of great use in studying these diurnal phenomena. The north pole of the globe must be elevated to an angle equal to the latitude of the observer, which can be done by means of the degrees marked on the metal meridian ring. It will then be seen at once what stars never set, which ones never rise, and during what part of the twenty-four hours any heavenly body at a known distance from the equator is above or below the horizon. For description of the celestial globe, see Appendix, Art. 400.

TIME.

Time is usually defined as "measured duration," and the standard unit of time has always been obtained in some way from the length of the day.

88. Apparent Solar Time. — The most natural way, since we are obliged to regulate our lives by the sun, is to reckon time by him; *i.e.*, to call it noon when the sun is on the meridian and highest, and to divide the day from one noon to another into its hours, minutes, and seconds. Time thus reckoned is called *apparent solar time* (see Appendix, Art. 422), which is the time shown by a correctly adjusted sundial. But because the sun's eastward motion in the sky is not uniform (owing to the oval form of the earth's orbit, and its inclination to the equator), these apparent solar days are not exactly of the same length. Thus, for instance, the interval from noon of Dec. 22d to noon of Dec. 23d is nearly a minute longer than the interval between the noons of Sept. 15th and 16th. As a consequence, it is only by very complicated and expensive machinery that a watch or clock can be made to keep time precisely with the sundial, and the attempt was long ago given up. Apparent solar time is now used only in communities where clocks and watches are rare, and sundials are the usual timepieces.

89. Mean Solar Time. — At present, for civil and business purposes, time is almost universally reckoned in days all of which have precisely the same length, and are just equal to the *average* apparent solar day; and this time, called *mean solar time* (Appendix, Art. 422), is that which is kept by all good timepieces.

Sundial time agrees with mean time four times a year; viz., upon April 15th, June 14th, Sept. 1st, and Dec. 24th. The greatest differences occur on Nov. 2d and Feb. 11th, when the sundial is respectively 16^m 20^s fast of the clock and 14^m 30^s slow. During the summer the difference never exceeds 6^m. This difference is called the *Equation of Time*, and is given in the almanac for every day in the year.

90. The Civil Day and the Astronomical Day. — The astronomical day begins at noon; the civil day at midnight, twelve hours earlier. Astronomical mean time is reckoned around through the whole twenty-four hours, instead of being counted in two series of twelve hours each. Thus 8 A.M. of Tuesday, Aug. 12th, civil reckoning, is Monday, Aug. 11th, 20^h, astronomical reckoning. Beginners need to bear this in mind in referring to the almanac.

91. Sidereal Time, or *Time reckoned by the Stars.* — As has been said (Art. 17), the sun is not fixed on the celestial sphere, but appears to creep completely around it once a year. The interval from noon to noon does not therefore correspond to the *true* diurnal revolution of the heavens. If we reckoned by the interval between two successive passages of any given star across the observer's meridian, we should find that the true day, the *sidereal* day, as it is called, is nearly 4^m shorter (3^m 56^s.9) than the ordinary solar day; the relation being such that *in a year the number of sidereal days exceeds that of solar by exactly one.* For many purposes, astronomers find it much more convenient to reckon by the stars than by the sun. They count the time, however, not by any real star, but from the *Vernal Equinox*, the sidereal clock being so set and regulated that it always shows zero hours, minutes, and seconds, at the moment when the Vernal Equinox is on the meridian (see Appendix, Art. 422). *Sidereal time*, of course, would not answer for business purposes, since its noon comes at all hours of the day and night at different seasons of the year. The almanac gives data by which sidereal time and mean solar time can be easily converted into each other.

92. The Determination of Time. — In practice, the problem always takes the shape of finding the *error* of a timepiece of some sort; *i.e.*, ascertaining how many seconds it is fast or slow. The instrument now ordinarily used for the purpose is the transit instrument, which is a small telescope mounted on an axis, placed exactly east and west, and level, so that as the telescope is turned it will follow the meridian; at least, the middle cross-wire in the field of view will do so. It is the

same as the meridian circle, except that it does not require
the costly graduated circle with its appendages. For description, see Appendix, Art. 416.

To determine with the transit the error of the sidereal clock
which is ordinarily used in connection with it, it is only necessary to observe the exact time indicated by the clock when
some star whose right ascension is known passes, or "transits," the middle wire of the instrument.

93. The right ascension of a star (Art. 18) is the number of
'hours' of arc (measured along the equator) by which the star
is east of the vernal equinox; and therefore when the star is
on the meridian, the right ascension also equals the number
of hours, minutes, and seconds since the transit of the vernal
equinox. In other words, we may say that the *right ascension
of a star is the sidereal time at the moment of its meridian transit.* (This is often called the observatory definition of right
ascension.) For instance, the right ascension of Vega (Alpha
Lyræ) is 18^h 33^m. If we observe its transit to occur at 18^h 40^m
by the clock, the clock is obviously 7^m fast.

With a good instrument, a skilled observer can thus determine the clock-error within about $\frac{1}{30}$ of a second of time.

To get *solar* time, we may observe the sun itself, the moment
of its transit being 'apparent noon.' But it is better, and it
is usual, to get the sidereal time first, and to deduce from that
the solar time by means of the necessary data which are furnished in the almanac.

The method by the transit instrument is most used, and is, on the
whole, the most convenient; but since the instrument requires to be
mounted upon a firm pier, it is not always available. When not, we
use some one of various other methods, for which reference must be
made to the General Astronomy. At sea, and by travellers on scientific expeditions, the time is usually determined by observing the altitude of the sun with a sextant some hours before or after noon. (See
Appendix, Art. 427.)

LONGITUDE.

94. The problem of finding the longitude is in many respects the most important of what may be called the "economic" problems of astronomy; *i.e.*, those of business utility to mankind. The great observatories of Greenwich and Paris were founded for the express purpose of furnishing the necessary data to enable the sailor to determine his longitude at sea; and the English government has given great prizes for the construction of clocks and chronometers fit to be used in such determinations.

The longitude of a place on the earth is defined as *the arc of the equator intercepted between the meridian which passes through the place and some meridian which is taken as the standard.*[1]

Now since the earth turns on its axis at a uniform rate, this arc is strictly proportional to, and may be measured by, the time intervening between the transits of any given star across the two meridians. The longitude of a place may therefore be defined as the *amount by which the time at Greenwich is earlier or later than the time at the station of the observer;* and this whether we reckon by solar or by sidereal time. Accordingly, terrestrial longitude is usually reckoned in hours, minutes, and seconds, rather than in degrees. Since the observer can easily find his own local time by the transit instrument, or by some of the many other methods, the *knot* of the problem is simply this : *To find the Greenwich time at any moment without going to Greenwich;* then we get the longitude at once by merely comparing it with our own time.

95. Methods of determining Longitude. — Incomparably the best method, whenever it is available, is to make a direct *tele-*

---

[1] As to the standard meridian, there is a variation of usage among different nations. The French reckon from the meridian of Paris, but most other nations use the meridian of Greenwich, at least at sea.

graphic comparison between the clock of the observer and that of some station, the longitude of which is known. The difference between the two clocks, duly corrected for their 'errors' (Art. 92), will be the true difference of longitude. The astronomical difference of longitude between the two places can thus be determined by four or five nights' observations within about 0.*02 — *i.e.*, within twenty feet or so, in the latitude of the United States. In many cases the telegraphic method, however, is not available ; never at sea, of course.

96. A second method is to use a *chronometer*, which is simply a very accurate watch. This is set to Greenwich time at some place whose longitude is known, and afterwards is supposed to keep that time wherever carried. The observer has only to compare his own local time, determined with the transit instrument or sextant, with the time shown by such a chronometer, and the difference is his longitude from Greenwich. This is the ordinary method at sea.

Practically, of course, no chronometer goes absolutely without gaining or losing; hence, it is always necessary to know and to allow for its gain or loss since the time it was last set. Moreover, it is never safe to trust a single chronometer, because of the liability of such instruments to change their rate in transportation. A number should be used, if possible.

Before the days of telegraphs and chronometers, astronomers were generally obliged to get their Greenwich time *from the moon*, which may be regarded as a clock-hand with the stars for dial figures ; but observations of this kind are troublesome, and the results inaccurate, as compared with those obtained by the telegraph and chronometer. (For further details, see General Astronomy, Arts. 109–116.)

97. Local and Standard Time. — Until recently it has been always customary to use *local* time, each station determining its own time by its own observations, and having, therefore,

a time differing from that of all other stations not on the same
meridian. Before the days of the telegraph, and while travelling was comparatively slow, this was best. At present there
are many reasons why it is better to give up the old system in
favor of a system of *standard time.* The change greatly facil-
itates all railway and telegraphic business, and makes it practically easy for everybody to have accurate time, since the
standard time can be daily wired from some headquarters to
every telegraph office.

According to the system now established in North America,
there are five such standard times in use, — the colonial, the
eastern, the central, the mountain, and the Pacific, — which
differ from Greenwich time by exactly four, five, six, seven,
and eight hours respectively, the minutes and seconds being
everywhere identical, and the same with those of the clock at
Greenwich. In order to determine the standard time by observation, it is only necessary to find the local time by one of the
methods given, and correct it according to the observer's longitude from Greenwich.

98. Where the Day begins. — It is clear that if a traveller
were to start from Greenwich on Monday noon, and travel
westward as fast as the earth turns to the east beneath his
feet, he would have the sun upon the meridian all day long,
and it would be continual noon. But what noon? It was
Monday when he started, and when he gets back to London
twenty-four hours later it will be Tuesday noon there, and yet
he has had no intervening night. When did Monday noon
become Tuesday noon?

It is agreed among mariners to make the change of date at
the 180th meridian from Greenwich. Ships crossing this line
from the east skip one day in so doing. If it is Monday after-
noon when a ship reaches the line, it becomes Tuesday after-
noon the moment she passes it, the intervening twenty-four
hours being dropped from the reckoning on the log-book.

Vice versa, when a vessel crosses the line from the western side, it counts the same day twice, passing from Tuesday back to Monday.

This 180th meridian passes mainly over the ocean, hardly touching land anywhere. There is some irregularity as to the date actually used on the different islands of the Pacific. Those which received their earliest European inhabitants *via* the Cape of Good Hope, have, for the most part, adopted the Asiatic date, even if they really lie east of the 180th meridian, while those which were first approached *via* Cape Horn have the American date. When Alaska was transferred from Russia to the United States, it was necessary to drop one day of the week from the official dates.

DETERMINATION OF THE POSITION OF A HEAVENLY BODY.

As the basis of our investigations in regard to the motions of the heavenly bodies, we require a knowledge of their places in the sky at known times. By the "place" of a body, we mean its right ascension and declination.

99. By the Meridian Circle (see Appendix, Art. 418). — If a body is bright enough to be seen by the telescope of the meridian circle, and comes to the meridian in the night-time, its right ascension and declination are best determined by that instrument. If the instrument·is in exact adjustment, the *right ascension* of the body is simply the sidereal time when it crosses the middle vertical wire of the reticle. The 'circle-reading,' on the other hand, corrected for refraction, gives the *declination*. A single complete observation with the meridian circle determines accurately both the right ascension and the declination of the object.

100. By the Equatorial. — If the body — a comet, for instance — is too faint to be observed by the telescope of the meridian circle, seldom very powerful, or comes to the meridian only in the daytime, we usually accomplish our object by using the equatorial (Appendix, Art. 414), and determine the position of the body by measuring with some kind of 'micrometer' the difference of right ascension and declination between it and a neighboring star whose place is given in some star-catalogue.

CHAPTER IV.

THE EARTH : ITS FORM AND DIMENSIONS ; ITS ROTATION,
MASS, AND DENSITY; ITS ORBITAL MOTION AND THE
SEASONS. — PRECESSION. — THE YEAR AND THE CAL-
ENDAR.

101. In a science which deals with the 'heavenly bodies,'
there might seem at first no place for the Earth. But certain
facts relating to the Earth, just such as we have to investigate with respect to her sister planets, are ascertained by astronomical methods, and a knowledge of them is essential as a
base of operations. In fact, Astronomy, like charity, "begins
at home," and it is impossible to go far in the study of the
bodies which are strictly "heavenly" until we have first acquired some accurate knowledge of the dimensions and motions
of the Earth itself.

102. The astronomical facts relating to the Earth are
broadly these : —

1. The earth is a great ball about 7920 miles in diameter.

2. It rotates on its axis once in twenty-four "sidereal"
hours.

3. It is not exactly spherical, but is slightly flattened at the
poles ; the polar diameter being nearly twenty-seven miles, or
about $\frac{1}{300}$ part less than the equatorial.

4. It has a mean density of about 5.6 times that of water,
and a mass represented in tons by 6 with twenty-one ciphers
following, (six thousand millions of millions of millions of
tons.)

5. It is flying through space in its orbital motion around
the sun, with a velocity of about eighteen and a half miles a

second; *i.e.*, about seventy-five times as swiftly as an ordinary cannon-ball.

103. The Earth's Approximate Form and Size. — It is not necessary to dwell on the ordinary proofs of the earth's globularity. We simply mention them.

1. It can be sailed around.

2. The appearance of vessels coming in from the sea indicates that the surface is everywhere *convex*.

3. The fact that as one goes from the equator towards the north the elevation of the pole increases in proportion to the distance from the equator, proves the same thing.

4. *The outline of the earth's shadow, as seen upon the moon during lunar eclipses, is such as only a sphere could cast.*

We may add, as to the smoothness and roundness of the earth, that if the earth be represented by an eighteen-inch globe, the difference between its greatest and least diameters would be only about one-sixteenth of an inch; the highest mountains would project only about one-seventieth of an inch, and the average elevation of continents and depths of the ocean would be hardly greater than a film of varnish. Relatively, the earth is really much smoother and rounder than most of the balls in a bowling-alley.

104. One of the simplest methods of showing the curvature of the earth is the following : —

In an expanse of still, shallow water (a long reach of canal, for instance), set a row of three poles about a mile apart, with their tops projecting to exactly the same height above the surface. On sighting across, it will then be found that the middle pole projects about eight inches (when refraction has been allowed for) above the line that joins the tops of the two end ones, and from this a rough estimate of the size of the earth can be made (see General Astronomy, Art. 134).

105. Measure of the Earth's Diameter. — The only accurate method of measuring the diameter of the earth is the following, the principle of which is very simple, and should be thoroughly mastered by the student: —

It consists in finding the *length in miles* of an arc of the earth's surface containing a *known number of degrees.* From this we get the length of one degree, and this gives the circumference of the earth (since it contains 360°), and from this the diameter is obtained by dividing it by 3.14159.

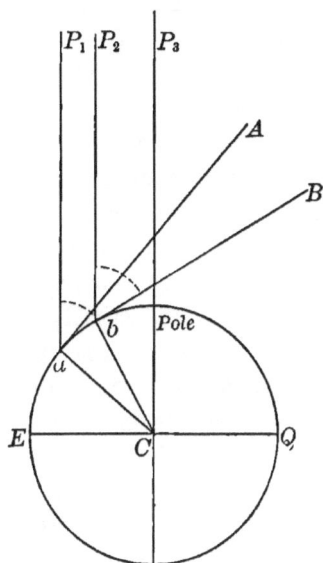

FIG. 8. — Measuring the Earth's Diameter.

To do this, we select two stations, *a* and *b* (Fig. 8), some hundreds of miles apart on the same meridian, and determine the latitude (or the altitude of the pole) at each station by astronomical observation. The difference of latitude (*i.e.*, *ECb* − *ECa*) is evidently the number of degrees in the arc *ab*, and the determination of this difference of latitude is the only *astronomical* operation necessary.

Next, the distance in miles between the two stations must be measured. This is *geodetic* work, and it is enough for our purpose here to say that it can be done with great precision by a process which is called 'triangulation.'

This measuring of arcs has been done on many parts of the earth's surface, and the result is that the *average* length of a degree is found to be a little more than sixty-nine miles, and the mean diameter of the earth about 7918 miles. The reason why we say average length and mean diameter is that the earth, as has been said, is not quite a globe, but is slightly flattened at its poles, so that the lengths of the degrees differ in different parts of the earth, as we shall soon see (Art. 110).

106. The Rotation of the Earth. — Ptolemy understood that the earth was *round*, but he and all his successors deliberately rejected the theory of its rotation. Though the idea that the earth might turn upon an axis was not unfamiliar, they considered that there were conclusive reasons against it. At the time when Copernicus of Thorn, in Poland (1473–1543), proposed his theory of the solar system, the only argument he could urge in favor of the earth's rotation[1] was that this hypothesis was much more *probable* than the older one that the heavens themselves revolve. All the phenomena then known would be sensibly the same on either supposition. The apparent daily motion of the heavenly bodies can be perfectly accounted for (within the limits of such observations as were then possible) either by supposing that they are actually attached to the celestial sphere, which turns daily, or that the earth itself spins upon an axis once in twenty-four hours; and for a long time the latter hypothesis did not seem to most people so reasonable as the older and more obvious one. A little later, after the telescope had been invented, *analogy* could be appealed to; for we can see with the telescope that the sun and moon and many of the planets really rotate upon axes. At present we can go still farther, and can absolutely demonstrate the earth's rotation by experiments, some of which even make it visible.

107. Foucault's Pendulum Experiment. — Among these experimental proofs, the most impressive is the "pendulum experiment" devised by Foucault in 1851. From the dome of the Pantheon, in Paris, he hung a heavy iron ball by a slender wire more than 200 feet long (Fig. 9). A circular rail,

---

[1] The word *rotation* denotes a spinning motion, like that of a wheel on its axis. The word *revolve* is more general, and may be used to describe such a spinning motion, or (and this is the more common use in Astronomy) to describe the motion of a body travelling around another, as when we say the earth 'revolves' around the sun.

with a little ridge of sand built upon it, was placed in such a way that a pin attached to the swinging ball would just scrape the sand and leave a mark at each vibration. To put the ball in motion, it was drawn aside by a cotton cord and left for some hours, until it came absolutely to rest. Then the cord was *burned off*, and the pendulum started to swing in a true plane.

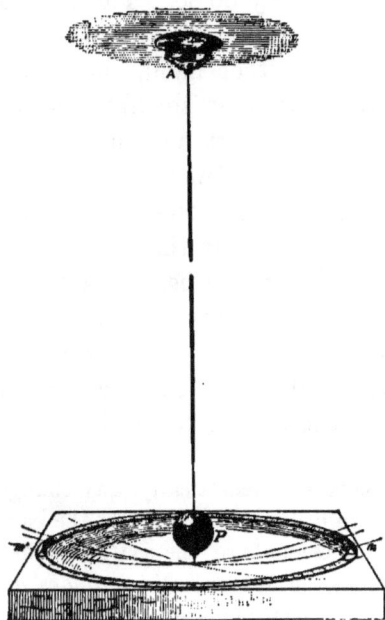

FIG. 9. — Foucault's Pendulum Experiment.

But this plane at once began to deviate slowly towards the right, so that the pin on the pendulum ball cut the sand ridge in a new place at each swing, shifting at a rate which would carry the line fully around in about thirty-two hours, if the pendulum did not first come to rest. In fact, the floor was actually and visibly turning under the plane defined by the swinging of the pendulum.

The experiment created great enthusiasm at the time, and has since been frequently performed. The pendulum used in such experiments must, in order to secure success, have a round ball, must be suspended by a round wire or on a point, and must be very heavy, very long, and very carefully protected against currents of wind. At the pole the plane of the pendulum will shift completely around once in twenty-four hours; at the equator, it will not turn at all; and in the intermediate regions, it will shift more or less rapidly according to the latitude of the place where the experiment is performed. (For fuller description, see General Astronomy, Arts. 140, 141.)

There are a number of other experimental proofs of the earth's rotation, which are really just as conclusive as the one above cited. (General Astronomy, Arts. 138–144.)

108. Invariability of the Earth's Rotation. — It is a question of great importance whether the day ever changes its length. Theoretically, it must almost necessarily do so. The friction of the tides and the fall of meteors upon the earth both tend to retard the rotation, while, on the other hand, the earth's loss of heat by radiation and the consequent shrinkage of the globe must tend to accelerate it, and to shorten the day. Then geological changes, the elevation and subsidence of continents, and the transportation of soil by rivers, act, some one way and some the other. At present we can only say that the change, if any change has occurred since Astronomy became accurate, has been too small to be detected. The day is certainly not longer or shorter by the $\frac{1}{100}$ part of a second than it was in the days of Ptolemy; probably it has not changed by the $\frac{1}{1000}$ part of a second, though of that we can hardly be sure.

109. Shiftings of the Earth's Axis. — Theoretically, any changes in the distribution of materials within or upon the globe of the earth ought to produce corresponding displacements of the axis, and these would principally show themselves as *variations in the latitudes and longitudes of observatories*. The actual variations are so minute, however, that it is only as recently as 1889 that they were first clearly detected by certain German observers, whose results have since been abundantly confirmed and extended. It is now beyond doubt that the earth really "wobbles" in whirling; and this causes each pole to describe an apparently irregular path around its mean position, never departing from it, however, by more than 40 or 50 feet. Dr. Chandler has shown that this motion is compounded of two, one circular, with a period of a year, the other oval, with a period of 428 days.

To explain certain geological phenomena it has been surmised that great and permanent displacements of the poles have occurred in the distant past. But of this, we have, as yet, no satisfactory evidence.

110. Effect of the Earth's Rotation on its Form. — The whirling of the earth on its axis tends to make the globe bulge at the equator and flatten at the poles, in the way illus-

trated by the well-known little apparatus shown in Fig. 10.
That the equator does really bulge in this way is shown by

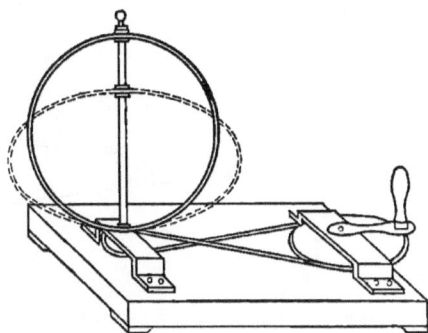

measuring *the length of a
degree of latitude on the various parts of the earth's
surface between the equator
and the pole*, in the manner
indicated a few pages back
(Art. 105). More than
twenty such arcs have been
measured, and it appears
that the length of the de-

FIG. 10.—Effect of Earth's Rotation on its Form.   grees increases regularly
from the equator towards the poles, as shown in the following
table : —

At the equator, one degree = 68.704 miles
At lat. 20°   "   "   = 68.786   "
"   "   40°   "   "   = 68.993   "
"   "   60°   "   "   = 69.230   "
"   "   80°   "   "   = 69.386   "
At the pole,   "   "   = 69.407   "

The difference between the equatorial and polar degree of
latitude is more than 0.7 of a mile, or over 3700 feet, while
the probable error of measurement cannot exceed a foot or
two to the degree.

From this table it can be calculated, by methods which
cannot be explained without assuming too much mathematical
knowledge in our readers, that the earth is orange-shaped, or
"an oblate spheroid," the diameter from pole to pole being
7899.74 miles, while the equatorial diameter is 7926.61 miles.
The difference, 26.87 miles, is about $\frac{1}{295}$ of the equatorial
diameter. This fraction, $\frac{1}{295}$, is called the *oblateness* or *ellipticity* of the earth.

Scholars are often puzzled by the fact that although the pole is
nearer the centre of the earth than the equator, yet the degrees of lat-

itude are *longest* at the pole. It is because the earth's surface there is more nearly flat than anywhere else, so that a person has to travel more miles to change the direction of his plumb-line one degree. Fig. 11 illustrates this. The angles *adb* and *fhg* are equal, but the arc *ab* is longer than *fy*.

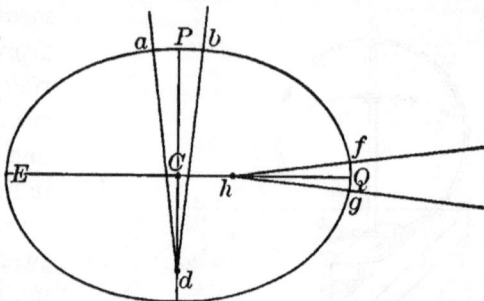

FIG. 11.—Length of Degrees in Different Latitudes.

111. Effect of the Earth's Rotation and Ellipticity upon the Force of Gravity. — For two reasons the force of gravity is less at the equator than at the poles. (1) The surface of the earth is there $13\frac{1}{2}$ miles farther from the centre, and this fact diminishes the gravity at the equator by about $\frac{1}{555}$. (2) The centrifugal force of the earth's rotation reduces the gravity at the equator by about $\frac{1}{289}$; the whole reduction, therefore ($\frac{1}{555}$ + $\frac{1}{289}$), is very nearly equal to $\frac{1}{190}$; *i.e.*, an object which weighs 190 pounds at the equator would weigh 191 pounds near the pole, — weighed by an accurate *spring-balance*. (In an ordinary balance, the loss of weight would not show, simply because the weights themselves would be affected as much as the body weighed, so that the balance would not be disturbed.) The effect of this variation of gravity from the pole to the equator is especially evident in the going of a pendulum clock. Such a clock, adjusted to keep accurate time at the equator, would gain 3ᵐ 37ˢ a day near the pole. In fact, one of the best ways of determining the form of the earth is by experiments with a pendulum at stations which differ considerably in latitude.

112. Surface and Volume of the Earth. — The earth is so nearly spherical that we can compute its surface and volume with sufficient accuracy by the formula for a perfect sphere, provided we put the earth's *mean* semi-diameter for the radius of the sphere. This mean semi-diameter is not the average of the polar and equatorial diameters, but is found by adding the polar diameter to twice the equatorial, and dividing by three. It comes out 7917.66 miles. From this we find the earth's surface to be, in round numbers, 197,000000 square miles, and its volume, or bulk, 260000,000000 cubic miles.

113. The Earth's Mass and Density. — The *volume* (or bulk) of a globe is simply the number of cubic miles of space which it contains. If the earth were all made of feathers or of lead, its volume would remain the same, as long as the diameter was not altered. The earth's *mass*, on the other hand, is the quantity of matter in it — the number of *tons* of rock and water which compose it, — and of course it makes a great difference with this whether the material be heavy or light. The *density* of the earth is the number of times its mass exceeds that of a sphere of pure water having the same dimensions.

The methods by which the mass of the earth can be measured depend upon a comparison between the attraction which the earth exerts upon a body at its surface and the attraction which is exerted upon the same body by another large body of known mass and at a known distance. The experiments are delicate and difficult, and we must refer for details to our larger book, General Astronomy, Arts. 164–179.

According to the best data at present available the earth's *density* is about 8.58, and its *mass* about 6000 millions of millions of millions of tons.

The most recent and valuable determinations are those made at Potsdam in 1888, and by Boys in England in 1893. Another is now in progress at Berlin.

114. Constitution of the Earth's Interior. — Since the average density of the earth's crust does not exceed three times that of water, while the mean density of the whole earth is about 5.58, it is clear that at the earth's centre the density must be very much greater than at the surface. Very likely it is as high as eight or ten times the density of water, and equal to that of the heavier metals.

There is nothing surprising in this. If the earth were once fluid, it is natural to suppose that the densest materials, in the process of solidification, would settle towards the centre.

Whether the centre of the earth is now solid or fluid, it is difficult to say with certainty. Certain tidal phenomena, to be mentioned hereafter, have led Lord Kelvin to conclude that the earth as a whole is probably solid throughout and "more rigid than glass," volcanic centres being mere "pustules," so to speak, in the general mass. To this most geologists demur, maintaining that at the depth of not many hundred miles the materials of the earth must be fluid, or at least semi-fluid. They infer this from the phenomena of volcanoes, and from the fact that the temperature continually increases with the depth, so far at least as we have yet been able to penetrate.

THE APPARENT MOTION OF THE SUN AND THE ORBITAL MOTION OF THE EARTH, AND THEIR IMMEDIATE CONSEQUENCES.

115. The Sun's Apparent Motion among the Stars. — The sun's apparent motion among the stars, which makes it describe the circuit of the heavens once a year, must have been among the earliest recognized astronomical phenomena, as it is one of the most important. The sun, starting in the spring, mounts northward in the sky each day at noon for three months, appears to stand still a few days at the summer solstice, and then descends towards the south, reaching in autumn the same noon-day elevation which it had in the spring. It keeps on its southward course to the winter solstice (in December), and then returns to its original height at the end of a year, by its course causing and marking the seasons.

Nor is this all. The sun's motion is not merely north and south, but it also *advances continually eastward* among the stars. It is true that we cannot see the stars near the sun in the

same way that we can those about the moon, so as to be able
directly to perceive this motion; but in the spring the stars
which are rising in the eastern horizon are different from
those which are found there in the summer or in the winter.
In March the most conspicuous of the eastern constellations
at sunset are Leo and Boötes. A little later Virgo appears; in
the summer Ophiuchus and Libra; still later Scorpio; while
in midwinter Orion and Taurus are ascending as the sun goes
down.

So far as the obvious appearances are concerned, it is quite
indifferent whether we suppose the earth to revolve around
the sun, or *vice versa*. That the earth really moves, however,
is absolutely demonstrated by two phenomena too minute and
delicate for observation without the telescope, but accessible
to modern methods. One of them is the *aberration of light*,
the other the *annual parallax of the fixed stars*. These can be
explained only by the actual motion of the earth, but we post-
pone their discussion for the present (see Art. 343, and Appendix, 435).

116. The Ecliptic; its Related Points and Circles. — By observing daily with the meridian circle the sun's declination
and the difference between its right ascension and that of
some standard star, we obtain a series of positions of the sun's
centre which can be plotted on the globe, and we can thus
mark out the path of the sun among the stars. It turns out
to be a great circle, as is shown by its cutting the celestial equator at two points just 180° apart (the so-called "equinoctial
points" or "equinoxes"), where it makes an angle with the
equator of approximately 23½° (23° 27' 14" in 1890).

This great circle is called the *Ecliptic*, because, as was early
discovered, eclipses happen only when the moon is crossing
it. Its position among the constellations is shown upon the
equatorial star-maps. It may be defined as the circle in which
the plane of the earth's orbit cuts the celestial sphere.

The angle which the ecliptic makes with the equator at the equinoctial points is called the *obliquity of the Ecliptic*. This obliquity is evidently equal to the sun's greatest distance from the equator; *i.e.*, its maximum declination, which is reached in December and June.

117. The two points in the ecliptic midway between the equinoxes are called the *solstices,* because at these points the sun "stands"; that is, ceases to move north or south. Two circles drawn through the solstices parallel to the equator are called the tropics, or " turning-lines," because there the sun turns from its northward motion to the southward, or *vice versa.* The two points in the heavens 90° distant from the ecliptic are called the poles of the ecliptic. The northern one is in the constellation of Draco, about midway between the stars Delta and Zeta Draconis, at a distance from the pole of the heavens equal to the obliquity of the ecliptic, or about $23\frac{1}{2}°$, and on the *Solstitial Colure,* the hour-circle which runs through the two solstices; the hour-circle which passes through the equinoxes being called the *Equinoctial Colure.* Great circles drawn through the poles of the ecliptic, and therefore perpendicular, or "secondaries," to the ecliptic, are known as circles of latitude. It will be remembered (Art. 20) that celestial longitude and latitude are measured with reference to the ecliptic, and not to the equator.

118. The Zodiac and its Signs. — A belt 16° wide (8° on each side of the ecliptic) is called the *Zodiac,* or zone of animals, the constellations in it, excepting Libra, being all figures of animals. It is taken of that particular width simply because the moon and all the principal planets always keep within it. It is divided into the so-called signs, each 30° in length, having the following names and symbols : —

Spring	Aries	♈		Autumn	Libra	♎
	Taurus	♉			Scorpio	♏
	Gemini	♊			Sagittarius	♐

Summer	Cancer	♋		Winter	Capricornus	♑
	Leo	♌			Aquarius	♒
	Virgo	♍			Pisces	♓

The symbols are for the most part conventionalized pictures of the objects. The symbol for Aquarius is the Egyptian character for water. The origin of the signs for Leo, Capricornus, and Virgo is not quite clear.

The zodiac is of extreme antiquity. In the zodiacs of the earliest history, the Fishes, Ram, Bull, Lion, and Scorpion appear precisely as now.

119. The Earth's Orbit. — The ecliptic must not be confounded with the earth's orbit. It is simply a great circle of the infinite celestial sphere, — the *trace* made upon that sphere by the plane of the earth's orbit, as was stated in its definition. The fact that the ecliptic is a great circle gives us no information about the earth's orbit itself, except that it *lies in one plane passing through the sun*. It tells us nothing as to the orbit's real form and size.

By reducing the observations of the sun's right ascension and declination through the year to longitude and latitude (the latitude would always be exactly zero except for some slight perturbations due chiefly to the moon's revolution around the earth), and combining these data with observations of the sun's apparent diameter, we can, however, ascertain the form of the earth's orbit and the law of its motion. The *size* of the earth's orbit, *i.e.*, its scale of miles, cannot be fixed until we find the sun's distance.

The result is that the orbit is found to be very nearly a circle, but not exactly so. It is an oval or ellipse, with the sun at one of its foci (as illustrated in Fig. 12), but is much more nearly circular than the oval there represented. Its eccentricity is only about $\frac{1}{60}$; that is to say, the distance from the centre of the sun to the middle of the ellipse is only about $\frac{1}{60}$ of the average distance of the sun from the earth.

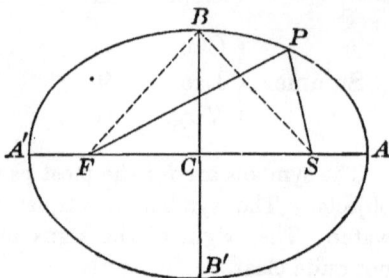

FIG. 12. — The Ellipse.

The method by which we proceed to ascertain the form of the orbit may be found in the Appendix, Art. 428. For a description of the ellipse, see Art. 429.

120. Definition of Terms. — The points where the earth is nearest to and most remote from the sun are called respectively the *Perihelion* and the *Aphelion* (Dec. 31st and June 30th), the line joining them being the *Major-axis* of the orbit. This line, indefinitely produced in both directions, is called the '*Line of Apsides*' (pronounced *Ap'-si-deez*), the major axis being a limited piece of it. A line drawn from the sun to the earth, or to any other planet at any point in its orbit, as SP in Fig. 12, is called the planet's *Radius Vector*.

The variations in the sun's apparent diameter due to our changing distance are too small to be detected without a telescope, so that the ancients failed to perceive them. Hipparchus, however, about 120 B.C., discovered that the earth is not in the centre[1] of the circular orbit which he supposed the sun to describe around it with uniform velocity.

Obviously the sun's apparent motion is not uniform, because it takes 186 days for the sun to pass from the vernal equinox,

---

[1] Hipparchus (and every one else until the time of Kepler, 1607) assumed on metaphysical grounds that the sun's orbit must necessarily be a circle, and described with a uniform motion, because (they said) the circle is the only *perfect* curve, and uniform motion is the only *perfect* motion proper for *heavenly* bodies.

March 20th, to the autumnal, Sept. 22d and only 179 days to
return.   Hipparchus explained this on the hypothesis that the
earth is out of the centre of the circle.

121. The Law of the Earth's Motion. — By combining the
measured apparent diameter of the sun with the differences of
longitude from day to day, we can deduce mathematically not

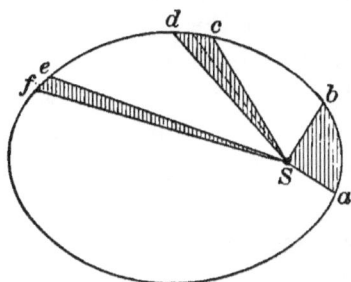

FIG. 13. — Equable Description of Areas.

only the form of the earth's orbit, but the law of her motion in
it.   It can be shown from the
comparison that the earth moves
in such a way that its *radius
vector describes areas proportional
to the time,* a law which Kepler
first brought to light in 1609;
that is to say, if *ab, cd, ef* (Fig.
13) be portions of the orbit de-
scribed by the earth in different weeks, the areas of the elliptical sectors *aSb, cSd,* and *eSf* are all equal.   A planet near
perihelion moves faster than at aphelion in just such proportion as to preserve this relation.

As Kepler left the matter, this is a mere fact of observation.
Newton afterwards proved that it is the necessary mechanical
consequence of the fact that the earth moves under the action
of a force always directed towards the sun.

It is true in every case of the elliptical motion of a heavenly body,
and enables us to find the position of the earth or of any planet, when
we once know the time of its orbital revolution (technically the
"period"), and the time when it was last at perihelion.   The solution
of the problem, first worked out by Kepler, lies, however, quite beyond
the scope of the present work.

122. Changes in the Earth's Orbit. — The orbit of the earth
changes slowly in form and position, though in the long run
it is absolutely unchangeable as regards the length of its
major axis and the duration of the year.

These so-called "secular changes" are due to "perturbations" caused by the action of the other planets upon the earth. Were it not for their attraction the earth would keep her orbit with reference to the sun and stars absolutely unaltered from age to age.

Besides these secular perturbations of the earth's *orbit*, the earth itself is also continually being slightly disturbed in its orbit. On account of its connection with the moon, it oscillates each month a few hundred miles above and below the true plane of the ecliptic, and by the action of the other planets is sometimes set backwards or forwards in its orbit to the extent of some thousands of miles. Of course every such displacement of the earth produces a corresponding slight change in the apparent position of the sun and of the nearer planets.

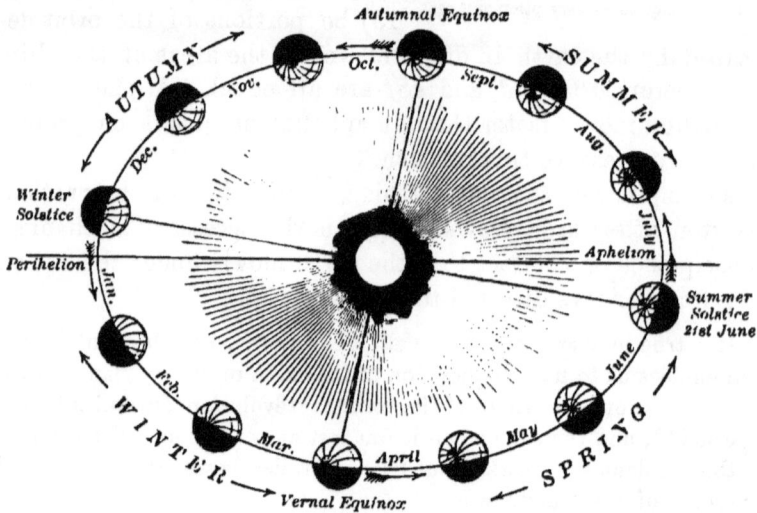

Fig. 14. — The Seasons.

123. The Seasons. — The earth in its motion around the sun always keeps its axis nearly parallel to itself during the whole year, for the mechanical reason that a spinning globe main-

tains the direction of its axis invariable, unless disturbed by
some outside force (very prettily illustrated by the gyroscope).   Fig. 14 shows the way in which the north pole of
the earth is tipped with reference to the sun at different
seasons of the year.   At the vernal equinox (March 20th) the
earth is situated so that the plane of its equator passes
through the sun.   At that time, therefore, the circle which
bounds the illuminated portion of the earth passes through
the two poles, as shown in Fig. 15, B, and day and night are
therefore equal, as implied by the
term 'equinox.' The same is again
true on the 22d of September.
About the 21st of June the earth is
so situated that its north pole is
inclined towards the sun by about
$23\frac{1}{2}°$, as shown in Fig. 15, A.   The
south pole is then in the unlighted
half of the earth's globe, while the north pole receives sunlight
all day long, and in all portions of the northern hemisphere
the day is longer than the night.   In the southern hemisphere, on the other hand, the reverse is true.

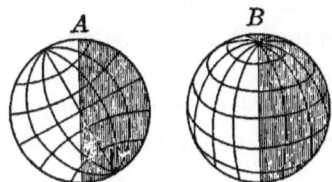

FIG. 15. — Position of Pole at Solstice
and Equinox.

At the time of the winter solstice the southern pole has perpetual sunshine, and the north pole is in the night.   At the
equator of the earth, day and night are equal at all times of
the year, and at that part of the earth there are no seasons in
the proper sense of the word.   Everywhere else the day and
night are unequal, except when the sun is at one of the
equinoxes.

In high latitudes the inequality between the lengths of the
day in summer and in winter is very great; and at places
within the polar circle there are always days in winter when
the sun does not rise at all, and others in the summer when it
does not set, but we have the phenomenon of the "midnight
sun," as it is called.   At the pole itself, the summer is one
perpetual day, six months in length, while the winter is a six-
months night.

Perhaps the student will get a better idea by thinking of the earth as a globe floating, just half immersed, on a sheet of still water, and so weighted that its poles dip at an angle of $23\frac{1}{2}°$, while it swims in a circle around the sun, a much larger globe, also floating on the same surface. The sheet of water corresponds to the ecliptic, while the plane of the equator is a circle on the globe itself, drawn square to the axis. If, now, the axis is kept pointing always the same way while the globe swims around, things will correspond to the motion of the earth around the sun.

124. Effects on Temperature. — The changes in the duration of *insolation* (exposure to sunshine) at any place involve changes of temperature, thus producing the seasons. It is clear that the surface of the soil at any place in the northern hemisphere will receive daily from the sun more than the average amount of heat whenever he is north of the celestial equator, and for two reasons: —

1. Sunshine lasts more than half the day.

2. The mean altitude of the sun during the day is greater than the average for the year, since he is higher at *noon* than at the time of the equinox, and in any case reaches the horizon at rising and setting.

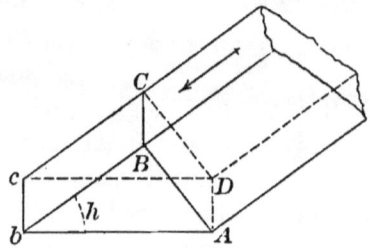

FIG. 16.

Effect of Sun's Elevation on Amount of Heat imparted to the Soil.

Now the more obliquely the rays strike, the less heat they bring to each square inch of surface, as is obvious from Fig. 16. A beam of sunshine which would cover the surface AC, if received squarely, will be spread over a much larger surface, Ac, if it falls at the angle h. The difference in favor of vertical rays is further exaggerated by the absorption of heat in our atmosphere, because the rays that are nearly horizontal have to traverse a much greater thickness of air before reaching the ground.

For these two reasons, therefore, the temperature rises

rapidly for a place in the northern hemisphere as the sun comes north of the equator. We, of course, receive the most heat in twenty-four hours at the time of the summer solstice; but this is not the hottest time of the summer. The weather is then getting hotter, and the maximum will not be reached until the increase ceases; *i.e.*, not until. the amount of heat lost in twenty-four hours equals that received in the same time. This maximum is reached in our latitude about the 1st of August. For similar reasons the minimum temperature in winter occurs about Feb. 1st.

125. Precession of the Equinoxes. — This is a slow westward motion of the equinoxes along the ecliptic. In explaining the seasons we have said (Art. 123) that the earth keeps its axis nearly parallel to itself during its annual revolution. It does not maintain strict parallelism, however, but owing to the attraction of the sun and moon on that portion of the mass of the earth which projects, like an equatorial ring, beyond the true spherical surface, the earth's axis continually but slowly shifts its place, keeping always nearly the same inclination to the plane of the ecliptic, so that its pole revolves in a small circle of $23\frac{1}{2}°$ radius around the pole of the ecliptic once in 25,800 years. Of course the celestial equator must move also, since it has to keep everywhere just 90° from the celestial pole; and, as a consequence, the equinoxes *move westward* on the ecliptic about $50''.2$ each year, as if to meet the sun. This motion of the equinox was called 'precession' by Hipparchus, who discovered [1] it about 125 B.C., but could not explain it. The explanation was not reached until the time of Newton, about 200 years ago.

126. Effect of Precession upon the Pole and the Zodiac. — At present the Pole-star, Alpha Ursæ Minoris, is about $1\frac{1}{4}°$

---

[1] He discovered it by finding that in his time the place of the equinox among the stars was no longer the same that it used to be in the days of Homer and Hesiod, several hundred years before.

from the pole, while in the time of Hipparchus the distance was fully 12°. During the next century the distance will diminish to about 30′, and then begin to increase.

If upon the celestial globe we trace a circle of 23½° radius, around the pole of the ecliptic as a centre, it will mark the track of the celestial pole among the stars.

It passes not very far from Alpha Lyræ (Vega), on the opposite side of the circle from the present Pole-star; about 12,000 years hence Vega will, therefore, be the Pole-star. Reckoning backwards, we find that some 4000 years ago Alpha Draconis (Thuban) was the Pole-star, and about 3½° from the pole.

Another effect of precession is that the *signs* of the zodiac do not now agree with the *constellations* which bear the same name. The *sign* of Ares is now in the constellation of Pisces, and so on; each sign having "backed" bodily, so to speak, into the constellation west of it.

The forces which cause precession do not act quite uniformly, and as a result the rapidity of the precession varies somewhat, and there is also a slight tipping or nodding of the earth's axis which is called *nutation*. (For a fuller account of the whole matter, see General Astronomy, Arts. 209–215.)

THE YEAR AND THE CALENDAR

127. Three different kinds of "year" are now recognized, — the *Sidereal*, the *Tropical* (or *Equinoctial*), and the *Anomalistic*.

The *sidereal* year, as its name implies, is the time occupied by the sun in apparently completing the circuit from a given star to the same star again. Its length is 365 days, 6 hours, 9 minutes, 9 seconds. From the mechanical point of view, this is the true year; *i.e.*, it is the time occupied by the earth in completing its revolution around the sun from a given direction in space to the same direction again.

The *tropical year* is the time included between two successive passages of the vernal equinox by the sun. Since the equinox moves yearly 50''.2 towards the west, the tropical year is shorter than the sidereal by about 20 minutes, its length being 365 days, 5 hours, 48 minutes, 46 seconds. *Since the seasons depend on the sun's place with respect to the equinox, the tropical year is the year of chronology and civil reckoning.*

The third kind of year is the *anomalistic* year, — the time between two successive passages of the perihelion by the earth. Since the line of apsides of the earth's orbit makes an eastward revolution once in about 108,000 years, this kind of year is nearly 5 minutes longer than the sidereal, its length being 365 days, 6 hours, 13 minutes, 48 seconds. It is but little used except in calculations relating to perturbations of the planets.

128. The Calendar. — The natural units of time are the day, the month and the year. The day is too short for convenience in dealing with considerable periods, such as the life of a man, for instance; and the same is true even of the month; so that for all chronological purposes the *tropical year* (the year of the seasons) has always been employed. At the same time, so many religious ideas and observations have been connected with the changes of the moon, that there has been a constant struggle to reconcile the month with the year. Since the two are incommensurable, no really satisfactory solution is possible, and the modern calendar of civilized nations entirely disregards the lunar phases. In early times the calendar was in the hands of the priesthood, and was mainly lunar, the seasons being either disregarded, or kept roughly in place by the occasional putting in or dropping of a month. The Mohammedans still use a purely lunar calendar, having a "year" of twelve months, which contains alternately 354 and 355 days. In their reckoning the seasons fall continually in different months, and their calendar gains on ours about one year in thirty-three.

129. The Julian Calendar. — When Julius Cæsar came into
power, he found the Roman calendar in a state of hopeless
confusion. He, therefore, with the advice of Sosigenes, the
astronomer, established (B.C. 45) what is known as the *Julian
Calendar*, which still, either untouched or with a trifling modification, continues in use among civilized nations. Sosigenes
discarded all reference to the moon's phases, and adopting
$365\frac{1}{4}$ days as the true length of the year, he ordained that
every fourth year should contain 366 days, — the extra day
being inserted by repeating the sixth day before the Calends
of March (whence such a year is called "Bissextile"). He
also transferred the beginning of the year, which before
Cæsar's time had been in March (as is indicated by the names
of several of the months, — December, the tenth month, for
instance), to Jan. 1st.

Cæsar also took possession of the month Quintilis, naming it July
after himself. His successor, Augustus, in a similar manner appropriated the next month, Sextilis, calling it August, and to vindicate his
dignity and make his month as long as his predecessor's he added to
it a day stolen from February.

The Julian calendar is still used unmodified in the Greek
Church, and also in many astronomical reckonings.

130. The Gregorian Calendar. — The true length of the
tropical year is not $365\frac{1}{4}$ days, but 365 days, 5 hours, 48 minutes, 46 seconds, leaving a difference of 11 minutes and 14
seconds by which the Julian year is too long. This difference
amounts to a little more than three days in 400 years. As a
consequence the date of the vernal equinox comes continually
earlier and earlier in the Julian calendar, and in 1582 it had
fallen back to the 11th of March instead of occurring on the
21st, as it did at the time of the Council of Nice (A.D. 325).
Pope Gregory, therefore, under the astronomical advice of
Clavius, ordered that the calendar should be restored by adding ten days, so that the day following Oct. 4th, 1582, should

be called the 15th instead of the 5th; further, to prevent any future displacement of the equinox, he decreed that thereafter only such *century years* should be leap years as are divisible by 400. Thus 1700, 1800, 1900, and 2100 are not leap years, but 1600 and 2000 are. The change was immediately adopted by all Catholic countries, but the Greek Church and most Protestant nations refused to recognize the Pope's authority. The new calendar was, however, at last adopted in England in 1752. At present (since the year 1800 was a leap year in the Julian calendar and not in the Gregorian) the difference between the two calendars is twelve days; but it will become thirteen in 1900, which will not be a leap year with us, though it will in Russia.

CHAPTER V.

THE MOON. — HER ORBITAL MOTION AND THE MONTH. —
DISTANCE, DIMENSIONS, MASS, DENSITY, AND FORCE OF
GRAVITY. — ROTATION AND LIBRATIONS. — PHASES. —
LIGHT AND HEAT. — PHYSICAL CONDITION. — TELE-
SCOPIC ASPECT AND PECULIARITIES OF THE LUNAR
SURFACE.

131. NEXT to the sun, the moon is the most conspicuous,
and to us the most important, of the heavenly bodies ; in fact,
she is the only one except the sun, which exerts the slightest
perceptible influence upon the interests of human life.　She
owes her conspicuousness and her importance, however, solely
to her nearness ; for she is really a very insignificant body as
compared with stars and planets.

132. The Moon's Apparent Motion; Definition of Terms, etc.
— One of the earliest observed of astronomical phenomena
must have been the eastward motion of the moon with reference to the sun and stars, and the accompanying change of
phase.　If, for instance, we note the moon to-night as very
near some conspicuous star, we shall find her to-morrow night
at a point considerably farther east, and the next night
farther yet ; she changes her place about 13° daily, and makes
the complete circuit of the heavens, from star to star again, in
about 27⅓ days.　In other words, she revolves around the
earth in that time, while she accompanies us in our annual
journey around the sun.　Since the moon moves eastward
among the stars so much faster than the sun (which takes a
year in going once around), she overtakes and passes him at

regular intervals ; and as her *phases* depend upon her apparent position with reference to the sun, this interval from new moon to new moon is specially noticeable, and is what we ordinarily understand as the "month."

The angular distance of the moon east or west of the sun at any time is called her *Elongation.* At new moon it is zero, and the moon is said to be in *Conjunction.* At full moon the elongation is 180°, and she is said to be in *Opposition.* In either case the moon is in *Syzygy.* (*Syzygy* means. "yoked together," the sun, moon, and earth being then nearly in line.) When the elongation is 90°, she is said to be in *Quadrature.*

133. Sidereal and Synodic Months. — The *sidereal* month is the time it takes the moon to make her revolution from a given star to the same star again; its length is $27\frac{1}{3}$ days (27 days, 7 hours, 43 minutes, 11.524 seconds). The mean daily motion, therefore, is 360° divided by this, or 13° 11' (nearly). The sidereal month is the true month from the mechanical point of view.

The *synodic* month is the time between two successive conjunctions or oppositions; *i.e.,* between two successive new or full moons. Its average length is about $29\frac{1}{2}$ days (29 days, 12 hours, 44 minutes, 2.841 seconds), but it varies considerably on account of the eccentricity of the moon's orbit.

If M be the length of the moon's sidereal period in days, E the length of the sidereal year, and S that of the synodic month, the three quantities are connected by a simple relation easily demonstrated. $\frac{1}{M}$ is the fraction of a circumference moved over by the moon in a day. Similarly, $\frac{1}{E}$ is the apparent daily motion of the sun. The difference is the amount which the moon *gains* on the sun daily. Now it gains a whole revolution in one synodic month of S days, and therefore must gain daily $\frac{1}{S}$ of a circumference. Hence we have the important equation

$$\frac{1}{M} - \frac{1}{E} = \frac{1}{S}$$

which is known as the *equation of synodic motion.* In a sidereal year the number of sidereal months is exactly one greater than the number of synodic months, the numbers being respectively 13.369 + and 12.369 +.

134. The Moon's Path among the Stars. — By observing the moon's right ascension and declination daily with suitable instruments, we can map out its apparent path, just as in the case of the sun (Art. 116). This path turns out to be (very nearly) a great circle, inclined to the ecliptic at an angle of 5° 8'. The two points where it cuts the ecliptic are called the "nodes," the *ascending* node being where the moon passes from the south side to the north side of the ecliptic, while the opposite node is called the descending node.

The moon at the end of the month never comes back *exactly* to the point of beginning among the stars, on account of the so-called "perturbations" of her orbit, due mostly to the attraction of the sun. One of the most important of these perturbations is the "regression of the nodes." These slide westward on the ecliptic just as the vernal equinox does (precession), but much faster, completing their circuit in about 19 years instead of 26,000.

135. Interval between the Moon's Successive Transits; Daily Retardation. — Owing to the eastward motion of the moon it comes to the meridian about 51 minutes later each day, on the average; but the retardation ranges all the way from 38 minutes to 66 minutes, on account of the variation in the rate of the moon's motion.

The *average* retardation of the moon's rising and setting is also, of course, the same 51 minutes; but the actual retardation is still more variable than that of the meridian transits, depending to some extent on the latitude of the observer as well as on the variations in the moon's motion. At New York the range is from 23 minutes to 1 hour and 17 minutes; that is to say, on some nights the rising of the moon is only 23 minutes later than on the preceding night. while at other times it is more than an hour and a quarter behindhand. In high latitudes the differences are still greater. In very high latitudes the moon, when it has its greatest possible declination, becomes circumpolar for a certain time each month, and remains visible without setting at all (like the midnight sun) for a greater or less number of days, according to the latitude of the observer.

136. Harvest and Hunter's Moon. — The full moon that occurs nearest the autumnal equinox is called the 'harvest moon'; the one next following, the 'hunter's moon.' At that time of the year the moon, while nearly full, rises for several consecutive nights almost at the same hour, so that the moonlight evenings last for an unusually long time. The phenomenon, however, is much more striking in Northern Europe and in Canada than in the United States.

137. Form of the Moon's Orbit. — By observation of the moon's apparent diameter in connection with observations of her place in the sky, we can determine the form of her orbit around the earth in the same way that the form of the earth's orbit around the sun was worked out (see Appendix, Art. 428). The moon's apparent diameter ranges from $33' 33''$, when as near the earth as possible, to $29' 24''$, when most remote; and her orbit turns out to be an ellipse like that of the earth around the sun, but of much greater eccentricity, averaging about $\frac{1}{18}$ (as against $\frac{1}{60}$). We say "averaging" because the actual eccentricity is variable, on account of perturbations.

The point of the moon's orbit nearest the earth is called the *Perigee*, that most remote the *Apogee*, and the indefinite line passing through these points the *Line of Apsides*, while the major axis is that portion of this line which lies between the perigee and apogee. This line of apsides is in continual motion, on account of perturbations (just as the line of nodes

is, Art. 134), but it moves eastward instead of westward, completing its revolution in about nine years.   In her revolution
about the earth, the moon observes the same law of equal
areas that the earth does in her orbit around the sun (Art.
121).

THE MOON'S DISTANCE.

138. In the case of any heavenly body, one of the first and
most fundamental inquiries relates to its distance from us:
until the distance has been somehow measured we can get no
knowledge of the real dimensions of its orbit, nor of the size,
mass, etc., of the body itself. The problem is usually solved
by measuring the apparent "parallactic" displacement of the
body, as seen by observers at widely separated stations.
Before proceeding farther, we must, therefore, say a few
words upon the subject of parallax.

139. Parallax. — In general the word "parallax" means
the difference between the directions of a heavenly body as
seen by the observer, and as seen from some standard point
of reference.   The *annual* or *heliocentric* parallax of a *star* is
the difference of the star's direction as seen from the *earth*
and from the *sun*.   The *diurnal* or *geocentric* parallax of the
sun, the moon, or a planet, is the difference between its direction as seen from the centre of the earth and from the observer's station on the earth's surface; or, what comes to the same
thing, *the geocentric parallax is the angle at the body made by
two lines drawn from it, one to the observer, the other to the centre
of the earth*.   (Stars have no geocentric parallax; the earth as
seen from them is a mere point.)

In Fig. 17, the parallax of the body P is the angle OPC.
Obviously this diurnal parallax is zero for a body directly overhead at Z, and is the greatest possible for a body on the horizon, as at P_h.

Moreover, and this is to be specially noted, this parallax of a body at the horizon — the "horizontal parallax" — is simply the *angular semi-diameter of the earth as seen from the body.* When, for instance, we say that the moon's horizontal parallax is 57′, it is equivalent to saying that seen from the moon the earth appears to have a diameter of 114′. In the same way, since the sun's parallax is 8″.8, the diameter of the earth as seen from the sun is 17″.6.

140. Relation between Parallax and Distance. — When the horizontal parallax of any heavenly body is ascertained, its distance follows at once through our knowledge of the earth's dimensions. If we know how large a ball of given size appears, we can tell how far away it is; if we know how large the earth looks from the moon, we can find the distance between them. Thus, when in the triangle CP_hO, Fig. 17, we know the angle at P_h, and the side CO, the radius of the earth, we can compute CP_h by a very easy trigonometrical

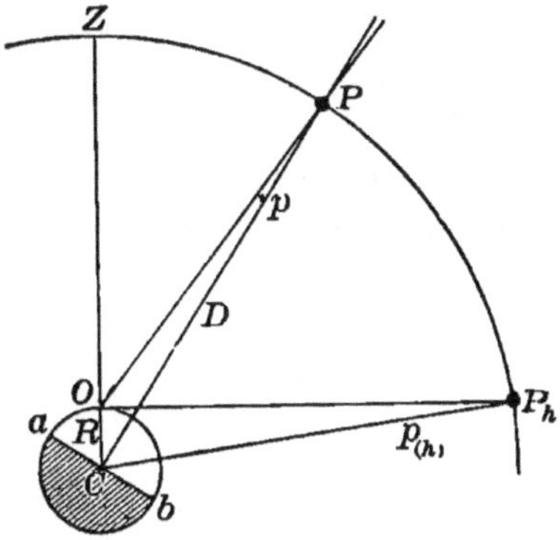

FIG. 17. — Diurnal Parallax.

calculation. Evidently the more remote the body, the smaller its parallax.

Since the radius of the earth varies slightly in different latitudes, we take the equatorial radius as a standard, and *the equatorial horizontal parallax* is the earth's equatorial semi-diameter as seen from the body. It is this which is usually meant when we speak simply of "the parallax" of the moon, of the sun, or of a planet without adding any qualification (but never when we speak of the parallax of a star; then we always mean the *annual* parallax).

141. Parallax, Distance, and Velocity of the Moon. — The moon's equatorial horizontal parallax found by corresponding observations made at different parts of the earth, is 3422″ (57′ 2″) according to Neison, but varies considerably on account of the eccentricity of the orbit. From this parallax we find that the moon's average distance from the earth is about 60.3 times the earth's equatorial radius, or 238,840 miles, with an uncertainty of perhaps 20 miles.

The maximum and minimum values of the moon's distance are given by Neison as 252,972 and 221,617 miles. It will be noted that the average distance is not the mean of the two extremes.

Knowing the size and form of the moon's orbit, the velocity of her motion is easily computed. It averages a little less than 2300 miles an hour, or about 3350 feet per second. Her mean apparent *angular* velocity among the stars is about 33′, which is just a little greater than the apparent diameter of the moon itself.

142. Diameter, Area, and Bulk of the Moon. — The mean apparent diameter of the moon is 31′ 7″. Knowing its distance, its real diameter comes out 2163 miles. This is 0.273 of the earth's diameter.

Since the surfaces of globes vary as the squares of their diameters, and their volumes as the cubes, this makes the surface area of the moon equal to about $\frac{1}{14}$ of the earth's, and the volume (or bulk) almost exactly $\frac{1}{49}$ of the earth's.

No other satellite is nearly as large as the moon in comparison with its primary planet. The earth and moon together, as seen from a distance, are really in many respects more like a double planet than like a planet and satellite of ordinary proportions. At a time, for instance, when Venus happens to be nearest the earth (at a distance of about twenty-five millions of miles), her inhabitants (if she has any) would see the earth considerably brighter than Venus herself at her best appears to us, and the moon would be about as bright as Sirius, oscillating backwards and forwards about half a degree each side of the earth, once a month.

143. Mass, Density, and Superficial Gravity of the Moon. —Her *mass* is about $\frac{1}{80}$ of the earth's mass (0.0125). The actual measurement of the moon's mass is an extremely difficult problem, and the methods pursued are quite beyond the scope of this book. Since the density is equal to $\frac{\text{Mass}}{\text{Volume}}$, the density of the moon as compared to that of the earth is found to be 0.613, or about 3.4 the density of water (the earth's density being 5.58). This is a little above the average density of the rocks which compose the crust of the earth.

The 'superficial gravity,' or the attraction of the moon for bodies at its surface, is about one-sixth that at the surface of the earth. This is a fact that must be borne in mind in connection with the enormous scale of the craters on the moon. Volcanic forces there would throw materials to a vastly greater distance than on the earth.

144. Rotation of the Moon. — The moon turns on its axis once a month, in exactly the time occupied by its revolution around the earth: its day and night are, therefore, each nearly a fortnight in length, and in the long run it keeps the same

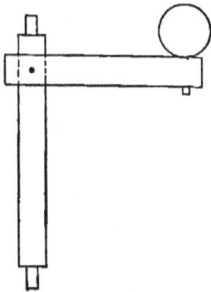

FIG. 18.

side always toward the earth. We see to-day precisely the same face of the moon which Galileo did when he first looked at it with his telescope. The opposite face has never been seen from the earth, and probably never will be.

It is difficult for some to see why a motion of this sort should be considered a rotation of the moon, since it is essentially like the motion of a ball carried on a revolving crank (Fig. 18). Such a ball, they say, "revolves around the shaft, but does not rotate on its own axis." It does rotate, however; for if we mark one side of the ball, we shall find the marked side presented successively to every point of the compass as the crank turns around, so that the ball turns on its own axis as really as if it were whirling upon a pin fastened to the table. By virtue of its connection with the crank, the ball has two distinct motions, — (1) the motion of translation, which carries its centre in a circle around the shaft; (2) an additional motion of rotation around a line drawn through its centre of gravity parallel to the shaft.

Rotation consists essentially in this : *A line connecting any two points in the rotating body, and produced to the celestial sphere, will sweep out a circle upon it.* In every rotating body, one line can be drawn through the centre of the body, however, so that the circle described by it in the sky will be infinitely small. This is the *axis* of the body.

145. Librations. — While in the long run the moon keeps the same face towards the earth, it is not so from day to day. With reference to the centre of the earth, it is continually oscillating a little, and these oscillations constitute what are called "Librations," of which we distinguish three; viz., (1) the libration in latitude, by which the north and south poles are alternately presented to the earth ; (2) the libration in longitude, by which the east and west sides of the moon are alternately tipped a little towards us; and (3) the diurnal libration, which enables us to look over whatever edge of the moon is uppermost when it is near the horizon. Owing to these librations we see considerably more than half of the moon's surface at one time and another. About 41 per cent of it is always visible; 41 per cent never visible, and a belt at the edge of the moon, covering about 18 per cent is rendered alternately visible and invisible by libration.

146. Phases of the Moon. — Since the moon is an opaque globe shining merely by reflected light, we can only see that hemisphere of her surface on which the sun is shining, and of the illuminated hemisphere only that portion which happens to be turned towards the earth.

When the moon is between the earth and the sun (new moon), the side presented to us is dark, and the moon is then invisible. A week later, at the end of the first quarter, half of the illuminated hemisphere is visible, and we have the

half-moon just as we do a week after the full. Between the
new moon and the half-moon, during the first and last quarters
of the lunation, we see less than half of the illuminated portion, and then have the "crescent" phase. Between half-
moon and the full moon, during the second and third quarters

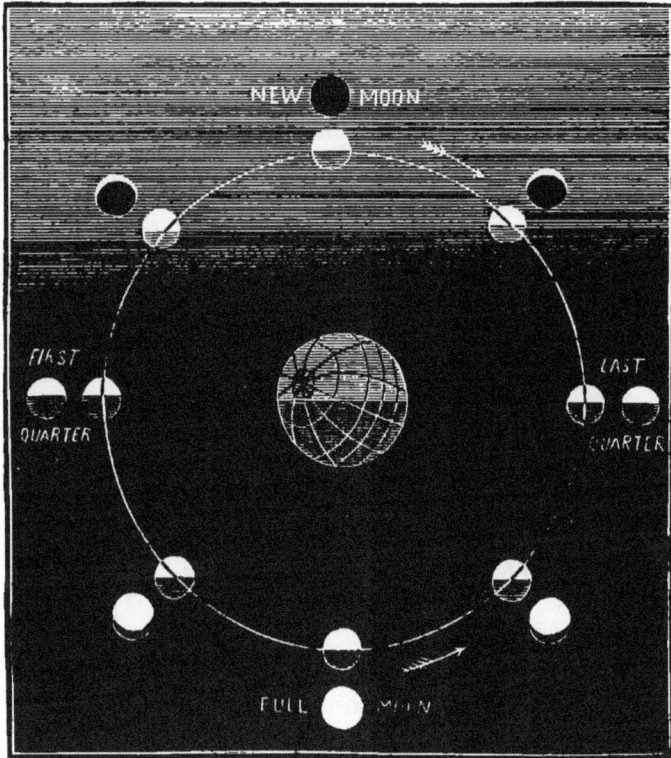

FIG. 19. — The Moon's Phases.

of the lunation, we see more than half of the moon's illuminated side, and we have then what is called the "gibbous"
phase.

Fig. 19 (in which the light is supposed to come from a point far
above the circle which represents the moon's orbit) shows the way in
which the phases are distributed through the month.

The line which separates the dark portion of the disc from the bright is called the *Terminator*, and is always a semi-ellipse, since it is a semicircle viewed obliquely, as shown by Fig. 20, *A*. Draughtsmen sometimes incorrectly represent the crescent form by a construction like Fig. 20, *B*, in which a smaller circle has a portion cut out of it by an arc of a larger one. It is to be noticed also that *ab*, the line which joins the "cusps" or points of the crescent, is always perpendicular to a line drawn from the moon to the sun, so that the horns are always turned directly away from the sun. The precise position in which they will stand

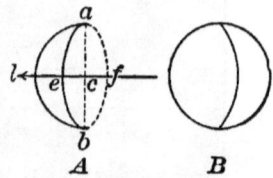

FIG. 20.

at any time is, therefore, perfectly predictable, and has nothing whatever to do with the weather. (Pupils have probably heard of the "wet moon" and "dry moon" superstition.)

147. Earth-shine on the Moon. — Near the time of new moon, the portion of the moon's disc which does not get the sunlight is easily visible, illuminated by a pale reddish light. This light is *earth-shine*, — the earth as seen from the moon being then nearly full. The red color is due to the fact that the light sent to the moon from the earth has passed twice through our atmosphere, and so has acquired the sunset tinge. Seen from the moon, the earth would be itself a magnificent moon about 2° in diameter, showing the same phases as the moon does itself.

Taking everything into account, the earth-shine is probably fifteen to twenty times as strong as the light of the moon at similar phases. Since the moon keeps always the same face towards the earth, the earth is visible only from that part of the moon which faces us, and remains nearly stationary in the lunar sky, neither rising nor setting. It is easy to see that she would be a very beautiful object, on account of the changes which would be continually going on upon her surface due to snow-storms, clouds, growth of vegetation, etc.

PHYSICAL CHARACTERISTICS OF THE MOON.

148. Absence of Air and Water. — The moon's atmosphere, if there is any, is extremely rare, its density at the moon's surface being probably not more than $\frac{1}{1000}$ part of that of our own atmosphere.

The evidence on the point is twofold: First, the telescopic appearance. There is no haze, shadows are perfectly black; there is no sensible twilight at the points of the crescent, and all outlines are visible sharply and without the least blurring such as would be due to the intervention of an atmosphere. Second, the absence of refraction when the moon intervenes between us and any distant body. When the moon 'occults' a star, for instance, there is no distortion or discoloration of the star-disc, but both the disappearance and the reappearance are practically instantaneous.

Of course if there is no air, there can be no *liquid* water, since the water would immediately evaporate and form an atmosphere of vapor if air were not present. It is not impossible, however, nor perhaps improbable, that *solid water* (ice and snow) may exist on the moon's surface. Although ice and snow liberate a certain amount of vapor, yet at a low temperature the quantity would be insufficient to make an atmosphere dense enough to be observed from the earth.

If the moon once formed a portion of the earth, as is likely, the absence of air and water requires explanation, and there have been many interesting speculations on the subject into which we cannot enter.

149. The Moon's Light. — In its quality moonlight is simply sunlight, showing a spectrum identical in every detail with that of the light coming from the sun itself, except as the intensity of different portions of the spectrum is slightly altered by its reflection from the lunar surface.

The brightness of full moonlight as compared with sunlight is about *one six-hundred-thousandth*. According to this, if the whole visible hemisphere were packed with full moons, we should receive from it only about one-eighth of the light of the sun.

The half-moon does not give nearly half as much light as the full moon. Near the full the brightness is suddenly and greatly increased, probably because at any time except the full the moon's visible surface is more or less darkened by shadows which disappear at the moment of full.

The average "albedo," or reflecting power, of the moon's surface is given by Zöllner as 0.174; i.e., the moon's surface reflects a little more than one-sixth of the light that falls upon it. There are, however, great differences in the brightness of the different portions of the moon's surface. Some spots are nearly as white as snow or salt, and others as dark as slate.

150. Heat of the Moon. — For a long time it was impossible to detect the moon's heat by observation. Even when concentrated by a large lens, it is too feeble to be shown by the most delicate thermometer. With modern apparatus, however, it is easy enough to perceive the heat of lunar radiation, though the *measurement* is extremely difficult. The total amount of heat sent by the full moon to the earth appears to be about $\frac{1}{170000}$ of that sent by the sun; i.e., the full moon in two days sends us about as much heat as the sun does in one second.

A considerable portion of the lunar heat seems to be simply reflected from the surface like light, while the rest, perhaps three-fourths of the whole, is "obscure heat"; i.e., heat which has first been absorbed by the moon's surface and then radiated, like the heat from a brick surface that has been warmed by the sunshine.

As to the *temperature* of the moon's surface, it is impossible to be very certain. During the long lunar night of fourteen days, the temperature must inevitably fall appallingly low, — perhaps 200° or 300° below zero. On the other hand, the

lunar rocks are exposed to the sun's rays in a cloudless sky
for fourteen days at a time, so that if they were protected by
air, like the rocks upon the earth, they would certainly become
intensely heated. But there is no air, and, on the whole, it is
probable that the temperature never rises much above the
freezing-point of water, since in the absence of air the heat
would be lost about as fast as it is received, and the condition
of things may be supposed to be somewhat like that on the
highest mountains of the earth (where there is perpetual snow
and ice), only more so.

151. Lunar Influences on the Earth. — The most important
effect produced upon the earth by the moon is the generation
of the tides in co-operation with the sun. There are also certain well-ascertained disturbances of the terrestrial magnetism
connected with the approach and recession of the moon in its
oval orbit; and this ends the chapter of proved lunar influences.

The multitude of current beliefs as to the controlling influence of the moon's phases and changes upon the weather and
the various conditions of life are mostly unfounded. It is
quite certain that if the moon has any influence at all of the
sort imagined, it is extremely slight; so slight that it has not
yet been demonstrated, though numerous investigations have
been made expressly for the purpose of detecting it. Different
workers continually come to contradictory results.

152. The Moon's Telescopic Appearance. — Even to the
naked eye the moon is a beautiful object, diversified with curious markings connected with numerous popular legends. In
a powerful telescope these naked-eye markings vanish, and
are replaced by a multitude of smaller details which make the
moon, on the whole, the most interesting of all telescopic
objects — especially to instruments of moderate size, say from
six to ten inches in diameter, which generally give a more

pleasing view than instruments either much larger or much smaller. An instrument of this size, with magnifying powers between 250 and 500, virtually brings the moon within a distance ranging from 1000 to 500 miles. Any object half a mile in diameter on the moon is distinctly visible. A long line or streak even less than a quarter of a mile across can easily be seen.

For most purposes the best time to look at the moon is when it is between six and ten days old: at the time of full moon few parts of the surface are well seen. It is evident that while with the telescope we should be able to see such objects as lakes, rivers, forests, and great cities, if they existed on the moon, it would be hopeless to expect to distinguish any of the minor indications of life, such as buildings or roads.

153. The Moon's Surface Structure. — The moon's surface for the most part is extremely broken. The earth's mountains are mainly in long ranges, like the Andes and Himalayas. On the moon the ranges are few in number; but, on the other hand, the surface is pitted all over with great craters, which resemble very closely the volcanic craters on the earth's surface, though on an immensely greater scale. The largest terrestrial craters

FIG. 21. — A Normal Lunar Crater (Nasmyth).

do not exceed six or seven miles in diameter; many of those on the moon are fifty or sixty miles across, and some have a diameter of more than a hundred miles, while smaller ones from five to twenty miles in diameter are counted by the hundred.

The normal lunar crater (Fig. 21) is nearly circular, surrounded by a mountain ring, which rises anywhere from 1000 to 20,000 feet above the neighboring country. The floor within the ring may be either above or below the outside level; some craters are deep, and some are filled nearly to the brim. Frequently, in the centre of the crater, there rises a group of peaks which attain the same elevation as the encircling ring, and these central peaks often show holes or minute craters in their summits.

On some portions of the moon these craters stand very thickly. This is especially the case near the moon's south pole. It is noticeable, also, that as on the earth the youngest mountains are generally the highest, so on the moon the most recent craters are generally deepest and most precipitous.

The height of a lunar mountain can be measured with considerable accuracy by means of its shadow.

The striking resemblance of these lunar craters to terrestrial volcanoes makes it natural to assume that they have a similar origin. This, however, is not quite certain, for there are considerable difficulties in the way of the volcanic theory, especially in the case of what are called the great "Bulwark Plains," so extensive that a person standing in the centre could not even see the summit of the surrounding ring at any point; and yet there is no line of distinction between them and the smaller craters, — the series is continuous. Moreover,

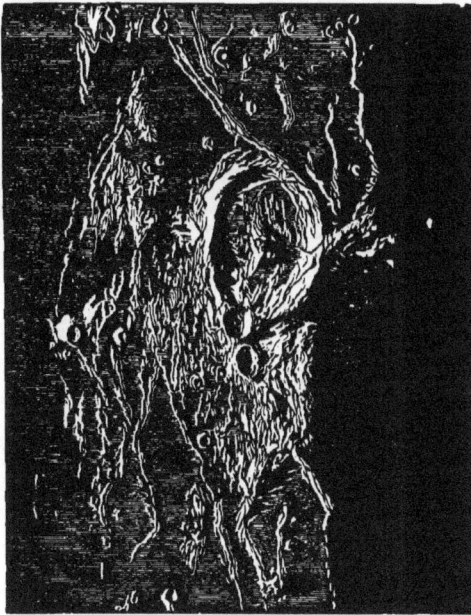

Fig. 22. — Gassendi (Nasmyth).

on the earth, volcanoes necessarily require the action of air and water, which do not now exist on the moon; so that if these lunar craters are really the result of volcanic eruptions, they must be ancient formations, for there is absolutely no evidence of any present volcanic activity. Fig. 22 represents one of the finest lunar craters, Gassendi, which is best seen about two days after the half moon.

154. Other Lunar Formations. — The craters and mountains are not the only interesting features on the moon's surface. There are many deep, narrow, crooked valleys which go by the name of "rills," and may once have been water-courses (see Fig. 23). Then there are many straight "clefts" half a mile or so wide, and of unknown depth, running in some cases several hundred miles straight through mountain and valley, without any apparent regard for the accidents of the surface.

Most curious of all are the light-colored streaks, or "rays," which radiate from certain of the craters, extending in some cases

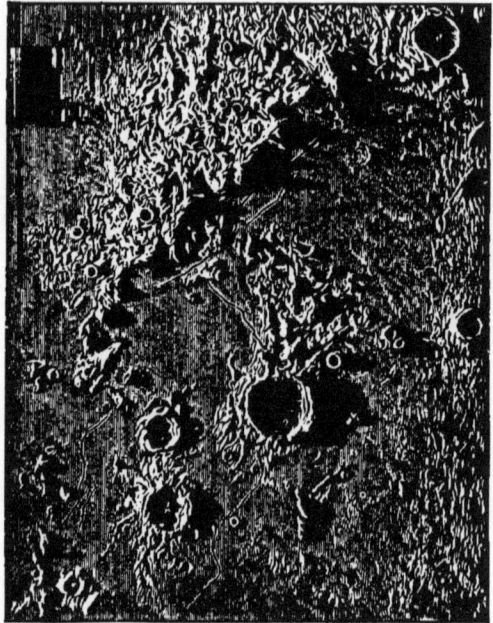

FIG. 23. — Archimedes and the Apennines (Nasmyth).

a distance of many hundred miles. They are usually from five to ten miles wide, and neither elevated nor depressed to any considerable extent with reference to the general surface. Like the clefts, they pass across valley and mountain, and sometimes straight through craters, without any change in width or color. No satisfactory explanation of them has yet

been given. The most remarkable of these "ray-systems" is
the one connected with the great crater Tycho, not very far
from the moon's south pole. The rays are not very conspicuous until within a few days of full moon, but at that time
they, and the crater from which they diverge, constitute by
far the most striking feature of the telescopic view.

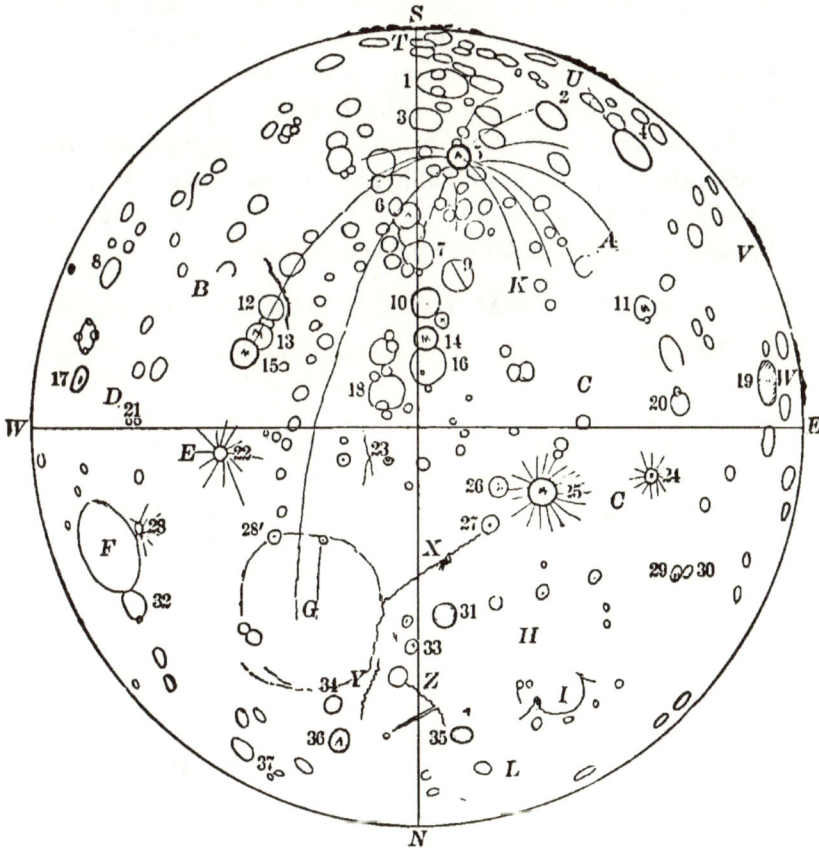

Fig. 24. — Map of the Moon, reduced from Neison.

155. Changes on the Moon. — It is certain that there are
no conspicuous changes on the moon's surface; no such transformations as would be presented by the earth viewed with a
telescope from the moon, — no clouds, no storms, no snow of

winter, and no spread of verdure in the spring. At the same time it is confidently maintained by some observers that here and there alterations do take place in the details of the lunar surface, while others as stoutly dispute it. The difficulty in settling the question arises from the great changes which take place in the appearance of a lunar object, according to the angle at which the sunlight strikes it. Other conditions also, such as the height of the moon above the horizon and the clearness and steadiness of the air, affect the appearance; and it is very difficult to secure a sufficient identity of conditions at different times of observation to be sure that apparent changes are real. It is probable that the question will finally be settled by photography. For further discussion of this subject, see General Astronomy, Art. 272.

KEY TO THE PRINCIPAL OBJECTS INDICATED IN FIG. 24.

A. Mare Humorum.

B. Mare Nectaris.

C. Oceanus Procellarum.

D. Mare Fecunditatis.

E. Mare Tranquilitatis.

F. Mare Crisium.

G. Mare Serenitatis.

H. Mare Imbrium.

I. Sinus Iridum.

K. Mare Nubium.

L. Mare Frigoris.

T. Leibnitz Mountains.

U. Doerfel Mountains.

V. Rook Mountains.

W. D'Alembert Mountains.

X. Apennines.

Y. Caucasus.

Z. Alps.

1. Clavius.	14. Alphonsus.	27. Eratosthenes.
2. Schiller.	15. Theophilus.	28. Proclus.
3. Maginus.	16. Ptolemy.	28'. Pliny.
4. Schickard.	17. Langrenus.	29. Aristarchus.
5. Tycho.	18. Hipparchus.	30. Herodotus.
6. Walther.	19. Grimaldi.	31. Archimedes.
7. Purbach.	20. Flamsteed.	32. Cleomedes.
8. Petavius.	21. Messier.	33. Aristillus.
9. "The Railway."	22. Maskelyne.	34. Eudoxus.
10. Arzachel.	23. Triesnecker.	35. Plato.
11. Gassendi.	24. Kepler.	36. Aristotle.
12. Catherina.	25. Copernicus.	37. Endymion.
13. Cyrillus.	26. Stadius.	

156. Lunar Maps and Nomenclature. — A number of maps of the moon have been constructed by different observers. The most recent and extensive is that by Schmidt of Athens, on a scale of seven feet in diameter; it was published by the Prussian government in 1878. Perhaps the best for ordinary observers is that given in Webb's "Celestial Objects for Common Telescopes." We present here (Fig. 24) a skeleton map, which indicates the position of about fifty of the leading objects.

As for the names of the lunar objects, the great plains upon the surface were called by Galileo "oceans," or "seas" (*Maria*), because he supposed that these grayish surfaces, which are visible to the naked eye and conspicuous in a small telescope, though not with a large one, were covered with water. Thus we have the "Oceanus Procellarum" (Sea of Storms), the "Mare Imbrium" (Sea of Showers), etc. The ten mountain ranges on the moon are mostly named for terrestrial mountains, as Caucasus, Alps, Apennines, though two or three bear the names of astronomers, like Leibnitz, Doerfel, etc. The conspicuous craters bear the names of ancient and mediæval astronomers and philosophers, as Plato, Archimedes, Tycho, Copernicus, Kepler, and Gassendi. This system of nomenclature seems to have originated with Riccioli, who made the first map of the moon in 1650.

156.* The first successful photographs of the moon were made by Rutherfurd of New York about 1866, and have remained unsurpassed until very recently.

At present the Paris and Lick observatories are taking the lead, and producing negatives unprecedented in size and clearness. The plates constitute a permanent and unimpeachable record of the state of the lunar surface, which will soon settle the question of changes upon it; and they supply the material for a much more perfect map of our satellite.

CHAPTER VI.

THE SUN. — ITS DISTANCE, DIMENSIONS, MASS, AND DEN-
SITY. — ITS ROTATION, SURFACE, AND SPOTS. — THE
SPECTROSCOPE AND THE CHEMICAL CONSTITUTION OF
THE SUN. — THE CHROMOSPHERE AND PROMINENCES.
— THE CORONA. — THE SUN'S LIGHT. — MEASUREMENT
AND INTENSITY OF THE SUN'S HEAT. — THEORY OF ITS
MAINTENANCE AND SPECULATIONS REGARDING THE
AGE OF THE SUN.

157. THE sun is a star, — the nearest of them; a hot, self-
luminous globe, enormous as compared with the earth and
moon, though probably only of medium size as a star; but to
the earth and the other planets which circle around it, it is
the grandest and most important of all the heavenly bodies.
Its attraction controls their motions, and its rays supply the
energy which maintains every form of activity upon their
surfaces.

158. The Sun's Distance. — The mean distance of the sun
from the earth (the *astronomical unit* of distance) is a little
less than 93,000000 miles. There are many methods of determining it, some of which depend on a knowledge of the Velocity of Light (Appendix, Arts. 434 and 436), while others
depend on finding the sun's horizontal parallax. (For a
resumé of the subject, see General Astronomy, Chap. XIV.)
The mean value of this parallax is very nearly 8″.8. In other
words, as seen from the sun, the earth has an apparent diameter of about 17″.6 (Art. 139). The distance is variable, to

the extent of about 1,500000 miles, on account of the eccen-tricity of the earth's orbit, the earth being almost 3,000000 miles nearer to the sun on Dec. 31st than on July 1st.

Knowing the distance of the earth from the sun, the earth's orbital velocity follows at once by dividing the circumference of the orbit by the number of seconds in a year. It comes out 18.5 miles per second. (Compare this with the velocity of a cannon-ball, which seldom exceeds 2500 feet per second.) *In travelling this* $18\frac{1}{2}$ *miles, the deflection of the earth's motion from a perfectly straight line amounts to less than one-ninth of an inch.*

159. The distance of the sun is of course enormous compared with any distance upon the earth's surface. Perhaps the simplest illustra-tion which will give us any conception of it is that drawn from the motion of a railway train, which, going a thousand miles a day (nearly forty-two miles an hour without stops) would take $254\frac{1}{2}$ years to make the journey. If sound were transmitted through interplan-etary space, and at the same rate as in our own air, it would make the passage in about fourteen years; *i.e.*, an explosion on the sun would be heard by us fourteen years after it actually occurred. Light trav-erses the distance in 499 seconds.

160. Dimensions of the Sun. — The sun's mean apparent diameter is 33' 4''. Since at its distance, 1'' equals 450.36 miles, its diameter is 866,500 miles, or $109\frac{1}{2}$ times that of the earth. If we suppose the sun to be hollowed out, and the earth placed at the centre of it, the sun's surface would be 433,000 miles away. Now since the distance of the moon from the earth is about 239,000 miles, she would be only a little more than half-way out from the earth to the inner surface of the hollow globe, which would thus form a very good background for the study of the lunar motions.

If we represent the sun by a globe two feet in diameter, the earth on the same scale would be 0.22 of an inch in diameter, the size of a very small pea. Its distance from the sun would be just about 220 feet, and the nearest star, still on the same scale, would be 8000 miles away, on the other side of the earth.

Since the surfaces of globes are proportional to the squares of their radii, the surface of the sun exceeds that of the earth in the ratio of $(109.5)^2 : 1$; i.e., the area of its surface is about 12,000 times the surface of the earth.

The volumes of spheres are proportional to the cubes of their radii, hence the sun's volume or bulk is $(109.5)^3$, or 1,300000 times that of the earth.

161. The Sun's Mass, Density, and Superficial Gravity. — The *mass* of the sun is nearly 332,000 times that of the earth. There are various ways of getting at this result, but they lie rather beyond the mathematical scope of this work.

Its *density*, as compared with that of the earth, is found by simply dividing its mass by its bulk (both as compared with the earth); i.e., the sun's density equals $\dfrac{332,000}{1,300000} = 0.255$, — a little more than a quarter of the earth's density.

To get its 'specific gravity' (i.e., its density compared with water), we must multiply this by the earth's mean specific gravity, 5.58. This gives 1.41. In other words, the sun's mean density is only about 1.4 times that of water, a very significant result as bearing on its physical condition, especially when we know that a considerable portion of its mass is composed of metals.

Of course this low density depends upon the fact that the temperature is enormously high, and the materials are mainly in a state of cloud, vapor, or gas.

The *superficial gravity* is about 27.6 as great as gravity on the earth; that is to say, a body which weighs one pound on the surface of the earth would there weigh 27.6 pounds, and a person who weighs 150 pounds here would there weigh nearly two tons. A body would fall 444 feet in the first second, and a pendulum which vibrates seconds on the earth would vibrate in less than a fifth of a second there.

162. The Sun's Rotation. — Dark spots are often visible upon the sun's surface, which pass across the disc from east to west and indicate an axial rotation. The average time occupied by a spot in passing around the sun and returning to the same apparent position, *as seen from the earth,* is 27.25 days. This interval, however, is not the true time of the sun's rotation, but the *synodic,* as is evident from Fig. 25. Suppose an observer on the earth at E sees a spot on the centre of the

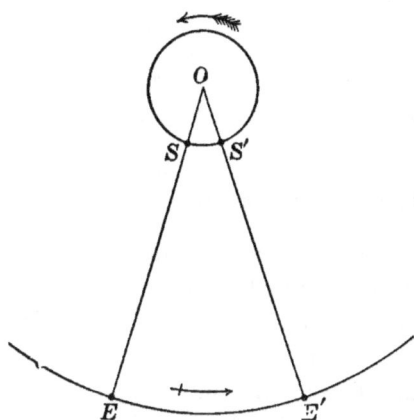

FIG. 25.
Synodic and Sidereal Revolutions of the Sun.

sun's disc at S; while the sun rotates, E will also move forward in its orbit, and the observer, the next time he sees the spot on the centre of the disc, will be at E', the spot having gone around the whole circumference plus the arc SS'.

The equation by which the true period is deduced from the synodic is the same as in the case of the moon; viz.,

$$\frac{1}{S} = \frac{1}{T} - \frac{1}{E},$$

T being the true period of the sun's rotation, E the length of the year, and S the observed synodic rotation. This gives $T = 25.35$. Different observers get slightly different results.

The paths of the spots across the sun's disc are usually more or less oval, showing that the sun's axis is inclined to the ecliptic, and so inclined that the north pole is tipped about $7\frac{1}{4}°$ towards the position which the earth occupies near the first of September. Twice a year the paths become straight, when the earth is in the plane of the sun's equator, June 3d and Dec. 5th (Fig. 26).

163. Peculiar Law of the Sun's Rotation. — It was noticed quite early that different spots give different results for the

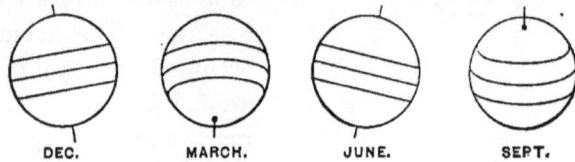

DEC.   MARCH.   JUNE.   SEPT.

FIG. 26. — Path of Spots across the Sun's Disc.

period of rotation, but the researches of Carrington, about forty years ago, first brought out the fact that the differences are systematic, so that at the solar equator the time of solar rotation is less than on either side of it. For spots near the sun's equator it is about 25 days; for solar latitude 30°, 26.5 days; and in solar latitude 40°, 27 days. The time of rotation of the sun's surface in latitude 45° is fully two days longer than at the equator; but we are unable to follow the law further towards the poles of the sun, because spots are almost never found beyond the parallel of 45°. No really satisfactory explanation of this strange acceleration of the spots at the sun's equator has yet been found.

164. Study of the Sun's Surface. — The heat and light of the sun are so intense that we cannot look directly at it with a telescope, as we do at the moon, and it is necessary, therefore, to provide either a special eye-piece with suitable shade-glass, or arrange the telescope, as in Fig. 27, so as to throw an image of the sun upon a screen.

In the study of the sun's surface, photography is for some purposes

FIG. 27. — Telescope and Screen.

very advantageous and much used. The instrument must, however, have lenses specially constructed for photographic operations, since an object-glass which would give admirable results for visual purposes would be worthless photograph-

FIG. 28. — The Great Sun Spot of September, 1870, and the Structure of the Photosphere. From a Drawing by Professor Langley. From the "New Astronomy," by permission of the Publishers.

ically. Since 1890, however, a few object glasses have been made with new kinds of glass, which are said to be good both for photography and for the eye. The exposure required to form a photographic picture is practically instantaneous. The negatives are usually from two inches up to eight or ten inches in diameter, and some of the best of them bear enlarging up to forty inches.

Photographs have the great advantage of freedom from prepossession on the part of the observer, and in an instant of time they secure a picture of the whole surface of the sun such as would require a skilful draughtsman hours to copy. But, on the other hand, they take no advantage of the instants of fine seeing, but represent the solar surface as it happened to appear at the moment when the plate was uncovered, affected by all the momentary distortions due to atmospheric disturbances.

165. The Photosphere. — The sun's surface seen with a telescope, under a medium magnifying power, appears to be of nearly uniform texture, though distinctly darker at the edges, and usually marked here and there with certain dark spots. With a higher power it is evident that the visible surface (called the photosphere) is by no means uniform, but is made up, as shown in Fig. 28, of a comparatively darkish background sprinkled over with grains, or "nodules," as Herschel calls them, of something more brilliant, — "like snowflakes on a gray cloth," according to Langley. These nodules or "rice-grains" are from 400 to 600 miles across, and, when the seeing is best, themselves break up into more minute "granules." For the most part, the nodules are about as broad as they are long, though of irregular form; but here and there, especially in the neighborhood of the spots, they are drawn out into long streaks, known as "filaments," "willow leaves," or "thatch straws."

Certain bright streaks called "faculæ" are also usually visible here and there upon the sun's surface, and though not very obvious near the centre of the disc, they become conspicuous near the "limb," or edge of the disc, especially in the neighborhood of the spots, as shown in Fig. 29. These faculæ are probably of the same material as the rest of the photosphere, but elevated above the general level and intensified in bright-

ness. When one of them passes off the edge of the disc, it is
sometimes seen as a little projection. The fact, however, that
their spectrum shows *bright lines* of calcium vapor, makes it
uncertain whether they may not be clouds of that substance
floating high above the photosphere.

In their nature, the photospheric "nodules" and faculæ are
in all probability *luminous clouds,* floating in a less luminous
atmosphere, just as a snow or rain-cloud, which has been

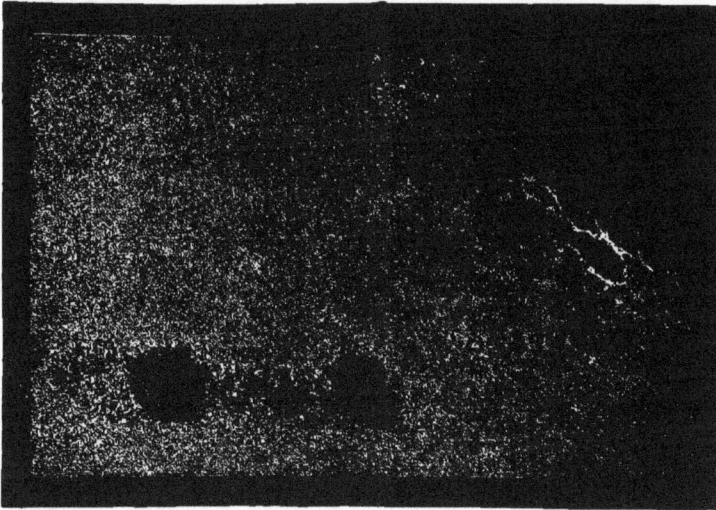

FIG. 29. — Faculæ at Edge of the Sun (De La Rue).

formed by the condensation of water-vapor, floats in the earth's
atmosphere. Such a cloud, while at a temperature even lower
than that of the surrounding gases, has a vastly greater power
of emitting light, and therefore appears very brilliant in comparison with the gas in which it floats, like the "mantle" of
a Welsbach gas-burner. There is considerable probability
that the principal element in the photosphere is *Carbon,*
though this cannot yet be regarded as proved.

166. Sun Spots. — Sun spots, whenever visible, are the most interesting and conspicuous objects upon the solar surface. The appearance of a normal sun spot (Fig. 30), fully formed and not yet beginning to break up, is that of a dark central "umbra," more or less circular, with a fringing "penumbra" composed of converging filaments. The umbra itself is not uniformly dark throughout, but is overlaid with filmy clouds,

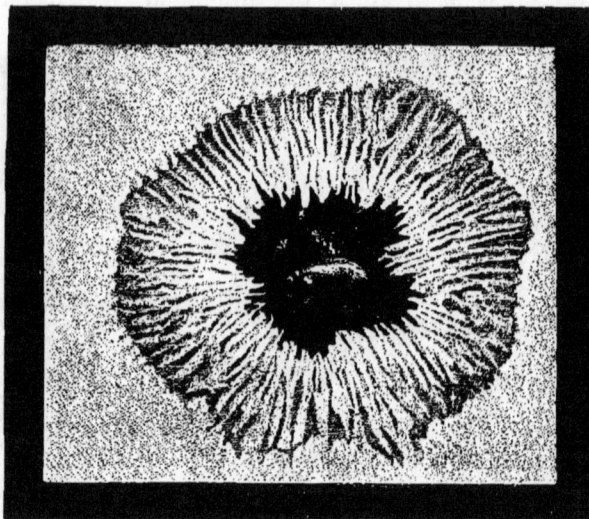

FIG. 30. — A Normal Sun Spot (Secchi; modified).

which usually are rather hard to see, but sometimes are conspicuous, as in the figure. Usually, also, within the umbra there are a number of round and very black spots, sometimes called "vortices," but often referred to as "Dawes's holes," after the name of their first discoverer.

Even the darkest portions of the umbra, however, are dark *only by contrast*. Photometric observations show that the nucleus of a spot gives about one per cent as much light as a corresponding area of the photosphere; the blackest portion of a sun spot is really more brilliant than a calcium light.

Very few spots are strictly normal. Frequently the umbra is out of the centre of the penumbra, or has a penumbra on one side only, and the penumbral filaments, instead of converging regularly towards the nucleus, are often distorted in every conceivable way. Spots are often gathered in groups within a common penumbra, separated from each other by brilliant "bridges," which extend across from the outside photosphere. Occasionally a spot has no penumbra at all, and sometimes we have what are called "veiled" spots, in which there seems to be a penumbra without any central nucleus.

167. Nature of Sun Spots.—The spots are probably shallow *depressions or hollows* in the photosphere filled with gases and vapors which are cooler than the surrounding regions, and therefore absorb a considerable portion of light, and make the spot look dark. The evidence that they are depressions consists in the change in their appearance as they approach the "limb," or edge of the disc. Here the penumbra becomes wider on the outer edge, and narrower on the inner edge, and just before the spot goes out of sight around the edge of the sun, the penumbra on the inner edge entirely disappears. The appearance is precisely such as would be shown by a

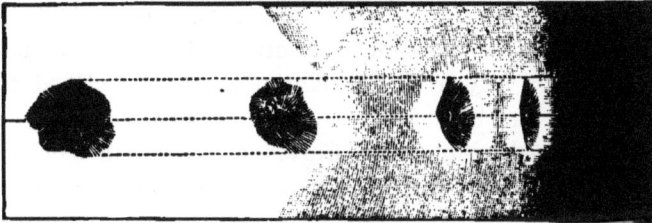

FIG. 31.   Sun Spots as Cavities.

saucer-shaped cavity in the surface of a globe, the bottom of the cavity being painted black to represent the umbra, and the sloping sides gray for the penumbra (see Fig. 31).

Observations upon a single spot would hardly be sufficient to prove this, because the spots are so irregular in their form ; but by observing the behavior of several hundred, the truth appears in the average result. Occasionally when a very .large spot passes off the sun's limb, a depression can be seen with the telescope. It is only fair to add, however, that some observers of great experience still dispute the received theory, and maintain that spots are dark clouds of some kind floating on, or just above, the photosphere.

That the nucleus of a spot is generally cooler as well as darker than the rest of the sun's surface, has been proved by several observers by direct experiments, though very near the edge of the sun the reverse has been found to be the case in some instances.

The penumbra is usually composed of "thatch straws," or long drawn out filaments, and these, as has been said, converge in a general way towards the centre of the spot. In the neighborhood of the spot, the surrounding photosphere is usually much disturbed and elevated into faculæ.

168. Dimensions of Sun Spots, etc. — The diameter of the *umbra* of a sun spot varies all the way from 500 miles, in the case of a very small one, to 50,000 miles in the case of a very large one. The *penumbra* surrounding a group of spots is sometimes 150,000 miles across, though that is an exceptional size. Quite frequently sun spots are large enough to be visible with the naked eye, and can actually be thus seen at sunset or through a fog, or by the help of a simple colored glass. The depth of the bottom of a spot is very difficult to determine, but according to Faye, Carrington, and some others, it seldom exceeds 2500 miles, and more often is between 500 and 1500.

The duration of sun spots is very various, but they are always short-lived phenomena from the astronomical point of view, sometimes lasting only for a few days, though more frequently for a month or two. In one instance a spot group attained the age of eighteen months.

Very little can be said as to their cause. Numerous theories, more or less satisfactory, have been proposed. On the whole, perhaps the most probable view is that they are the effect of eruptions. It is not likely, however, that they are the holes or craters through which the eruptions break out, as Secchi at one time thought, and as Mr. Proctor maintained to the very last: it is more probable, in accordance with Secchi's later views, that when an eruption takes place, a hollow, or *sink*, results in the photospheric cloud-surface somewhere near it, in which hollow the cooler gases and vapors collect. It is almost universally admitted that in some way they are due to matter descending from above upon the photosphere, but there is wide difference of opinion as to the nature and source of the falling substance, — whether it is meteoric, or was formed by condensation in the upper regions of the solar atmosphere; or thrown up from the photosphere by eruption, as the text suggests.

169. Distribution of Spots, and their Periodicity. — It is a significant fact that the spots are confined mostly to two zones of the sun's surface between 5° and 40° of north and south solar latitude. Practically none are ever found beyond the latitude of 45°, but at the time when spots are most numerous, a few are found near the equator. In 1843 Schwabe of Dessau, by the comparison of an extensive series of observations covering nearly twenty years, showed that the sun spots are probably periodic, being at some times much more numerous than at others, with a roughly regular recurrence every ten or eleven years. A few years later he fully established this remarkable result. Wolf of Zurich has collected all the observations discoverable, and has obtained a pretty complete record back to 1610, when Galileo first discovered these objects. The average period is 11.1 years, but the maxima are somewhat irregular, both in time and as to the extent of the surface covered by spots. The last maximum occurred in 1892-3. During the maximum the sun is

never free from spots, from 25 to 50 being frequently visible
at once. During the minimum, on the contrary, weeks and
even months pass without the appearance of a single one.
The cause of this periodicity is not yet known.

Another curious and important fact has recently been brought out
by Spoerer, though not yet explained. Speaking broadly, the disturbance which produces the spots of a given period first manifests
itself in two belts, about 30° north and south of the sun's equator.
These belts then draw in towards the equator, and the spot-maximum
occurs when their latitude is about 16°; while the disturbance finally
dies out at a latitude of from 5° to 10°, about twelve or fourteen
years after its first outbreak. Two or three years before this disappearance, however, two new zones of disturbance show themselves.
Thus at the spot-minimum there are usually four well marked spot-
belts; two near the sun's equator, due to the expiring disturbance,
and two in high latitudes, due to the newly beginning outbreak.

170. Terrestrial Influence of Sun Spots. — One influence of
sun spots on the earth is perfectly demonstrated. When the
spots are numerous, magnetic disturbances (*magnetic storms*)
are most numerous and most violent upon the earth — a fact
not to be wondered at, since notable disturbances upon the
sun's surface have been immediately followed by magnetic
storms with brilliant exhibitions of the Aurora Borealis, as in
1859 and 1883. But no one has yet been able to explain the
nature of the connection by which disturbances upon the sun's
surface affect the magnetic condition of the earth, though the
fact is beyond doubt.

It has been attempted, also, to show that the periodical disturbance
of the sun's surface is accompanied by effects upon the earth's meteorology, — upon its temperature, barometric pressure, storminess, and
the amount of rain-fall. On the whole, it can only be said that while
it is possible that *real* effects are produced, they must be very slight,
and are almost entirely covered up by the effect of purely terrestrial
causes. The results obtained thus far in attempting to co-ordinate
sun-spot phenomena with meteorological phenomena are unsatisfac-

tory and often contradictory. We may add that the spots cannot produce any sensible effect by their direct action in diminishing the light and heat of the sun. They do not *directly* alter the amount of solar radiation at any time by so much as one part in a thousand.

THE SOLAR SPECTRUM AND ITS REVELATIONS.

About 1860 the spectroscope appeared in the field as a new and powerful instrument for astronomical research, resolving at a glance many problems which before did not seem even open to investigation.

171. Principle of the Spectroscope. — The *essential* part of the apparatus is either a prism or a train of prisms, or else a diffraction "grating,"[1] which is capable of performing the same office of "dispersing" (*i.e.*, of spreading and sending in different directions) the light rays of different colors.

If with such a "dispersion piece," as we may call it (either prism or grating), one looks at a distant *point* of light, he will see instead of a point a long, bright streak, red at one end and violet at the other. If the object looked at is a *line* of light, parallel to the edge of the prism or to the lines of the grating, then instead of a colored streak without width, he gets a colored band or ribbon of light, the *spectrum*, which may show markings which will give him much valuable information. It is usual to form this line of light by admitting the rays through a narrow "slit" placed at one end of a tube, which carries at the other end an achromatic object-glass having the slit in the principal focus. This tube, with slit and lens, constitutes the "collimator." Instead of looking at the spectrum with the naked eye, it is better also in most cases to use a small "view telescope," so called to distinguish it from the large telescope to which the spectroscope is often attached.

---

[1] The "grating" is merely a piece of glass or speculum metal, ruled with many thousand straight, equidistant lines, from 5000 to 20,000 in the inch.

172. Construction of the Spectroscope. — The instrument, therefore, as usually constructed, and shown in Fig. 32, consists of three parts, — collimator, dispersion-piece, and view

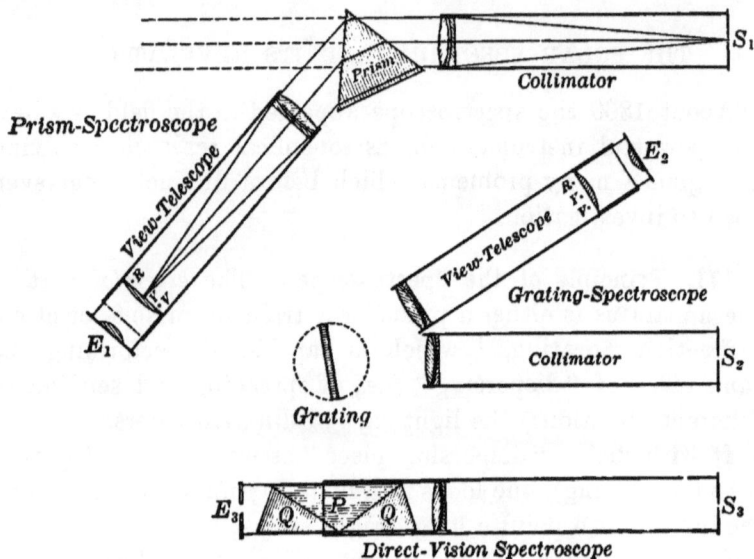

Fig. 32. — Different Forms of Spectroscope.

telescope, — although in the "direct-vision" spectroscope, shown in the figure, the view telescope is omitted. If the slit S be illuminated by strictly homogeneous light (*i.e.*, light all of one color), say *yellow*, the "real image" of the slit will be found at Y. If, at the same time, light of a different color — *red* for instance — be also admitted, a second image will be formed at R, and the observer will then see a spectrum with two bright lines, *the lines being really nothing more than images of the slit.*

If violet light be admitted, a third image will be formed at V, and there will be three bright lines. If light from a candle be admitted, there will be an infinite number of these slit-images close together, like the pickets in a fence, without interval or break, and we then get what is called a 'continuous' spectrum. If, however, we look at sunlight or moonlight or· the light of a star, we shall find a spectrum continuous in the main, but crossed by numerous dark lines, or missing slit-images (as if some of the fence-pickets had been knocked off, leaving gaps).

173. Principles upon which Spectrum Analysis depends. — These, substantially, as announced by Kirchhoff in 1858, are the three following : —

1st. A *continuous spectrum* is given by bodies which are so dense that the molecules interfere with each other in such a way as to prevent their free vibration ; *i.e.,* by bodies which are either solid or liquid, or, if gaseous, are under pressure.

2d. The spectrum of a luminous gas under low pressure is *discontinuous,* that is, it is made up of *bright lines or bands,* and these lines are characteristic. The same substance *under similar conditions* always gives the same set of lines, and usually it does so even under conditions which differ rather widely ; but when the circumstances differ too much, it may give two or more different spectra.

3d (and most important for our purpose just now). A gas or vapor *absorbs* from a beam of white light passing through it *precisely those rays of which its own spectrum consists;* so that the spectrum of white light which has been transmitted through such a vapor, if the vapor is cooler than the original source of light, exhibits a "reversed" spectrum of the gas ; *i.e.,* we get a spectrum which shows *dark* lines in place of the characteristic *bright* lines.

We therefore infer that the sun is covered by an envelope. of gases, not so hot as the luminous clouds which form the photosphere, and that these gases by their absorption produce the dark lines which we see.

174. Experiment illustrating Reversal of Spectrum. — The principle of reversal is illustrated by Fig. 33. Suppose that in front of the spectroscope we place a spirit lamp with a little

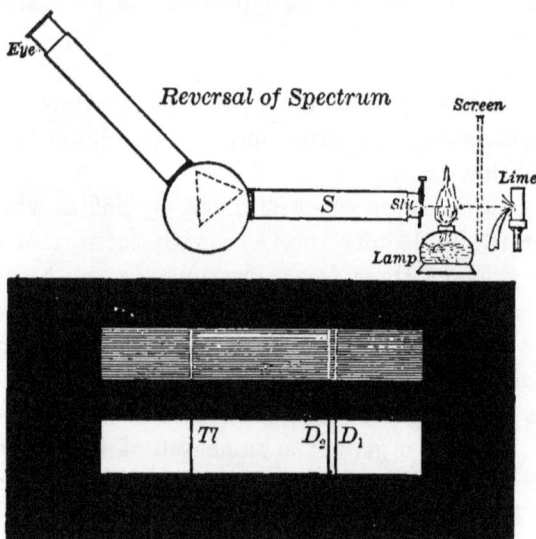

FIG. 33. — Reversal of the Spectrum.

carbonate of soda and some salt of thallium upon the wick. We shall then get a spectrum showing the two yellow lines of sodium and the green line of thallium, all bright, as in the upper of the two spectra. If now the lime light be started behind the flame, we shall at once get the effect shown in the lower figure, — a continuous spectrum crossed by three black lines which exactly replace the bright ones. Thrust a screen between the lamp flame and the lime, and the dark lines instantly turn bright again.

The dark lines which appear when the screen is removed are dark only *relatively to the background :* when the screen is taken away they really *brighten* a little (say 2 or 3 per cent); but the brightness of the background increases hundreds of times, and so far exceeds that of the lines themselves that they look black.

175. Chemical Constituents of the Solar Atmosphere. — By taking advantage of these principles, we can detect a large number of well-known terrestrial elements in the sun. The solar spectrum is crossed by dark lines,[1] which with an instrument of high power number several thousand.

By proper arrangements it is possible to identify among these lines many which are due to the presence in the sun's atmosphere of known terrestrial elements in the state of vapor. To effect the comparison necessary for this purpose, the spectroscope must be so arranged that the observer can confront the spectrum of sunlight with that of the substance to be tested. In order to do this, half of the slit is covered by a little reflector or "comparison prism," which reflects into the tube the light of the sun, while the other half of the slit receives directly the light of some flame or electric spark. On looking into the spectroscope, the observer will then see a spectrum, the lower half of which, for instance, is

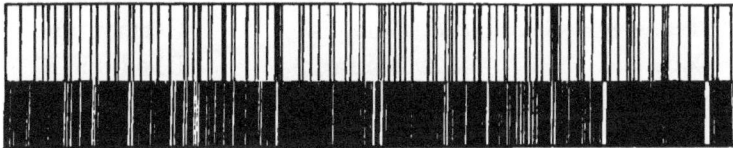

FIG. 34. — Comparison of the Spectrum of Iron with the Solar Spectrum. From a Negative by Professor Trowbridge.

made by sunlight, while the upper half is made by light coming from an electric spark between two metal points, say of iron. This latter spectrum will show the bright lines of iron vapor, and the observer can then easily see whether they do or do not correspond exactly with the dark lines of the solar spectrum.

---

[1] They are generally referred to as Fraunhofer's lines, because Fraunhofer was the first to map them. To some of the principal ones he assigned letters of the alphabet, which are still retained; thus A is a strong red line at the extreme end of the spectrum; C, one in the scarlet; D, one in the yellow; and H, one in the violet.

In such comparisons photography may be most effectively used instead of the eye. Fig. 34 is a rather unsatisfactory reproduction, on a reduced scale, of a negative made by Professor Trowbridge of Cambridge. The lower half is the violet portion of the sun's spectrum, and the upper half that of an electric arc charged with the vapor of iron.[1] The reader can see for himself with what absolute certainty such a photograph indicates the presence of iron in the solar atmosphere. A few of the lines in the photograph which do not show corresponding lines in the solar spectrum are due to impurities in the carbons of the electric arc, and not to iron.

176. Elements known to exist in the Sun. — As the result of such comparisons, we have the following list of thirty-six elements which are now (1895) known to exist in the sun : —

* Calcium, 11.	* Strontium, 23.	Copper, 30.
* Iron, 1.	Vanadium, 8.	Zinc, 29.
* Hydrogen, 22.	* Barium, 24.	Cadmium, 26.
* Sodium, 20.	Carbon, 7.	* Cerium, 10.
* Nickel, 2.	Scandium, 12.	Glucinum, 33.
* Magnesium, 19.	Yttrium, 15.	Germanium, 32.
* Cobalt, 6.	Zirconium, 9.	Rhodium, 27.
Silicon, 21.	Molybdenum, 17.	Silver, 31.
Aluminium, 25.	Lanthanum, 14.	Tin, 34.
* Titanium, 3.	Niobium, 16.	Lead. 35.
* Chromium, 5.	Palladium, 18.	Erbium, 28.
* Manganese, 4.	Neodymium, 13.	· Potassium, 36.

The substances are arranged according to the *intensity* of the dark lines by which they are represented in the solar spectrum, while the numbers appended indicate the rank which each would hold if the arrangement had been based upon the *number* of lines. An asterisk denotes that the lines of the element often or always appear as bright lines in the spectrum of the chromosphere. (Art. 180.)

In the atmosphere of the sun these bodies must be, of course, in the condition of vapor, which is somewhat cooler than

---

[1] Of course, in the negative, dark lines show bright, and *vice versa*.

the clouds which form the photosphere. It will be noticed that all of them, carbon alone excepted, are *metals* (chemically hydrogen is as much a metal as any of the others), and that a number of the elements which are among the most important in the constitution of the earth fail to present themselves. Thus far oxygen, nitrogen, chlorine, bromine, iodine, sulphur, phosphorus, and mercury all appear to be missing.

We must be cautious, however, in drawing negative conclusions. It is quite possible that the spectra of these bodies under solar conditions may be so different from their spectra as presented in our laboratories, that we cannot easily recognize them : many substances, under different conditions, give two or more widely different spectra.

177. The Reversing Layer. — According to Kirchhoff's theory the dark lines are formed by the passing of light emitted by minute solid or liquid particles of photospheric clouds through the somewhat cooler vapors which compose the substances that we recognize by the dark lines in the spectrum. If this is so, the spectrum of the gaseous envelope, which by its absorption forms the dark lines, ought to show a spectrum of corresponding bright lines when seen by itself. The opportunities are rare when it is possible to obtain a spectrum of this gaseous envelope separate from that of the photosphere; but at the time of a total eclipse, at the moment when the sun's disc has just been obscured by the moon, and the sun's atmosphere is still visible beyond the moon's limb, the observer ought to see this bright-line spectrum, if the slit of the spectroscope be carefully directed to the proper point; and the observation has actually been made. The lines of the solar spectrum, which up to the time of the total obscuration of the sun remain dark as usual, are suddenly *reversed,* and the whole field of the spectroscope is filled with brilliant colored lines, which flash out quickly, and then gradually fade away, disappearing in about two seconds.

The natural interpretation of this phenomenon is that the formation of the dark lines in the solar spectrum is, mainly at least, produced by a *very thin* stratum closely covering the photosphere, since the moon's motion in two seconds would correspond to a thickness of only 500 miles.

There are reasons, however, to doubt whether the lines are *all* produced in such a thin layer. According to Mr. Lockyer, the solar atmosphere is very extensive, and certain lines of the spectrum appear to be formed only in the regions of lower temperature high up above the surface of the photosphere. It is probable also that many lines originate *within* the photosphere and not above it, being caused by the vapors which lie between the cloud-masses that give the brilliant light.

178. Sun-Spot Spectrum. — The spectrum of a sun spot differs from the general solar spectrum not only in its diminished brilliancy, but in the great widening of certain lines, the thinning of others, and the change of some (especially the lines of hydrogen) to bright lines on some occasions. The majority of the Fraunhofer lines, however, are not much affected either way.

In the green and blue portions of the spectrum the darkest part of a sun-spot spectrum is found to be composed of fine dark lines close packed. This shows that the darkening is due to the absorption of light by *gases* and *vapors;* not by mist or smoke, for then the spectrum would be continuous.

Sometimes, in connection with sun spots, certain lines of the spectrum are bent and broken, as shown in Fig. 35. These distortions are explained by the swift motion towards or from the observer of the gaseous matter, which by its absorption produces the line in question. In the case illustrated in the figure, hydrogen was the substance.

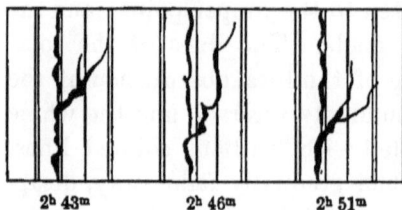

2ʰ 43ᵐ   2ʰ 46ᵐ   2ʰ 51ᵐ
FIG. 35. — The C line in the Spectrum of a Sun Spot.

and its motion was towards the observer, nearly at the rate of
300 miles a second at one point.

179. Doppler's Principle. — The principle upon which the
explanation of this displacement and distortion of lines depends was first enunciated by Doppler in 1842. It is this:
*when the distance between us and a body which is emitting regular
vibrations, either of sound or of light, is de'creasing, then the
number of pulsations received by us in each second is increased,
and the length of the waves is correspondingly diminished.* Thus
the pitch of a musical tone rises in the case supposed, and in
the same way the *refrangibility* of a light wave, which depends
upon its wave length, is increased, so that it will fall nearer
the violet end of the spectrum. This principle finds numerous
applications in modern astronomical spectroscopy, and it is of
extreme importance that the student should clearly understand it.

180. The Chromosphere. — Outside the photosphere, or shining surface of the sun, lies the so-called chromosphere, of which
the stratum of gases that produce the dark lines in the solar
spectrum is the hottest and densest portion. The word is
derived from the Greek, *chroma* (color), and means "color-
sphere." It is so-called because it is brilliantly scarlet, owing
this color to the hydrogen gas which is its most conspicuous
component. In structure, it is like a sea of flame, covering
the photosphere to a depth of from 5000 to 10,000 miles, and
as seen through a telescope at the time of a total eclipse, it
has been well described as looking like a "prairie on fire."
There is, however, no real burning in the case; *i.e.*, no heat-
producing combination of hydrogen with oxygen, or with any
other element.

Under ordinary circumstances the chromosphere is invisible,
drowned in the light of the photosphere. It can be seen with
the telescope only for a few seconds at a time, during the fleet-

ing moments of a total eclipse; but with the spectroscope it can be studied at other times, as we shall see.

181. Prominences. — The prominences, or protuberances, are scarlet clouds which are seen during a total eclipse, projecting from behind the edge of the moon. They are simply extensions of the chromosphere, or isolated clouds of the same gaseous substances, chiefly hydrogen. Their true nature was established at an eclipse in 1868, when their spectrum was first satisfactorily made out. The spectrum is composed of numerous bright lines, conspicuous among which are the lines of hydrogen, together with a brilliant yellow line (sometimes called D_3 because near the two so-called D lines) and the so-called H and K lines of calcium, with a number of others that are always present though more difficult to observe. At times also when the solar forces are peculiarly energetic hundreds of other lines appear, especially those of iron, titanium, magnesium and sodium. For a long time the D_3 line remained entirely unidentified, and the name of *helium*, or "sun-metal," was proposed and accepted for the hypothetical element to which it is due. In 1895, however, Dr. Ramsay, one of the discoverers of "*argon*," found the D_3 line in the spectrum a gas disengaged by heating and pumping from a rare mineral known as *uraninite*, and very soon it was found by him and other observers in various other minerals and in meteoric iron. Along with the D_3 line, were also found several other unidentified lines of the chromosphere spectrum, which, with D_3 and the hydrogen lines, are also found in the spectra of certain nebulæ and variable stars. It was a great triumph thus to "run helium to earth," though as yet very little is known as to its nature and properties except that, next to hydrogen, it is the lightest of all known gases, and in chemical inertness appears to resemble argon itself.

182. Spectroscopic Observations of the Prominences and Chromosphere. — Since the spectrum of these objects is composed of a small number of brilliant lines, it is possible to observe them with a spectroscope in full daylight. The explanation of the way in which the spectroscope effects this lies rather beyond our limitations; but it is sufficient for our purpose to say that by attaching a spectroscope to a good telescope the prominences can now be studied at leisure any clear day. They are wonderfully interesting and beautiful objects. Some of them, the so-called "quiescent" prominences, are of enormous size, 50,000 or even 100,000 miles in height, faint and diffuse, remaining almost unchanged for days. Others are much more brilliant and active, especially those that are associated with sun spots, as many of them are. These "eruptive" prominences often alter their appearance very rapidly, — so fast that one can sometimes actually see the motion: velocities from 50 to 200 miles a second are frequently met with. As a rule the eruptive prominences are not so large as the quiescent ones, but occasionally they surpass them, and a few have been observed to attain elevations of more than 200,000 miles. Fig. 36 gives specimens of both kinds.

182.* Photography of Prominences. — Quite recently it has become possible to photograph these objects at any time by utilizing the H and K lines in their spectrum. An explanation of the method lies quite beyond our scope, but Professor Hale, the director of the new Yerkes Observatory, and Deslandres in Paris, have been specially successful in this line, and have both constructed spectroscopic apparatus with which, at a single operation, they obtain a picture of the entire chromosphere and its prominences, surrounding an image of the sun itself with its spots and faculous regions. The solar image is really only a picture of those parts of the disc where the calcium lines are bright, and is by no means so perfect a picture as photographs made in the usual way: but it is sufficient to show how the prominences stand related to the solar surface, and its comparison with an ordinary photograph brings out many interesting peculiarities. The new method is a great step in the study of solar physics.

183. The Corona.— Probably the most beautiful and impressive of all natural phenomena is the corona, the "glory" of light which surrounds the sun at a total eclipse. The por-

Quiescent Prominences.

Flames.   Jets and Spikes near Sun's Limb, Oct. 5, 1871.
Eruptive Prominences.
Fig. 36.

tion of it near the sun is dazzlingly bright and of a pearly lustre, contrasting beautifully with the scarlet prominences, which stud it like rubies. It seems to be mainly composed of projecting filaments of light, which near the sun are pretty well defined, but at a little distance fade out and melt into the general radiance. Near the poles of the sun the corona does not usually extend very far and has a pretty definite outline, but in the spot regions and near the sun's equator faint streams sometimes extend to a distance of several degrees; and at the distance of the sun every degree means more than a million of miles.

A very striking and perplexing feature is the existence of perfectly straight dark rays or rifts, which reach clear down to the very edge of the sun.

The corona varies very greatly in brightness at different eclipses, according to the apparent diameter of the moon at the time. The portion of the corona nearest the sun is so much brighter than the outer regions that a little increase of the moon's diameter cuts off a very large proportion of the light. The total light of the corona is usually at least two or three times as great as that of the full moon.

Fig. 37 represents the corona as seen in the eclipse of 1882.

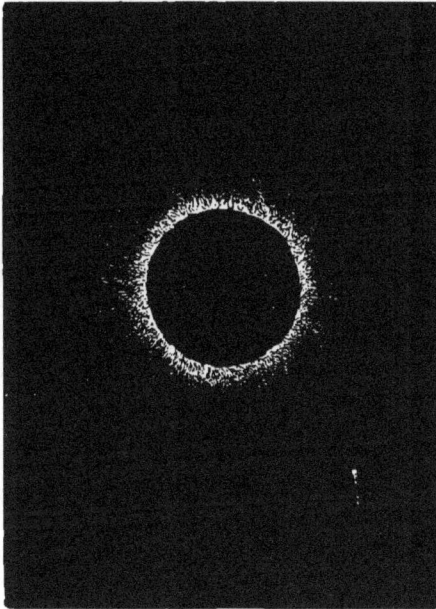

FIG. 37.
Corona of the Egyptian Eclipse, 1882.

184. Spectrum of the Corona. — A characteristic feature of its spectrum is a bright green line, generally known as the "1474" line.[1] This line was at first supposed to be due to iron, and the coincidence was for a long time puzzling (since the vapor of iron is a very improbable substance to be found at an elevation above the hydrogen of the chromosphere), until

[1] So-called because it coincides with a dark line on Kirchhoff's map of the solar spectrum, which was the chart in use when the line was first discovered, in 1869.

it was discovered that the line is really a close double. One of the two components of the dark line is due to iron, while the other, the true corona line, is due to some still unknown gaseous element (probably lighter than hydrogen), which has been called *coronium*, after the analogy of helium. It is to be hoped that before very long this substance also may be "run to earth" as helium has been.

Besides this conspicuous green line, the hydrogen lines are also faintly visible in the corona spectrum; and by means of photography it has been found that the violet and ultra-violet portions of the spectrum are also rich in bright lines, the two wide lines or bands, known as H and K in the ordinary solar spectrum, being especially bright and conspicuous.

185. The corona is proved to be a true appendage of the sun, and not, as has been at times supposed, a mere optical phenomenon, nor one due to the atmosphere of the earth or moon, by two established facts : —

1st. That its spectrum is not that of reflected sunlight, but of a self-luminous gas; and

2d. Because photographs of the corona, made at widely different stations along the track of an eclipse, agree exactly in details.

Its real nature and relation to the sun is very difficult to explain. It is a gaseous envelope, at least mainly gaseous, as our atmosphere is, but it does not stand in any such relations to the globe beneath as does the air. Its phenomena are not yet satisfactorily explained, and remind us far more of auroral streamers and of comets' tails than of anything that occurs in the lower regions of the earth's atmosphere. The material of the corona is of excessive rarity, as is shown by the fact that in a number of cases comets have passed directly through it (as, for instance, in 1882) without the slightest perceptible disturbance. Its density, therefore, must be almost inconceivably less than that of the best vacuum which we are able to produce.

SUN'S LIGHT AND HEAT.

186. The Sun's Light. — By photometric measures, which
we cannot explain here, it is found that the sun gives us 1575
billions of billions (1575 followed by 24 ciphers) times as
much light as a standard candle [1] would do at that distance.

The amount of light received from the sun is about 600,000
times that given by the full moon, about 7,000,000000 times
that of Sirius, the brightest of the fixed stars, and fully
200,000,000000 times that of the Pole-star. As to the intensity of sunlight, or the intrinsic brightness of the sun's surface, we find that it is about 190,000 times as bright as that
of the candle flame, and fully 150 times as bright as the lime
of a calcium light; so that even the darkest part of a sun spot
outshines the lime light. The brightest part of an electric arc-
light comes nearer sunlight in intensity than anything else we
know of, being from a half to a quarter as bright as the solar
surface itself.

The sun's disc is brightest near the centre, but the variation
is slight until we get pretty near the edge, where the light
falls off rapidly. Just at the sun's limb, the brightness is not
much more than a third as great as at the centre. The color
also is there modified, becoming a sort of an orange-red. This
darkening and change of color are due to the general absorption of light by the lower portions of the sun's atmosphere.
According to Langley, if this atmosphere were suddenly removed the surface would shine out somewhere from two to
five times as brightly as now, and its tint would become
strongly blue, like the color of an electric arc.

187. The Quantity of Solar Heat; the Solar Constant. — The
"solar constant" is the number of heat units which a square
unit of the earth's surface, unprotected by any atmosphere and

---

[1] The standard candle is a sperm candle weighing one-sixth of a pound
and burning 120 grains an hour. An ordinary gas-burner usually gives a
light equivalent to from ten to fifteen candles.

squarely exposed to the sun's rays, would receive from the sun in a unit of time. The heat-unit most used at present is the "calory,"[1] which is the quantity of heat required to raise the temperature of one kilogram of water 1° C.; and as the result of the best observations thus far made (Langley's) it appears that the "Solar Constant" is approximately thirty of these calories to a square metre in a minute. At the earth's surface a square metre, owing to the absorption of a large percentage of heat by the air, would, however, seldom actually receive more than from ten to fifteen calories in a minute.

The method of determining the solar constant is simple, as far as the principle goes, but the practical difficulties are serious, and thus far have prevented our obtaining all the accuracy desirable. The determination is made by allowing a beam of sunlight of known diameter to fall upon a known quantity of water for a known time, and measuring how much the water rises in temperature. The principal difficulty lies in determining the proper allowance to be made for absorption of the sun's heat in passing through the air. Besides this it is necessary to measure, and allow for, the heat which is received by the water from other sources than the sun.

188. Solar Heat at the Earth's Surface. — Since it requires about eighty calories of heat to melt one kilogram of ice, it follows that, taking the solar constant at thirty, the heat received from the sun when overhead would melt in an hour a sheet of ice about nine-tenths of an inch thick. From this it is easily computed that the amount of heat received by the earth from the sun in a year would melt a shell of ice 165 feet thick all over the earth's surface.

"Solar engines" have been constructed within the last few years, in which the heat received upon a large reflector is made to evaporate water in a suitable boiler and to drive a

---

[1] A "small calory" is also used, one thousandth as large as this : viz. the quantity of heat which will raise the temperature of one *gram* of water 1° C.

steam engine. It is found that the heat received upon a reflector ten feet square can be made to give practically about
one horse-power.

189. Radiation from the Sun's Surface. — If we attempt to
estimate the intensity of the radiation from the surface of the
sun itself, we reach results which are simply amazing. We
must multiply the solar constant observed at the earth by the
square of the ratio between the earth's distance from the sun
and the distance of the sun's surface from its own centre ; *i.e.*,
by the square of $\left(\dfrac{93,000000}{433,250}\right)$, or about 46,000: in other words,
the amount of heat emitted in a minute by a square foot of
the sun's surface is about 46,000 times as great as that received
by a square foot of surface at the distance of the earth. Carrying out the figures, we find that if the sun were frozen
over completely to a depth of over sixty feet, the heat it emits
would be sufficient to melt the ice in one minute ; that if a
bridge of ice could be formed from the earth to the sun by an
ice-column 2½ miles square, and if in some way the entire solar
radiation could be concentrated upon it, it would be melted in
one second, and in seven more would be dissipated in vapor.

Expressing it in terms of energy, we find that the solar radiation is more than 120,000 horse-power continuously, for each
square metre of the sun's surface.

So far as we can now see, only a very small fraction of this whole
radiation ever reaches a resting-place. The earth intercepts about
$\frac{1}{2300000000}$ and the other planets of the solar system receive in all
perhaps from ten to twenty times as much. Something like $\frac{1}{100000000}$
seems to be utilized within the limits of the solar system.

190. The Sun's Temperature. — We can determine with some
accuracy the *amount of heat* which the sun gives ; to find its
temperature is a very different thing, and we really have very
little knowledge about it, except that it must be extremely

high, — far higher than that of any terrestrial source of heat now known. The difficulty is that our laboratory experiments do not give the necessary data from which we can determine what temperature substances like those of which the sun is composed must have, in order to enable them to send out heat at the rate which we observe. Of two bodies at precisely the same temperature, one may send out heat a hundred times more rapidly than the other.

The estimates as to the temperature of the photosphere run all the way from the very low ones of some of the French physicists (who set it at about 2500° C.) to those of Secchi and Ericsson, who put the figure among the millions. The prevailing opinion sets it between 5000° and 10,000° C., or from 9000° to 18,000° F. [1]

A very impressive demonstration of the intensity of the sun's heat is found in the fact that in the focus of a powerful burning lens all known substances melt and vaporize ; and yet it can be shown that at the focus of the lens the temperature can never even nearly equal that of the source from which the heat is derived.

191. Constancy of the Sun's Heat. — It is still a question whether the total amount of the sun's radiation does or does not vary from time to time. There may be considerable fluctuations in the hourly or daily quantity of heat, without our being able to detect them with our present means of observation.

As to any steady *progressive* increase or decrease in the amount of heat received from the sun, it is quite certain that no considerable change has occurred for the past 2000 years, because the distribution of plants and animals on the earth's surface is practically the same as in the earliest days of history. It is, however, rather probable than otherwise that the great changes of climate, which Geology indicates as having formerly taken place on the earth, may ultimately be traced to changes in the condition of the sun.

---

[1] The determination of Wilson and Gray in 1893–5, makes it 8000° C., or a little more than 14,000° F.

192. Maintenance of the Solar Heat. — We cannot here discuss the subject fully, but must content ourselves with saying,
first, *negatively*, that this maintenance cannot be accounted
for on the supposition that the sun is a hot body, solid or
liquid, simply cooling; nor by combustion; nor (adequately)
by the fall of meteors on the sun's surface, though this cause
undoubtedly operates to a limited extent.   Second, we can say
positively that the solar radiation *can* be accounted for on the
hypothesis first proposed by Helmholtz, that the sun is mainly
gaseous, and shrinking slowly but continuously.   While we
cannot see any such shrinkage, because it is too slow, it is a
matter of demonstration that if the sun's diameter should contract about 300 feet a year, heat enough would be generated
to keep up its radiation without any lowering of its temperature.   If the shrinkage were more than about 300 feet,
the sun would be hotter at the end of the year than it was at
the beginning.

We can only say that while no other theory meets the conditions of the problem, this appears to do so perfectly, and
therefore has probability in its favor.

193. Age and Duration of the Sun. — Of course if this
theory is correct, the sun's heat must ultimately come to an
end; and looking backward it must have had a beginning.   If
the sun keeps up its present rate of radiation, it must, on this
hypothesis, shrink to about half its diameter in some 5,000000
years at the longest.   It will then be eight times as dense as
now, and can hardly continue to be mainly gaseous, so that
the temperature must begin to fall quite sensibly.   It is not,
therefore, likely, in the opinion of Professor Newcomb, that
the sun will continue to give heat sufficient to support the
present conditions upon the earth for much more than
10,000000 years, if so long.

On the other hand, it is certain that the shrinkage of the
sun to its present dimensions from a diameter larger than that

of the orbit of Neptune, the remotest of the planets, would produce about 18,000000 times as much heat as the sun now throws out in a year; hence, IF the sun's heat has been, and still is, wholly due to the contraction of its mass, it cannot have been emitting heat at the *present rate*, on this shrinkage hypothesis, for more than 18,000000 years. But notice the 'if.' It is quite possible that the solar system may have received in the past supplies of heat other than that due to the contraction of the sun's mass. If so, it may be much older.

194. Constitution of the Sun. — To sum up: The received opinion as to the constitution of the sun is that the *central mass,* or nucleus, is probably *gaseous,* under enormous pressure, and at an enormous temperature.

The *photosphere* is probably a sheet of *luminous clouds,* constituted mechanically like terrestrial clouds, that is, of small, solid, or liquid particles, very likely of carbon, floating in gas.

These photospheric clouds float in an atmosphere composed of those gases which do not condense into solid or liquid particles at the temperature of the solar surface. This atmosphere is laden, of course, with the vapors out of which the clouds have been condensed, and constitutes the *reversing layer* which produces the dark lines of the solar spectrum.

The *chromosphere* and *prominences* appear to be composed of permanent gases, mainly hydrogen and helium, which are mingled with the vapors in the region of the photosphere, but rise to far greater elevations. For the most part the prominences appear to be formed by jets of hydrogen, ascending through the interstices between the photospheric clouds, like flames playing over a coal fire.

As to the *corona,* it is as yet impossible to give any satisfactory explanation of all the phenomena that it presents, and since thus far it has been possible to observe it only during the brief moments of total eclipses, progress in its study has been necessarily slow.

CHAPTER VII.

ECLIPSES AND THE TIDES. — FORM AND DIMENSIONS OF
SHADOWS. — ECLIPSES OF THE MOON. — SOLAR ECLIPSES,
— TOTAL, ANNULAR, AND PARTIAL. — NUMBER OF
ECLIPSES IN A YEAR. — RECURRENCE OF ECLIPSES
AND THE SAROS. — OCCULTATIONS. — THE TIDES.

195. The word *Eclipse* (literally a 'swoon') is a term applied to the sudden darkening of a heavenly body, especially of the sun or moon. An eclipse of the moon is caused by its passing through the shadow of the earth; an eclipse of the sun by the moon's passing between the sun and the observer, or, what comes to the same thing, by the passage of the moon's shadow over the observer. The 'Shadow,' in Astronomy, is the *space* from which sunlight is excluded by an intervening body; speaking geometrically, it is a *solid*, not a *surface*. If we regard the sun and the other heavenly bodies as spherical, which, of course, they are very nearly, these shadows are *cones* with their axes in the line which joins the centres of the sun and the shadow-casting body, the point being always directed away from the sun. If interplanetary space were a little hazy, we should see every planet accompanied by its shadow, like a black tail behind it.

ECLIPSES OF THE MOON.

196. Dimensions of the Earth's Shadow. — The length of the shadow is easily found. In Fig. 38, O is the centre of the sun and E the centre of the earth, and aCb is the shadow of the earth cast by the sun. It is readily shown by Geometry that if we call EC, the length of the shadow, L, and OE, the distance of the earth from the sun, D, then

$$L = D \times \left(\frac{r}{R - r} \right),$$ R being OA the radius of the sun, and r

the radius of the earth Ea. This fraction, $\left(\frac{r}{R - r} \right)$, is about

$\frac{1}{108.5}$, so that $L = \frac{1}{108.5} D.$

This gives 857,000 miles for the average length of the earth's shadow. The length varies about 14,000 miles on each side

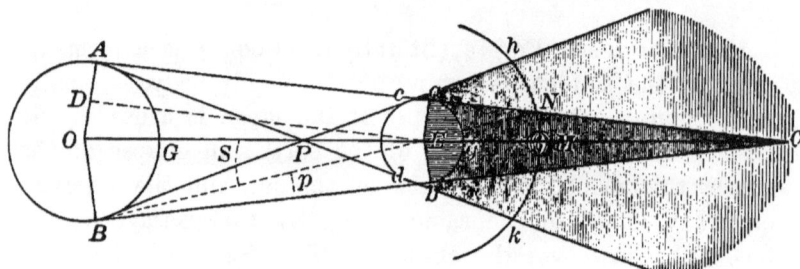

FIG. 38. — The Earth's Shadow.

of the mean, in consequence of the variation of the earth's distance from the sun at different times of the year.

From the cone aCb all sunlight is excluded, or would be were it not for the fact that the atmosphere of the earth bends some of the rays which pass near the earth's surface into its shadow. The effect of this atmospheric refraction is to increase the diameter of the shadow about two per cent, but to make it less perfectly dark.

If we draw the lines, Bc and Ad, crossing at P, between the earth and the sun, they will bound the *penumbra*, within which a part, but not the whole, of the sunlight is cut off: an observer outside of the shadow, but within this partly shaded space, would see the earth as a black body encroaching on the sun's disc, but not covering it.

197. Lunar Eclipses. — The axis, or central line, of the earth's shadow is always directed to a point directly opposite the sun. If, then, at the time of full moon the moon happens to be near the ecliptic, *i.e.*, not far from one of the nodes (the points where her orbit cuts the ecliptic), she will pass through the shadow and be eclipsed. Since, however, the moon's orbit is inclined 5° 8' to the ecliptic, lunar eclipses do not happen very frequently, seldom more than twice a year; because the moon at the full usually passes north or south of the shadow, without touching it.

Lunar eclipses are of two kinds, partial and total; total when she passes completely into the shadow; partial when she only partly enters it, going so far to the north or south of the centre that only a portion of the disc is obscured. An eclipse of the moon when central (*i.e.*, when the moon crosses the centre of the shadow) may continue total for about two hours, the interval from the first to the last contact being about two hours more. This depends upon the facts that the moon's hourly motion is nearly equal to its own diameter, and that the diameter of the earth's shadow where the moon crosses it is between two and three times the diameter of the moon itself. The duration of an eclipse that is not central varies of course with the part of the shadow traversed by the moon.

198. Phenomena of Total Eclipses of the Moon. — Half an hour or so before the moon reaches the shadow, its edge begins to be sensibly darkened by the penumbra, and the edge of the shadow itself, when it first touches the moon, appears nearly black by contrast with the bright parts of the moon's surface. To the naked eye the outline of the shadow looks fairly sharp, but even with a small telescope it appears indefinite, and with a large telescope of high magnifying power the edge of the shadow becomes entirely indistinguishable, so that it is impossible to determine within half a minute or so the time when it reaches any particular point.

After the moon has wholly entered the shadow, her disc is usually distinctly visible, illuminated with a dull copper-colored light, which is sunlight deflected around the earth into the shadow by the refraction of our atmosphere, as illustrated by Fig. 39. The brightness of the moon's disc during a total eclipse of the moon differs greatly at different times, according

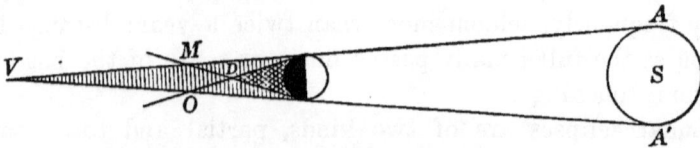

FIG. 39. — Light bent into Earth's Shadow by Refraction.

to the condition of the weather on the parts of the earth which happen to lie at the edges of the earth's disc as seen from the moon. If it is cloudy and stormy there, little light will reach the moon; if it happens to be clear, the quantity of light deflected into the shadow may be very considerable. In the lunar eclipse of 1884, the moon was for a time absolutely invisible to the naked eye, a very unusual circumstance.

During the eclipse of Jan. 28th, 1888, although the moon was pretty bright to the eye, Pickering found that its photographic power, when centrally eclipsed, was only about $\frac{1}{1400000}$ of what it had been before the shadow covered it.

199. Computation of a Lunar Eclipse. — The computation of a lunar eclipse is not at all complicated, though we do not propose to enter into it. Since all its phases are seen everywhere at the same absolute instant wherever the moon is above the horizon, it follows that a single calculation giving the Greenwich times of the different phenomena is all that is needed. Such computations are made and published in the Nautical Almanac. The observer needs only to correct the predicted time by simply adding or subtracting his longitude from Greenwich, in order to get the true local time. With an eclipse of the sun the case is very different.

ECLIPSES OF THE SUN.

200. The Length of the Moon's Shadow is very nearly
$\frac{1}{400}$ of its distance from the sun, and averages 232,150 miles.
It varies not quite 4000 miles, ranging from 236,050 to
238,300.

Since the mean length of the shadow is less than the mean
distance from the earth (238,800 miles), it is evident that *on
the average* the shadow will fall short of the earth. The eccentricity of the moon's orbit, however, is so great that she is
sometimes more than 30,000 miles nearer than at others. If
when the moon is nearest the earth, the shadow happens to
have at the same time its greatest possible length, its point
may reach nearly 18,400 miles beyond the earth's surface. In

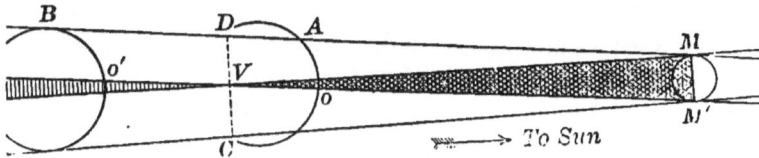

FIG. 40. — The Moon's Shadow on the Earth.

this case the "cross-section" of the shadow, where the earth's
surface cuts it (at *o* in Fig. 40) will be about 168 miles in
diameter, which is the largest value possible. On the other
hand, when the moon is farthest from the earth, we may have
the state of things indicated by placing the earth at *B*, in Fig.
40. The vertex, *V*, of the shadow will then fall 24,700 miles
short of the earth's surface, and the cross-section of the
"shadow produced" will have a diameter of 206 miles at *o'*,
where the earth's surface cuts it.

201. Total and Annular Eclipses. — To an observer within
the shadow-cone (*i.e.*, between *V* and the moon, Fig. 40), the
sun will be totally eclipsed. An observer in the "produced"
cone, beyond *V*, will see the moon apparently smaller than the

sun, leaving a ring of the sun uneclipsed; this is w hat is called an "*annular*" eclipse. These annular eclipses are considerably more frequent than the total, and now and then an eclipse is annular in part of its course across the earth, and total in part. This is when the point of the moon's shadow extends beyond the surface of the earth, but does not reach as far as its centre.

The track of the eclipse across the earth will of course be a narrow stripe having its width equal to the cross-section of the shadow, and extending across the hemisphere which is turned towards the moon at the time, though not necessarily passing the centre of that hemisphere. Its course is always from the west towards the east, but usually with considerable motion toward the north or south.

202. The Penumbra and Partial Eclipses. — The penumbra can easily be shown to have a diameter on the line *CD* (Fig. 40) a little more than twice the diameter of the moon, or over 4000 miles. An observer situated within this penumbra has a partial eclipse. If he is near to the cone of the shadow, the sun will be mostly covered by the moon; if near the outer edge of the penumbra, the moon will but slightly encroach on the sun's disc. While, therefore, a total or annular eclipse is visible as such only by observers within the narrow path traversed by the shadow-spot, the same eclipse will be visible as a partial one anywhere within 2000 miles on each side of the path; and the 2000 miles must be reckoned square to the axis of the shadow, and may correspond to a much greater distance when reckoned around upon the spherical surface of the earth.

203. Velocity of the Shadow, and Duration of an Eclipse. — Were it not for the earth's rotation, the moon's shadow would pass the observer at the rate of about 2100 miles an hour. The earth, however, is rotating towards the east in the same general direction as that in which the shadow moves, so that the relative velocity is usually much less.

A total eclipse of the sun observed at a station near the equator, under the most favorable conditions possible, may continue total for 7^m 58^s. In latitude 40° the duration can barely equal $6\frac{1}{4}^m$. At the equator an annular eclipse may last for 12^m 24^s, the maximum width of the ring of the sun visible around the moon being $1'$ $37''$.

In the observation of an eclipse, four contacts are recognized: the *first*, when the edge of the moon first touches the edge of the sun ; the *second*, when the eclipse becomes total or annular; the *third*, at the cessation of the total or annular phase; and the *fourth*, when the moon finally leaves the solar disc. From the first contact to the fourth the time may be a little over two hours. In a partial eclipse, only the first and fourth are observable, and the interval between them may be very small when the moon just grazes the edge of the sun.

The magnitude of an eclipse is usually reckoned in " digits," the digit being $\frac{1}{12}$ of the sun's diameter. An eclipse of nine digits is one in which the disc of the moon covers three-fourths of the sun's diameter at the middle of the eclipse.

204. Phenomena of a Solar Eclipse. — There is nothing of special interest until the sun is mostly covered, though before that time the shadows cast by the foliage begin to be peculiar.

The light shining through every small interstice among the leaves, instead of forming as usual a circle on the ground, makes a little crescent — an image of the partly covered sun.

About ten minutes before totality the darkness begins to be felt, and the remaining light, coming as it does from the edge of the sun, is not only faint but yellowish, more like that of a calcium light than sunshine. Animals are perplexed and birds go to roost. The temperature falls, and dew appears. In a few moments, if the observer is so situated that his view commands the distant western horizon, the moon's shadow is seen coming, much like a heavy thunder shower, and advancing with almost terrifying swiftness. As soon as the shadow arrives, and sometimes a little before, the corona and prominences become visible, while the brighter planets and stars of the first three magnitudes make their appearance.

The suddenness with which the darkness pounces upon the observer is startling. The sun is so brilliant that even the small portion which remains visible up to the moment of total obscuration so dazzles the eye that it is unprepared for the sudden transition. In a few moments, however, the eye adjusts itself, and it is found that the darkness is really not very intense. If the totality is of short duration, say not more than two minutes, there is not much difficulty in reading an ordinary watch-face. In an eclipse of long duration (four or five minutes) it is much darker, and lanterns become necessary.

205. Calculation of a Solar Eclipse. — A solar eclipse cannot be dealt with in any such summary way as a lunar eclipse, because the times of contact and the phenomena are different at every different station. The path which the shadow of a total eclipse will describe upon the earth is roughly mapped out in the Nautical Almanacs several years beforehand, and with the chart are published the data necessary to enable one to calculate with accuracy the phenomena for any given station; but the computation is rather long and somewhat complicated.

Oppolzer, a Viennese astronomer, published a few years ago a remarkable book entitled "The Canon of Eclipses," containing the elements of all eclipses (8000 solar and 5200 lunar) occurring between the year 1207 B.C. and 2162 A.D., with maps showing the approximate tracks of all the solar eclipses.

206. Frequency of Eclipses and Number in a Year. — The least possible number in a year is two, both of the sun; the largest seven, five solar and two lunar: the most usual number is four. The eclipses of a given year always take place at two opposite seasons, which may be called the "eclipse months" of the year, near the times when the sun crosses the nodes of the moon's orbit. Since the nodes move westward around the ecliptic once in about nineteen years (Art. 134), the time oc-

cupied by the sun in passing from a node to the same node
again is only 346.62 days, which is sometimes called the
" eclipse year."

Taking the whole earth into account, the solar eclipses are
the more numerous, nearly in the ratio of 3 : 2. *It is not so,
however, with those that are visible at a given place.* A solar
eclipse can be seen only by persons who happen to be on the
track described by the moon's shadow in its passage across
the globe, while a lunar eclipse is visible over considerably
more than half the earth, either at its beginning or end,
if not throughout its whole duration, — and this more than
reverses the proportion; *i.e.*, at any given place lunar eclipses
are considerably more frequent than solar. Solar eclipses that
are total somewhere or other on the earth's surface are not
very rare, averaging about one for every year and a half. But
at any given place a total eclipse happens only once in about
360 years in the long run.

During the 19th century, six shadow-tracks have already traversed
the United States, and one more will do so on May 27th, 1900, the
path in this case running from Texas to Virginia.

207. Recurrence of Eclipses; the Saros. — It was known to
the Egyptians, even in prehistoric times, that eclipses occur
at regular intervals of 18 years and $11\frac{1}{3}$ days ($10\frac{1}{3}$ days if there
happen to be five leap years in the interval). They named
this period the "Saros." It consists of 223 synodic months,
containing 6585.32 days, while 19 "eclipse years" contain
6585.78. The difference is only about 11 hours, in which time
the sun moves on the ecliptic about 28'. If, therefore, a solar
eclipse should occur to-day with the sun exactly at one of the
moon's nodes, at the end of 223 months the new moon will
find the sun again close to the node (only 28' west of it), and
a very similar eclipse will occur again; but the track of this
new eclipse will lie about 8 hours of longitude *farther west* on
the earth, on account of the odd .32 of a day in the Saros,

The usual number of eclipses in a Saros is a little over 70, varying two or three one way or the other.

In the Saros closing Dec. 22d, 1889, the total number was 72, — 29 lunar and 43 solar. Of the latter, 29 were central (13 total, 16 annular), and 14 were only partial.

THE TIDES.

208. Cause of the Tides. — Since the tides depend upon the action of the sun and of the moon upon the waters of the earth, they may properly be considered here before we deal with the planetary system. We do not propose to go into the mathematical theory of the phenomena at all, as it lies far beyond our limitations; but any person can see that a liquid globe falling freely towards an attracting body, which attracts the nearer portions more powerfully than the more remote, will be drawn

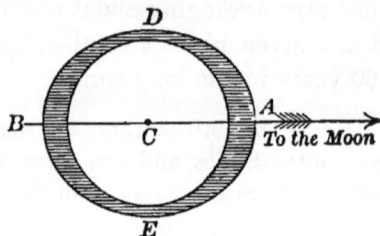

Fig. 41. — The Tides.

out into an elongated lemon-shaped form, as illustrated in Fig. 41; and if the globe, instead of being liquid, is mainly solid, but has large quantities of liquid on its surface, substantially the same result will follow. Now the earth is free in space, and though it has other motions, it is also *falling* towards the moon and towards the sun, and is affected precisely as it would be if its other motions did not exist. The consequence is that at any time there is a tendency to elongate those diameters of the earth which are pointed towards the moon and towards the sun. The sun is so much farther away than the moon that its effect in thus deforming the surface of the earth is only about two-fifths as great as that of the moon.

209. The tides consist in a regular rise and fall of the ocean surface, the average interval between corresponding high waters on successive days at any given place being twenty-four hours and fifty-one minutes, which is precisely the same as the average interval between two successive passages of the moon across the meridian; and since this coincidence is maintained indefinitely, it of itself makes it certain that there must be some causal connection between the moon and the tides. Some one has said that the odd fifty-one minutes is the moon's "ear mark."

That the moon is largely responsible for the tides is also shown by the fact that when the moon is in perigee, at the nearest point to the earth, the tides are nearly twenty per cent higher than when she is in apogee.

210. Definitions. — While the water is rising, it is *flood* tide; while falling, it is *ebb* tide. It is *high water* at the moment when the water level is highest, and *low water* when it is lowest. The *spring tides* are the largest tides of the month, which occur near the times of new and full moon, while the *neap tides* are the smallest, and occur at half moon, the relative heights of spring and neap tides being about as 7 : 3. At the time of the spring tides, the interval between the corresponding tides of successive days is less than the average, being only about 24 hours, 38 minutes (instead of 24 hours, 51 minutes), and then the tides are said to *prime*. At the neap tides, the interval is greater than the mean, — about 25 hours, 6 minutes, and the tide *lags*. The *establishment* of a port is the mean interval between the time of high water at that port and the next preceding passage of the moon across the meridian. The "establishment" of New York, for instance, is 8 hours, 13 minutes. The actual interval between the moon's transit and high water varies, however, nearly half an hour on each side of this mean value at different times of the month, and under varying conditions of the weather.

211. Motion of the Tides. — If the earth were wholly composed of water, and if it kept always the same face towards the moon, as the moon does towards the earth, then (leaving out of account the sun's action for the present) a permanent tide would be raised upon the earth, as indicated in Fig. 41. The difference between the water level at A and D would be a little less than two feet. Suppose, now, the earth put in rotation. It is easy to see that the two tidal waves A and B would move over the earth's surface, following the moon at a certain angle dependent on the inertia of the water, and tending to move with a westward velocity equal to the earth's eastward rotation, — about a thousand miles an hour at the equator. The sun's action would produce similar tides superposed upon the moon's tide, and about two-fifths as large, and at different times of the month these two pairs of tides would sometimes conspire and sometimes be opposed.

If the earth were entirely covered with deep water, the tide waves would run around the globe regularly; and if the depth of the water were not less than thirteen miles, the tide crests, as can be shown (though we do not undertake it here), *would follow the moon at an angle of just* 90°: it would be high water just where it might at first be supposed we should get low water, the place of high water being shifted 90° by the rotation of the earth.

If the depth of the water were, as it really is, much less than thirteen miles, the tide wave in the ocean could not keep up with the moon, and this would complicate the results. Moreover, the continents of North and South America, with the southern Antarctic continent, make a barrier almost from pole to pole, leaving only a narrow passage at Cape Horn. As a consequence it is quite impossible to determine *by theory* what the course and character of tide waves must be. We have to depend upon observations, and observations are more or less inadequate, because, with the exception of a few islands, our only possible tide stations are on the shores of continents where local circumstances largely control the phenomena.

212. Free and Forced Oscillations. — If the water of the ocean is suddenly disturbed, as, for instance, by an earth- quake, and then left to itself, a "free wave" is formed, which, if the horizontal dimensions of the wave are large as compared with the depth of the water (*i.e.*, if it is many hundred miles in length), will travel at a rate which depends simply on the depth of the water.

Its velocity is equal, as can be proved, to the velocity acquired by a body in falling through half the depth of the ocean. Observations upon waves caused by certain earthquakes in South America and Japan have thus informed us that between the coasts of those coun· tries the Pacific averages between two and one-half and three miles in depth.

Now as the moon in its apparent diurnal motion passes across the American continent each day and comes over the Pacific Ocean, it starts such a "parent" wave in the Pacific, and a second one is produced twelve hours later. These waves, once started, move on nearly (but not exactly) like a free earth- quake wave: not exactly, because the velocity of the earth's rotation being about 1040 miles at the equator, the moon moves (relatively) westward faster than the wave can natu- rally follow it; and so for a while the moon slightly acceler- ates the wave. The tidal wave is thus, in its origin, a "forced oscillation": in its subsequent travel it is very nearly, but not entirely, "free."

Of course as the moon passes on over the Indian and Atlan- tic oceans, it starts waves in them also, which combine with the parent wave coming in from the Pacific.

213. Course of Travel of the Tide Wave. — The parent wave appears to start twice a day in the Pacific Ocean, off Callao, on the coast of South America. From this point the wave travels northwest through the deep water of the Pacific, at the rate of about 850 miles an hour, reaching Kamtchatka in ten hours. Through the shallow water to the west and southwest the velocity is only from 400 to 600 miles an hour, so that the wave is six hours old when it reaches New Zealand. Passing on by Australia and combining with the small wave which the moon starts in the Indian Ocean, the resultant tide crest reaches the Cape of Good Hope in about twenty-nine hours, and enters the Atlantic. Here it combines with a smaller tide wave, twelve hours younger, which has "backed" into the Atlantic around Cape Horn, and it is also modified by the direct tide produced by the moon's action upon the Atlantic. The tide resulting from the combination of these three then travels northward through the Atlantic at the rate of about 700 miles an hour. It is about forty hours old when it first reaches the coast of the United States in Florida; and our coast lies in such a direction that it arrives at all the principal ports within two or three hours of the same time. It is forty-one or forty-two hours old when it reaches New York and Boston. To reach London, it has to travel around the northern end of Scotland and through the North Sea, and is nearly sixty hours old when it arrives at that port.

In the great oceans there are three or four such tide crests, following nearly in the same track, but with continual minor changes.

214. Height of the Tides. — In mid-ocean the difference between high and low water is usually between two and three feet, as observed on isolated islands in the deep water. On the continental shores the height is ordinarily much greater.

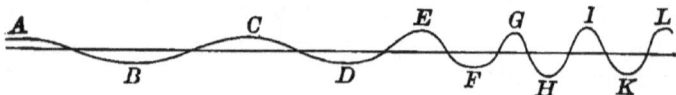

FIG. 42. — Increase in Height of Tide on approaching the Shore.

As soon as the tide wave "touches bottom," so to speak, the velocity is diminished, the tide crests are crowded more closely together, and the height of the tide is very much increased, as indicated in Fig. 42.

Theoretically it varies inversely as the fourth root of the depth ; *i.e.*, where the water is 100 feet deep, the tide wave should be twice as high as at the depth of 1600 feet.

Where the configuration of the shore forces the tide into a corner, it sometimes rises very high. At Minas Basin on the Bay of Fundy, tides of seventy feet are not uncommon, and an altitude of 100 feet is said to occur sometimes. At Bristol in the English Channel, tides of forty or fifty feet are reached ; at the same time on the coast of Ireland, just opposite, the tide is very small.

215. Tides in Rivers. — The tide wave ascends a river at a rate which depends upon the depth of the water, the amount of friction, and the swiftness of the stream. It may, and generally does, ascend until it comes to a rapid where the velocity of the current is greater than that of the wave. In shallow streams, however, it dies out earlier. Contrary to what is usually supposed, it often ascends to an elevation far above that of the highest crest of the tide wave at the river's mouth. In the La Plata and Amazon, the tide goes up to an elevation of at least 100 feet above the sea-level. The velocity of a tide wave in a river seldom exceeds ten or twenty miles an hour, and is ordinarily much less.

CHAPTER VIII.

THE PLANETARY SYSTEM.

THE PLANETS IN GENERAL. — THEIR NUMBER, CLASSI-
FICATION, AND ARRANGEMENT. — BODE'S LAW. — THEIR
ORBITS. — KEPLER'S LAWS AND GRAVITATION. — AP-
PARENT MOTIONS AND THE SYSTEMS OF PTOLEMY
AND COPERNICUS. — DETERMINATION OF DATA RELAT-
ING TO THE PLANETS, THEIR DIAMETER, MASS, ETC. —
HERSCHEL'S ILLUSTRATION OF THE SOLAR SYSTEM. —
DESCRIPTION OF THE TERRESTRIAL PLANETS, MERCURY,
VENUS, AND MARS.

216. THE earth is one of a number of bodies called planets
which revolve around the sun in oval orbits that are nearly
circular and lie nearly in one plane or level. There are eight
of them which are of considerable size, besides a group of several hundred minute bodies called the asteroids, which seem
to represent in some way a ninth planet, either broken to
pieces or somehow ruined in the making.

217. Classification of the Planets. — The four inner ones
have been called by Humboldt the *terrestrial* planets, because
the earth is one of them, and the others resemble it in size
and density. In the order of distance from the sun they are
Mercury, Venus, the earth, and Mars. The four outer ones
Humboldt calls the *major* planets, because they are much
larger and move in larger orbits. They seem to be bodies of
a different sort from the earth, very much less dense and

probably of higher temperature. They are Jupiter, Saturn, Uranus, and Neptune. The asteroids (from the Greek *astēr-eidos, i.e.,* star-like planets), called by some planetoids, or minor planets, all lie in the vacant space between Mars and Jupiter, and appear to contain in the aggregate about as much material as would make a planet not far from the size of Mars. All of the planets except Mercury and Venus have satellites. The earth has one, Mars two, Jupiter four, Saturn eight, Uranus four, Neptune one, — twenty in all.

218. The following little table contains in round numbers the principal numerical facts as to the planets: —

Name.	Distance in Astronomical Units.	Period.	Diameter.
Mercury	0.4	3 months	3000 miles
Venus	0.7	7½ months	7700 "
Earth	1.0	1 year	7913 "
Mars	1.5	1 yr. 10 mos.	4200 "
Asteroids	3.0 ±	3 years to 9 years	500 to 10 miles
Jupiter.	5.2	11.9 years	86,000 miles
Saturn.	9.5	29.5 "	73,000 "
Uranus	19.2	84.0 "	32,000 "
Neptune	30.1	164.8 "	35,000 "

This table should be learned by heart. More accurate data will be given hereafter, but the round numbers are quite sufficient for all ordinary purposes, and are much more easily remembered.

219. Bode's Law. — If we set down a row of 4's, to the second 4 add 3, to the third 6, to the fourth 12, etc., a series of numbers will result which, divided by 10, will represent the planetary distances very nearly, except in the case of Neptune, whose distance is only 30 instead of 38, as the rule would make it. Thus —

4	4	4	4	4	4	4	4	4
3	6	12	24	48	96	192	384
4	7	10	16	[28]	52	100	196	388
☿	♀	⊕	♂	①	♃	♄	♅	♆

(The characters below the numbers are the *symbols* of the planets, used in almanacs instead of their names.)

This law seems to have been first noticed by Titius of Wittenberg, but bears the name of Bode, Director of the Observatory of Berlin, who first secured general attention to it.

No logical reason can yet be given for it. It may be a mere convenient coincidence, or it may be the result of the process of development which brought the solar system into its present state.

220. Kepler's Laws. — Three famous laws discovered by Kepler (1607–1620) govern the motions of the planets: —

I. The orbit of each planet is an ellipse with the sun in one of its foci. (See Appendix, Art. 429, for a description of the ellipse.)

II. In the motion of each planet around the sun, the radius vector describes equal areas in equal times. (See Art. 121, Fig. 13, for illustration.)

III. The squares of the periods of the planets are proportional to the cubes of their mean distances from the sun. This is known as the *Harmonic* Law. Stated as a proportion it reads: $P_1^2 : P_2^2 :: A_1^3 : A_2^3$, or in words: The *square* of the *period* of planet No. 1: *square* of the *period* of planet No. 2:: *cube* of the *mean distance* of planet No. 1: *cube* of the *mean distance* of planet No. 2. Planets No. 1 and No. 2 are any pair of planets selected at pleasure. (For fuller illustration, see Appendix, Art. 430.)

It was the discovery of this law which so filled Kepler with enthusiasm that he wrote, " If God has waited 6000 years for a discoverer, I can wait as long for a reader."

221. Gravitation. — When Kepler discovered these three laws he could give no reason for them — no more than we can now for Bode's law; — but some sixty years later Newton discovered that they all follow necessarily as the consequence of the law of gravitation, which he had discovered; namely, that *"every particle of matter in the universe attracts every other particle with a force that varies directly as the masses of the particles, and inversely as the square of the distance between them."* It would take us far beyond our limits to attempt to show *how* Kepler's laws follow from this, but they do. The only mystery in the case is the mystery of the *"attraction"* itself; for this word *"attraction"* is to be taken as simply describing an effect without in the least explaining it.

Things take place *as if* the atoms had in themselves intelligence to recognize each other's positions, and power to join hands in some way, and pull upon each other through the intervening space, whether it be great or small. But neither Newton nor any one else supposes that atoms are really endowed with any such power, and the explanation of gravity remains to be found: very probably it is somehow involved in that constitution of the material universe which makes possible the transmission through space of light and heat, and electric and magnetic forces.

222. Sufficiency of Gravitation to explain the Planetary Motions. — We wish to impress as distinctly as possible upon the student one idea; this namely, that given a planet once in motion, nothing further than gravitation is required to explain perfectly all its motions forever after. Many half-educated people have an idea that some other force or mechanism must act to keep the planets going. This is not so: not a single motion in the whole planetary system has ever yet been detected for which gravitation fails to account.

223. Map of the Orbits. — Fig. 43 shows the smaller orbits of the system (including the orbit of Jupiter) drawn to scale, the radius of the earth's orbit being taken as four-tenths of an inch.

On this scale, the diameter of Saturn's orbit would be 7.4 inches, that of Uranus would be 13.4 inches, and that of Neptune about two

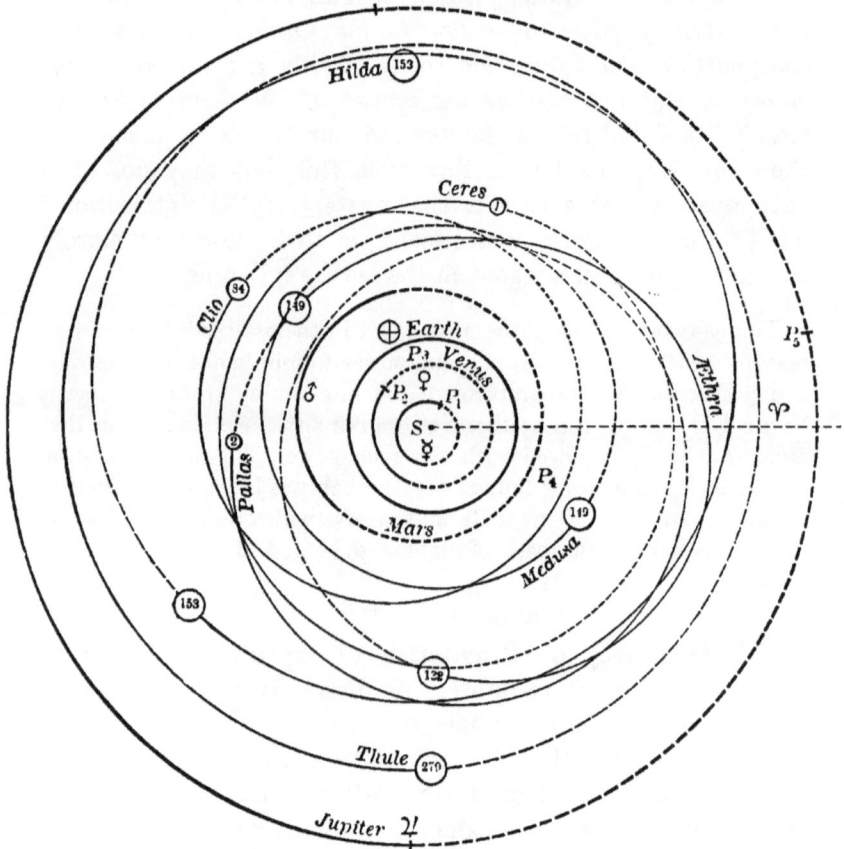

FIG. 43. — Plan of the Smaller Planetary Orbits.

feet. The nearest fixed star, on the same scale, would be a mile and a quarter away.

It will be seen that the orbits of Mercury, Mars, Jupiter, and several of the asteroids are quite distinctly "out of centre " with respect to the sun. The orbits are so nearly circular that there is no noticeable difference between their length and their breadth, but the eccentricity shows plainly in the position of the sun.

224. Inclination of the Orbits. — The orbits are drawn as if they all lay on the plane of the ecliptic; *i.e.*, on the surface of the paper. This is not quite correct. The orbit of the asteroid Pallas should be really tipped up at an angle of nearly 30°, and that of Mercury, which is more inclined to the ecliptic than the orbit of any other of the principal planets, is sloped at an angle of 7°. The inclinations, however, are so small (excepting the asteroids) that they may be neglected for ordinary purposes. On the scale of the diagram, Neptune, which rises and falls the most of all with reference to the plane of the

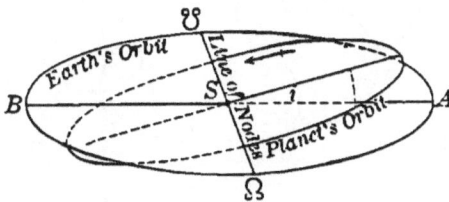

FIG. 44. — Inclination and Line of Nodes.

ecliptic, would never be more than a third of an inch above or below the level of the paper.

The line in which the plane of a planet's orbit cuts the plane of the earth's orbit at the ecliptic is called the *Line of Nodes*. Fig. 44 shows how the line of nodes and the inclination of the two orbits are related.

225. Geocentric Motions of the Planets; *i.e.,* **their motions with respect to the earth regarded as the centre of observation.**

While the planets revolve regularly in nearly circular orbits around the sun, with velocities[1] which depend upon their distance from it, the motions relative to the earth are very different, being made up of the planet's real motion combined

---

[1] A planet's velocity in miles per second equals very nearly $\dfrac{18.5}{\sqrt{\text{Distance}}}$.

with the apparent motion due to that of the earth in her own orbit.

If, for instance, we keep up observations, for a long time, of the direction of Jupiter as seen from the earth, at the same time watching the changes of its distance by measuring the alterations of the planet's apparent size as seen in the telescope, and then plot the results to get the form of the orbit of Jupiter with reference to the earth, we get a path like that shown in Fig. 45, which represents his motion relative to the earth during a term of about twelve years. The appearances are all the same as if the earth were really at rest while the planet moved in this odd way.

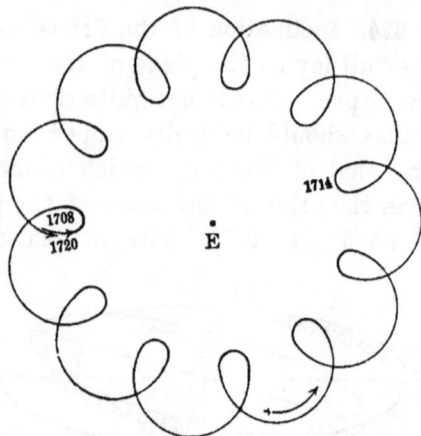

FIG. 45.
Apparent Geocentric Motion of Jupiter.

The procedure for finding this relative orbit of Jupiter is the same as that indicated in Appendix, Art. 428, for finding the form of the earth's orbit around the sun.

226. Direct and Retrograde Motion. — With the eye alone the changes in a planet's diameter would not be visible, and we should notice only the alternating direct (eastward) and retrograde (westward) motion of the planet among the stars. If we watch one of the planets (say Mars) for a few weeks, beginning at the time when it rises at sunset, we shall find that each night it has travelled some little distance to the west; and it will keep up this westward or retrograde motion for some weeks, when it will stop or become "stationary," and will then reverse its motion and begin to move eastward. If

we watch long enough (*i.e.*, for several years) we shall find
that it keeps up this oscillating motion all the time, the length
of its eastward swing being always greater than that of the
corresponding westward one.   All the planets, without exception, behave alike in this respect, as to their alternate direct
and retrograde motion among the stars.

227. Elongation and Conjunction. — The visibility of a
planet does not, however, depend upon its position among

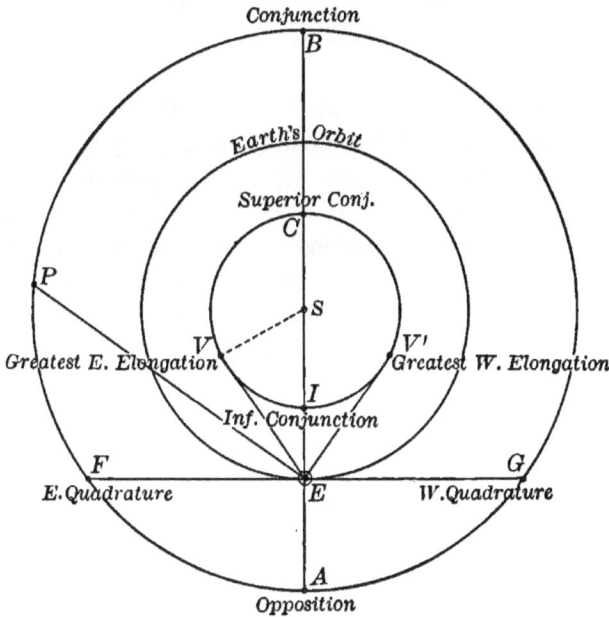

FIG. 46. — Planetary Configurations.

the stars, but upon its position in the sky with reference to
the sun's place.   If it is very near the sun, it will be above
the horizon only by day, and generally we cannot see it.   The
Elongation of a planet is the apparent distance from the sun
in degrees, as seen from the earth, of course.   In Fig. 46, for
the planet *P*, it is the angle *PES.*   When the planet is in line

with the sun as seen from the earth, at *B, C,* or *I* in the figure, the elongation is zero, and the planet is said to be in *conjunction; inferior conjunction,* if the planet is between the earth and the sun, as at *I; superior,* if beyond the sun, as at *B* or *C.* When the elongation is 180°, as at *A,* the planet is said to be in *opposition.* When the planet is at an elongation of 90°, as at *F* or *G,* it is in quadrature. Evidently only those planets which *lie within* the earth's orbit, and are called 'inferior' planets, can have an inferior conjunction; and only those which are outside the earth's orbit (the superior planets) can come to quadrature or opposition.

228. Synodic Period. — The synodic period of a planet is the time occupied by it in passing from conjunction to conjunction again, or from opposition to opposition; so called because the word "synod" is derived from two Greek words which mean 'a coming together.' The relation of the synodic period to the sidereal is the same for planets as in the case of the moon. If *E* is the length of the true (sidereal) year, and *P* the planet's period, *S* being the length of the synodic period, then

$$\frac{1}{S} = \frac{1}{E} - \frac{1}{P}$$

(The difference between $\frac{1}{E}$ and $\frac{1}{P}$ is to be taken without regard to which of the two is the larger.)

229. The Synodic Motion, or Apparent Motion of a Planet with respect to 'Elongation' or to the Sun's Place in the Sky. — In this respect there is a marked difference between the superior and inferior planets.

(*a*) The inferior planets are never seen very far from the sun, but appear to oscillate back and forth in front of and behind him. Venus, for instance, starting at superior conjunction at *C* (Fig. 46), seems to come out eastward from the sun as an evening star, until, at the point V, she reaches her greatest *eastern* elongation, about 47° from the sun. Then she begins to diminish her elongation, and approaches the sun, until she comes to inferior conjunction, at I. From there she continues to move westward as morning star, until she comes to V', her greatest *western* elongation, and there she begins to diminish her western elongation until, at the end of the synodic period, she is back at superior conjunction. The time taken to move from V' to V through C is, in her case, more than three times that required to slide back from V to V' through I. The gain of eastern elongation is up-hill work, as she is then, so to speak, pursuing the sun, which itself moves eastward nearly a whole degree every day along the ecliptic.

(*b*) The superior planets may be found at all elongations, and do not oscillate back and forth with reference to the apparent place of the sun, but continually increase their western elongation or decrease their eastern. *They always come to the meridian earlier on each successive night*, though the difference is not uniform.

230. Ptolemaic and Copernican Systems. — Until the time of Copernicus (about 1540) the Ptolemaic System prevailed unchallenged. It rejected the idea of the earth's rotation (though Ptolemy accepted the *rotundity* of the earth), placing her at the centre of things and teaching that the apparent motions of the stars and planets were real ones. It taught that the celestial sphere revolves daily around the earth, carrying the stars and planets with it, and that besides this diurnal motion, the moon, the sun, and all the planets revolve around the earth within the sphere, the two former steadily, but the planets with the peculiar looped motion shown in Fig. 45.

Copernicus put the sun at the centre, and made the earth revolve on its axis and travel around the sun, and showed that it was possible in this simple way to account for all the otherwise hopelessly complicated phenomena of the planetary and diurnal motions, so far as then known.   It was not until after the invention of the telescope, and the introduction of new methods of observation, that the facts which absolutely demonstrate the orbital motion of the earth were brought to light; viz., Aberration of Light (Appendix, Art. 435) and Stellar Parallax (Art. 433).

THE PLANETS THEMSELVES.

231. In studying the planetary system we meet a number of inquiries which refer to the planet itself and not to its orbit; relating, for instance, to its *magnitude;* its *mass, density,* and *surface-gravity;* its *diurnal rotation* and *oblateness;* its *brightness, phases,* and *reflecting power,* or *"albedo";* the *peculiarities of its spectrum;* its *atmosphere;* its *surface-markings* and *topography;* and, finally, its *satellite system.*

232. Magnitude. — The size of a planet is found by measuring its apparent diameter (in seconds of arc) with some form of "micrometer" (see Appendix, Art. 415).   Since we can find the distance of a planet from the earth at any moment when we know its orbit, this micrometric measure will give us the means of finding at once the planet's diameter in miles.

If we take r to represent the number of times by which the planet's semi-diameter exceeds that of the earth, then the area of the planet's surface compared with that of the earth equals r^2, and its volume or bulk equals r^3.   The nearer the planet, other things being equal, the more accurately r and the quantities to be derived from it can be determined.   An error of $0''.1$ in measuring the apparent diameter of Venus, when nearest us, counts for less than thirteen miles; while in Neptune's case, the same error would correspond to more than 1300 miles.

233. Mass, Density, and Gravity. — If the planet has a satellite, its mass is very easily and accurately found from the following proportion, which we simply state without demonstration (see General Astronomy, Arts. 536, 539); viz.: —

$$Mass\ of\ Sun : mass\ of\ Planet :: \frac{A^3}{T^2} : \frac{a^3}{t^2},$$

in which A is the mean distance of the planet from the sun and T its sidereal period of revolution, while a is the distance of the satellite from the planet, and t its sidereal period; whence

$$Mass\ of\ Planet = Sun \times \left(\frac{a^3}{t^2} \times \frac{T^2}{A^3} \right).$$

Substantially the same proportion may be used to compare the planet with the earth; viz.: —

$$(Earth + Moon) : (Planet + Satellite) :: \frac{a_1^3}{t_1^2} : \frac{a_2^3}{t_2^2},$$

a_1 and t_1 being here the period and distance of the moon, and a_2 and t_2 those of the planet's satellite.

If the planet has no satellite, the determination of its mass is a difficult matter, depending upon perturbations produced by it in the motions of the other planets.

Having the planet's mass compared with the earth, we get its *density* by dividing the mass by the volume, and the superficial gravity is found by dividing by r^2 the mass of the planet compared with that of the earth.

234. The Rotation Period and Data connected with it. — The length of the planet's day, when it can be determined at all, is ascertained by observing with the telescope some spot on the planet's disc and noting the interval between its returns to the same apparent position. The inclination of the planet's equator to the plane of its orbit, and the position of its equinoxes, are deduced from the same observations that give the planet's diurnal rotation; we have to observe the path pursued by a spot in its motion across the disc. Only Mars, Jupiter, and Saturn permit us to find these elements with any considerable accuracy.

The ellipticity or oblateness of the planet, due to its rotation, is found by taking measures of its polar and equatorial diameters.

235. Data relating to the Planet's Light. — A planet's brightness and its reflecting power, or "albedo," are determined by photometric observations, and the spectrum of the planet's light is of course studied with the spectroscope. The question of the planet's atmosphere is investigated by means of various effects upon the planet's appearance and light, and by the phenomena that occur when the planet comes very near to a star or to some other heavenly body which lies beyond. The planet's surface-markings and topography are studied directly with the telescope, by making careful drawings of the appearances noted at different times. Photography, also, is beginning to be used for the purpose. If the planet has any well-marked and characteristic spots upon its surface by which the time of rotation can be found, then it soon becomes easy to identify such as are really permanent, and after a time we can chart them more or less perfectly; but we add at once that Mars is the only planet of which, so far, we have been able to make anything which can be fairly called a map.

236. Satellite System. — The principal data to be ascertained are the distances and periods of the satellites. These are obtained by micrometric measures of the apparent distance and direction of each satellite from the planet, followed up for a considerable time. In a few cases it is possible to make observations by which we can determine the diameters of the satellites, and when there are a number of satellites together their masses may sometimes be ascertained from their mutual perturbations. With the exception of our moon and Iapetus, the outer satellite of Saturn, all the satellites of the solar system *move almost exactly in the plane of the equator* of the planet to which they belong; at least, so far as known, for we do not know with certainty the position of the equators of Uranus and Neptune. Moreover, all the satellites, except the moon and Hyperion, the seventh satellite of Saturn, move in orbits that are practically *circular*.

237. Tables of Planetary. Data. — In the Appendix we present tables of the different numerical data of the solar system, derived from the best authorities and calculated for a solar parallax of 8''.80, the sun's mean distance being, therefore, taken as 92,897,000 miles. These tabulated numbers, however, differ widely in accuracy. The periods of the planets and their distances in 'astronomical units' are very accurately known; probably the last decimal in the table may be trusted. Next in certainty come the masses of such planets as have satellites, expressed in terms of the sun's mass. The masses of Venus and Mercury are much more uncertain.

The distances of the planets *in miles*, their masses *in terms of the earth's mass*, and their diameter *in miles*, all involve the solar parallax, and are affected by the slight uncertainty in its amount. For the remoter planets, diameters, volumes, and densities are all subject to a very considerable percentage of error. The student need not be surprised, therefore, at finding serious discrepancies between the values given in these tables and those given in others, amounting in some cases to ten or twenty per cent, or even more. Such differences merely indicate the actual uncertainty of our knowledge. Fig. 47 gives an idea of the relative sizes of the planets.

The sun, on the scale of the figure, would be about a foot in diameter.

238. Sir John Herschel's Illustration of the Dimensions of the Solar System. — In his "Outlines of Astronomy," Herschel gives the following illustration of the relative magnitudes and distances of the members of our system : —

" Choose any well-levelled field. On it place a globe two feet in diameter. This will represent the sun. *Mercury* will be represented by a grain of mustard seed on the circumference of a circle 164 feet in diameter for its orbit; *Venus*, a pea, on a circle of 284 feet in diameter; the *Earth*, also a pea, on a circle of 430 feet; *Mars*, a rather large pin's head, on a circle of 654 feet; the *asteroids*, grains of sand, on orbits having a diameter of 1000 to 1200 feet; *Jupiter*, a moderate-sized orange, on a circle nearly half a mile across; *Saturn*, a small orange, on a circle of four-fifths

FIG. 47. — Relative Size of the Planets.

of a mile; *Uranus*, a full-sized cherry or small plum, upon a circumference of a circle more than a mile in diameter; and, finally, *Neptune*, a good-sized plum, on a circle about $2\frac{1}{2}$ miles in diameter."

We may add that, on this scale, the nearest star would be on the opposite side of the globe, 8000 miles away.

THE TERRESTRIAL PLANETS, — MERCURY, VENUS, AND MARS.

239. Mercury has been known from the remotest antiquity, and among the Greeks it had for a time two names, — Apollo when it was morning star, and Mercury when it was evening star. It is so near the sun that it is comparatively seldom seen with the naked eye, but when near its greatest elongation it is easily enough visible as a brilliant reddish star of the first magnitude, low down in the twilight. It is best seen in the evening at such eastern elongations as occur in the spring. When it is morning star, it is best seen in the autumn.

It is exceptional in the solar system in various ways. It is the *nearest* planet to the sun, *receives the most light and heat*, is the *swiftest* in its movement, and (excepting some of the asteroids) *has the most eccentric orbit*, with the *greatest inclination to the ecliptic*. It is also the *smallest in diameter* (again excepting the asteroids), has the *least mass*, and (probably) the *greatest density* of all the planets.

240. Its Orbit. — The planet's mean distance from the sun is 36,000000 miles, but the eccentricity of its orbit is so great (0.205) that the sun is 7,500000 miles out of the centre, and the distance ranges all the way from $28\frac{1}{2}$ millions to $43\frac{1}{2}$ millions, while the planet's velocity in the different parts of its orbit varies from 36 miles a second to only 23. A given area upon its surface receives on the average nearly seven times as much light and heat as it would on the earth; but the heat received when the planet is at perihelion is $2\frac{1}{4}$ times greater than at aphelion. For this reason there must be at least two seasons in its year, due to the changing distance of the planet from the sun, whatever may be the position of its equator or the length of its day. The sidereal period is 88 days, and the synodic period (or time from conjunction to conjunction) is 116 days. The greatest elongation ranges from 18° to 28°, and occurs about 22 days before and after the inferior conjunction. The inclination of the orbit to the ecliptic is about 7°.

241. Planet's Magnitude, Mass, etc. — The *apparent diameter* of Mercury varies from 5″ to about 13″, according to its distance from us; and its real diameter is very near 3000

miles.   This makes its surface about one-seventh that of the
earth, and its bulk or volume one-eighteenth.   The planet's
mass is very difficult to determine, since it has no satellite,
and it is not accurately known.   Probably it is about one-
twenty-seventh of the earth's mass; it is certainly smaller
than that of any other planet (asteroids excepted).   Our uncertainty as to the mass prevents us from assigning certain
values to its density or superficial gravity, though it is probably two-thirds as dense as the earth, and the force of gravity
upon it is about one-quarter what it is upon the earth.

242. Telescopic Appearances, Phases, etc. — Seen through
the telescope the planet looks like a little moon, showing

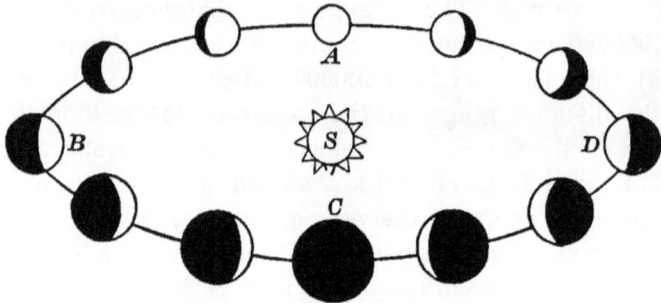

FIG. 48. — Phases of Mercury and Venus.

phases precisely similar to those of our satellite.   At inferior
conjunction the dark side is towards us; at superior conjunction the illuminated surface.   At greatest elongation, the
planet appears as a half-moon.   It is gibbous between superior
conjunction and greatest elongation, while between inferior
conjunction and greatest elongation it is crescent.   Fig. 48
illustrates these phases.

The atmosphere of the planet cannot be as dense as that of
the earth or Venus, because at a transit it shows no encircling
ring of light, as Venus does (Art. 248); both Huggins and
Vogel, however, report that the spectrum of the planet, in

addition to the ordinary dark lines belonging to the spectrum of reflected sunlight, shows certain bands known to be due to water-vapor, thus indicating that water exists in the planet's atmosphere.

Generally, Mercury is so near the sun that it can be observed only by day; but when proper precautions are taken to screen the object-glass of the telescope from direct sunlight, the observation is not especially difficult. The surface presents very little of interest. The disc is brighter at the edge than at the centre, but the markings are not well enough defined to give us any really satisfactory information as to its topography.

The *albedo*, or reflecting power, of the planet is very low, only 0.13, somewhat inferior to that of the moon and very much below that of any other of the planets. No satellite is known, and there is no reason to suppose that it has any.

243. Diurnal Rotation of the Planet. — In 1889, Schiaparelli, the Italian astronomer, announced that he had discovered certain markings upon the planet, and that they showed that the planet rotates on its axis *only once during its orbital period of eighty-eight days*, thus keeping the same face always turned towards the sun, in the same way that the moon behaves with respect to the earth. Owing to the eccentricity of the planet's orbit, however, it must have a large *libration* (Art. 145), amounting to about 23½° on each side of the mean; *i.e.*, seen from a favorable station on the planet's surface, the sun, instead of rising and setting, as with us, would seem to oscillate back and forth through an arc of 47° once in 88 days.

This asserted discovery is very important and has excited great interest. Schiaparelli is probably correct, but it may be well to wait for confirmation of his observations by others before absolutely accepting the conclusion.

244. Transits of Mercury. — At the time of inferior conjunction, the planet usually passes north or south of the sun, the inclination of its orbit being 7°; but if the conjunction occurs when the planet is very near its node (Art. 224), it crosses the sun's disc and becomes visible upon it as a small black spot; not, however, large enough to be seen without a telescope, as Venus can under similar circumstances. Since the earth passes the planet's line of nodes on May 7th and Nov. 9th, transits can occur only near those days.

The transits of the last half of the present century are as follows: May Transits. — May 6th, 1878 ; May 9th, 1891. November Transits. — Nov. 12th, 1861 ; Nov. 5th, 1868 ; Nov. 7th, 1881 ; Nov. 10th, 1894. The next transits will occur in Nov. 1907, and 1914 ; and in May, 1924.

Transits of Mercury are of no particular astronomical importance, except as furnishing accurate determinations of the planet's place in the sky at a given time.

VENUS.

245. The second planet in order from the sun is Venus, the brightest and most conspicuous of all. It is so brilliant that at times it casts a shadow, and is easily seen by the naked eye in the daytime. Like Mercury it had two names among the Greeks, — Phosphorus as morning star, and Hesperus as evening star.

Its mean distance from the sun is 67,200000 miles, and its distance from the earth ranges from 26,000000 miles (93 − 67) to 160,000000 (93 + 67). No other body ever comes so near the earth except the moon, and occasionally a comet. The eccentricity of the orbit of Venus is the smallest in the planetary system, only 0.007, so that the greatest and least distances of the planet from the sun differ from the mean less than 500,000 miles. Its sidereal period is 225 days, or seven and a half months ; and its synodic period 584 days, — a year and seven months. From inferior conjunction to greatest elongation is only 71 days. The inclination of its orbit is not quite $3\frac{1}{2}°$, — less than half that of Mercury.

246. Magnitude, Mass, Density, etc. — The apparent diameter of the planet varies from 67″, at the time of inferior conjunction, to only 11″, at superior; the great difference arising from the enormous variation in the distance of the planet from the earth. The real diameter of the planet in miles is about 7700. Its surface compared with that of the earth is $\frac{95}{100}$; its volume, $\frac{92}{100}$. By means of the perturbations she produces upon the earth, the *mass* of Venus is found to be a little less than four-fifths of the earth's mass, so that her mean density is a little less than the earth's. In magnitude she is the earth's twin sister.

247. General Telescopic Appearance; Phases, etc. — The general telescopic appearance of Venus is striking on account of her great brilliancy, but exceedingly unsatisfactory because nothing is distinctly outlined upon the disc.

When about midway between greatest elongation and inferior conjunction, the planet has an apparent diameter of 40″, so that with a magnifying power of only 45 she looks exactly like the moon four days old, and of the same apparent size. (Very few persons, however, would think so on the first view through the telescope; the novice always underrates the apparent size of a telescopic object.)

The phases of Venus were first discovered by Galileo in 1610, and afforded important evidence as to the truth of the Copernican system as against the Ptolemaic. Fig. 49 represents the planet's disc as seen at five points in its orbit. 1, 3, and 5 are taken at superior conjunction, greatest elongation, and near inferior conjunction, respectively; while 2 and 4 are at intermediate points. (No. 2 is badly engraved, however; the sharp corners are impossible.)

The planet attains its maximum brightness when its apparent area is at a maximum, about thirty-six days before and after inferior conjunction. According to Zöllner. the 'albedo' of the planet is 0.50; *i.e.*, it reflects about half the light which falls upon it, the reflecting power being about three times that of the moon, and almost four times that of Mercury. It is, however, slightly exceeded by the reflecting power of Uranus and Jupiter, while that of Saturn is about

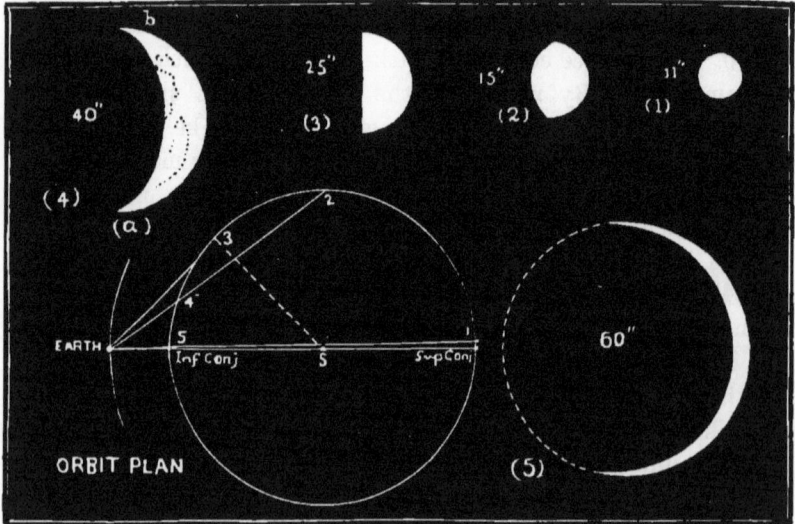

FIG. 49. — Telescopic Appearances of Venus.

the same. This high albedo probably indicates a surface mostly covered with clouds, since few rocks or soils could match its brightness. Like Mercury, Mars, and the moon, the disc of Venus is brightest at the edge, — in contrast with the appearance of Jupiter and Saturn.

248. Atmosphere of the Planet. — When the planet is near inferior conjunction, the horns of the crescent extend notably beyond the diameter; and when *very* near conjunction, a thin line of light has been seen by some observers to complete the whole circumference of the disc. This is due to the refraction of sunlight bent around the planet's globe by its atmosphere, a phenomenon still better seen when the planet is entering upon the sun's disc at a transit. The black disc is

then encircled by a delicate luminous ring, as illustrated by
Fig. 50. The planet's atmosphere is probably from one and

one-half to two times as
dense as our own, and
the spectrum shows
evidence of water-va-
por in it. Many ob-
servers have also re-
ported faint lights as
visible at times on the
dark portions of the
planet's disc. These
cannot be accounted
for by any mere reflec-
tion or refraction of
sunlight, but must orig-
inate on the planet it-
self. They recall the
Aurora Borealis and
other electrical mani-

FIG. 50. — Atmosphere of Venus as seen during a
Transit. (Vogel, 1882.)

festations on the earth, though it is impossible to give a certain
explanation of them as yet.

249. Surface-markings, Rotation, etc. — As has been said,
Venus is a very unsatisfactory telescopic object. She presents no obvious surface-markings, — nothing but occasional
indefinite shadings: sometimes, however, when in the crescent
phase, intensely bright spots have been reported near the
points of the crescent, which may perhaps be "ice-caps" like
those which are seen on Mars. The darkish shadings may
possibly be continents and oceans, dimly visible, or they may
be atmospheric objects; observations do not yet decide. From
certain irregularities occasionally observed upon the "terminator" (Art. 146), various observers have concluded that there
are high mountains upon the planet.

As to the rotation-period of the planet, nothing is yet certainly known. The length of its day has been set, on very insufficient grounds, at about 23 hours and 21 minutes; but the recent work of Schiaparelli makes it almost certain that this result cannot be trusted, and renders it rather probable that Venus behaves like Mercury in its diurnal rotation, the length of its sidereal day being equal to the time of its orbital revolution. The planet's disc shows no sensible oblateness.

No satellite has ever been discovered; not, however, for want of earnest searching.

250. Transits. — Occasionally Venus passes between the earth and the sun at inferior conjunction, giving us a so-called "transit." She is then visible, even to the naked eye, as a black spot on the sun's disc, crossing it from east to west. When the transit is central it occupies about eight hours, but when the track lies near the edge of the disc the duration is correspondingly shortened. Since the earth passes the nodes of the orbit on June 5th and Dec. 7th, all the transits occur near these days, but they are extremely rare phenomena. Their special interest consists in their availability for the purpose of finding the sun's parallax (see Appendix, Art. 437, and General Astronomy, Chap. XVI.).

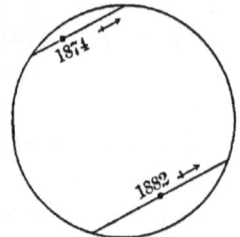

The first observed transit in 1639 was seen by only two persons, — Horrox and Crabtree in England, — but the four which have occurred since then have been observed in all parts of the world by scientific expeditions sent out for the purpose by the different governments. The transits of 1769 and 1882 were visible in the United States. Transits of Venus have occurred or will occur at the following dates : —

FIG. 51.
Transit of Venus's Tracks.

$$\begin{cases} \text{Dec. 7th, 1631.} \\ \text{Dec. 4th, 1639.} \end{cases} \qquad \begin{cases} \text{June 5th, 1761.} \\ \text{June 3d, 1769.} \end{cases}$$

{ Dec. 9th, 1874.   { June 8th, 2004.
{ Dec. 6th, 1882.   { June 6th, 2012.

Fig. 51 shows the tracks of Venus across the sun's disc in the transits of 1874 and 1882.

MARS.

251. This planet, also, has always been known. It is so conspicuous on account of its fiery red color and brightness, as well as the rapidity and apparent capriciousness of its movement among the stars, that it could not have escaped the notice of the very earliest observers.

Its mean distance from the sun is a little more than one and a half times that of the earth (141,500000 miles), and the eccentricity of its orbit is so considerable (0.093) that its radius vector varies more than 26,000000 miles. At opposition the planet's *average* distance from the earth is 48,600000 miles; but when opposition occurs near the planet's perihelion, this distance is reduced to less than 36,000000 miles, while near aphelion it is over 61,000000 miles. At conjunction the average distance from the earth is 234,000000 miles.

The apparent diameter and brightness of the planet of course vary enormously with these great changes of distance. At a favorable opposition (when the planet's distance from us is the least possible) it is more than fifty times as bright as at conjunction, and fairly rivals Jupiter; when most remote, it is hardly as bright as the Pole-star.

The favorable oppositions occur always in the latter part of August, and at intervals of fifteen or seventeen years. The last such opposition was in 1892, and the next will be in 1907.

The inclination of the orbit is small, 1° 51'. The planet's *sidereal* period is 687 days (one year, ten and a half months); its *synodic* period is much the longest in the planetary system, being 780 days, or nearly two years and two months. During

710 of these 780 days it moves towards the east, and retro-
grades during 70.

252. Magnitude, Mass, etc. — The apparent diameter of the
planet ranges from 3″.6, at conjunction, to 25″ at a favorable
opposition. Its real *diameter* is approximately 4300 miles,
with an error of perhaps 50 miles one way or the other.
This makes its *surface* about two-sevenths, and its *volume*
one-seventh of the earth's. Its *mass* is a little less than one-
ninth of the earth's mass, its *density* 0.73, and its *superficial
gravity* 0.38 ; *i.e.*, a body which here weighs 100 pounds would
have a weight of only 38 pounds on the surface of Mars.

253. General Telescopic Aspect, Phases, etc. — When the
planet is nearest, it is more favorably situated for telescopic
observation than any other heavenly body, the moon alone
excepted. It then shows a ruddy disc,
which, with a magnifying power of 75,
is as large as the moon. Since its orbit
is outside the earth's, it never exhibits
the crescent phases like Mercury and
Venus ; but at quadrature it appears
distinctly gibbous, as in Fig. 52, about
like the moon three days from full. Like
Mercury, Venus, and the moon, its disc
is brightest at the limb (*i.e.*, at its circular edge) ; but at the "terminator," or

FIG. 52.
Mars at Quadrature.

boundary between day and night upon the planet's surface,
there is a slight shading, which, taken in connection with certain other phenomena, indicates the presence of an atmosphere.
This atmosphere, however, is probably much less dense than
that at the earth, as indicated by the infrequency of clouds
and of other atmospheric phenomena familiar to us on the
earth. Huggins and Vogel have reported that the planet's
spectrum shows the lines of water-vapor, but the recent obser-

vations of Campbell at the Lick Observatory do not confirm
this, and go to show that whatever atmosphere exists must be
very rare indeed, probably not more than one-fourth as dense
as our own.

Zöllner gives the albedo of Mars as 0.26, just double that of
Mercury, and much higher than that of the moon, but only
about half that of Venus and the major planets. Near opposition the brightness of the planet suddenly increases in the
same way as that of the moon near the full (Art. 149).

254. Rotation, etc. — The spots upon the planet's disc
enable us to determine its period of rotation with great precision. Its sidereal day is found to be 24 hours, 37 minutes,
22.67 seconds, with a probable error not to exceed one-fiftieth
of a second. It is the only one of the planets which has the
length of its day determined with any such accuracy. The
exactness is obtained by comparing the drawings of the planet
made two hundred years ago with others made recently.

The inclination of the planet's equator to the plane of its
orbit is very nearly 24° 50' (26° 21' to the *ecliptic*). So far,
therefore, as depends upon that circumstance, Mars should
have seasons substantially the same as our own, and certain
phenomena make it evident that such is the case.

The planet's rotation causes a slight flattening of the poles,
— hardly sensible to observation, but probably about $\frac{1}{200}$.
(Larger values, now known to be erroneous, are given in
many text-books.)

255. Surface and Topography. — With even a small telescope, not more than four or five inches in diameter, the planet
is a very beautiful object, showing a surface diversified with
markings, light and dark, which for the most part are found to
be permanent. Occasionally, however, we see others of a temporary character, supposed to be clouds; but these are surprisingly rare, compared with clouds upon the earth. The

permanent markings are broadly divisible into three classes : —

First, the white patches, two of which are specially conspicuous near the planet's poles, and are generally supposed to be masses of snow or ice, since they behave just as would be expected if such were the case. The northern one dwindles away during the northern summer, when the north pole is turned towards the sun, while the southern one grows rapidly larger; and *vice versa* during the southern summer.

Second, patches of bluish gray or greenish shade, covering about three-eighths of the planet's surface, and generally supposed to be bodies of water, though this is very far from certain.

Third, extensive regions of various shades of orange and

FIG. 53. — Telescopic Views of Mars.

yellow, covering nearly five-eighths of the surface, and interpreted as land.

These markings, of course, are best seen when near the centre of the planet's disc; near the limb they are lost in the brilliant light which there prevails, and at the terminator they fade out in the shade. Fig. 53 gives an idea of the planet's general appearance, though without pretending to minute accuracy.

256. Recent Discoveries. The Canals and their Gemination.
— In addition to these three classes of markings, the Italian astronomer Schiaparelli, in 1877 and 1879, announced the discovery of a great number of fine, straight lines, or " canals," as he called them, crossing the ruddy portions of the planet's

surface in various directions ; and in 1881 he announced that many of them become *double* at times. For several years there was a suspicion that he was perhaps misled by some illusion, because other observers, with telescopes more powerful than his, were unable to make out anything of the sort. More recently, however, his results have been abundantly confirmed both in Europe, and in this country at the Lick Observatory, and the observatory of Mr. Lowell in Arizona. It appears that the power of the telescope is not so important in the observation of these objects as steadiness of the air and keenness of the observer's eye : nor are they usually best seen when Mars is nearest, but their visibility depends upon the *season* on the planet ; and this is especially the case with their "gemination."

As to their real nature there is wide difference of opinion, and it is doubtful if the true explanation has yet been proposed. According to Mr. Lowell, the polar caps are really snow masses, which melt in the (Martian) spring, and the water makes its way towards the equator, over the planet's mountainless plains, for several weeks obscuring the well-known markings which are visible at other times. For him the dark portions of the planet's surface are not seas, but land covered with vegetation of some sort, while the ruddy portions are rocky deserts intersected with the "canals," which, in his view, are really irrigating water-courses ; and on account of their straightness he is disposed to accept them as *artificial.* When the waters reach these canals, vegetation springs up along their banks on either side, and these streaks of vegetation are what we see. Where the water-courses cross each other there are dark, round "lakes," as they have been called, which he interprets as *oases.*

Of course the difficulties of the theory are obvious : for instance, the almost absolute levelness of the planet's surface which it assumes, and especially the fact that at Mars the solar radiation is only half as intense as upon the earth. This, recalling the low density of his atmosphere, would naturally lead to the supposition that the temperature even at his equator must be lower than that at the summits of our highest mountains, and far below the freezing point of water. It was this consideration that has led some astronomers to suggest that the polar caps are not *ice*-sheets at all, but formed of congealed carbon-dioxide ($C O_2$), or some substance of similar properties.

But whatever the explanation may be, there is no longer any doubt that the "canals" are real, and that they, and the surface in general, undergo noticeable changes of appearance with the progress of the planet's seasons. / At the same time Professor Holden of the Lick Observatory says that during the years 1888–1895 nothing has been observed there, so far as he knows, which goes to confirm Mr. Lowell's "very positive and striking conclusions." The day may perhaps come when photography will be able to lend its aid to the solution of the problem, or some heat-measurer may be contrived sensitive enough to give us positive information as to the planet's temperature. If the polar caps are really snow caps, Mars must obtain surface heat from some still unexplained supply.

257. Maps of the Planet. — A number of maps of Mars have been constructed by different observers since the first one was made by Maedler in 1830. Fig. 54 is reduced from one which was published in 1888 by Schiaparelli, and shows most of his "canals" and their "geminations." While there may be some doubt as to the accuracy of the minor details, there can be no doubt that the main features of the planet's surface are substantially correct. The nomenclature, however, is in a very unsettled state. Schiaparelli has taken his names mostly from ancient geography, while the English areographers,[1] following the analogy of the lunar maps, have mainly used the names of astronomers who have contributed to our knowledge of the planet's surface.

---

[1] The Greek name of Mars is Ares; hence "Areography" is the description of the surface of Mars.

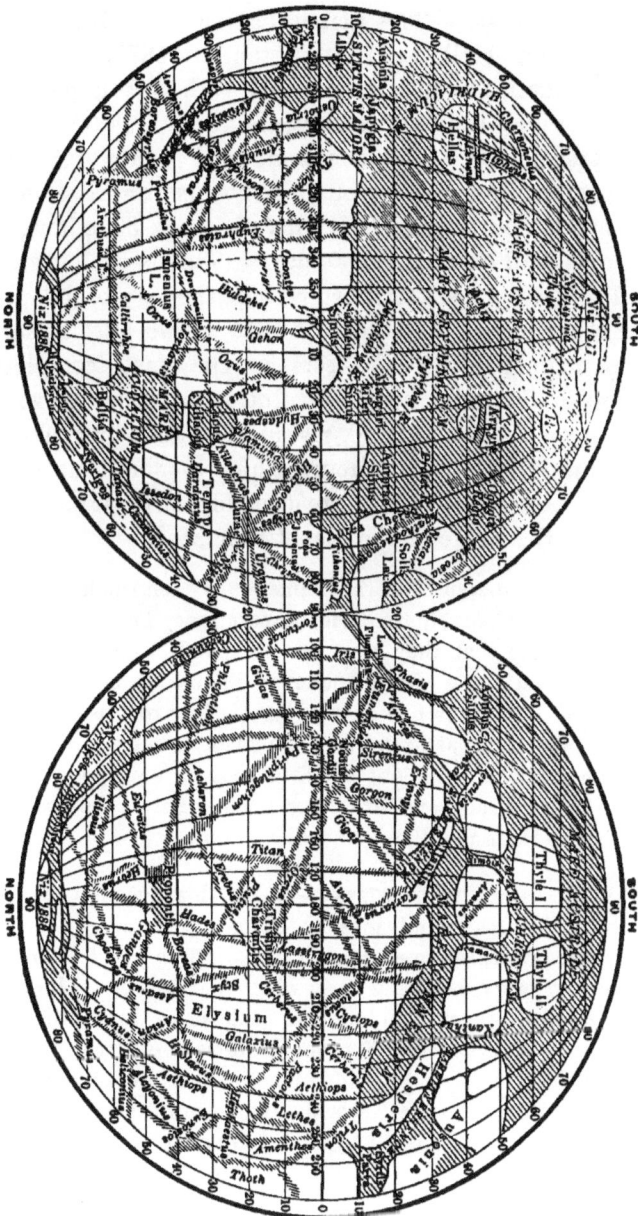

Fig. 54. — Chart of Mars as observed from 1877 to 1888. (Schiaparelli.)

258. Satellites. — The planet has two satellites, discovered by Professor Hall, at Washington, in 1877. They are extremely small, and observable only with very large telescopes. The outer one, Deimos, is at a distance of 14,600 miles from the planet's centre, and has a sidereal period of 30 hours, 18 minutes; while the inner one, Phobos, is at a distance of only 5800 miles, and its period is only 7 hours, 39 minutes, — less than one-third of the planet's day. (This is the only case of a satellite with a period shorter than the day of its primary.) Owing to this circumstance, it *rises in the west*, as seen from the planet's surface, and sets in the east, completing its strange, backward, diurnal revolution in about eleven hours. Deimos, on the other hand, rises in the east, but takes nearly 132 hours in its diurnal circuit, which is more than four of its months. Both the orbits are sensibly circular, and lie very closely in the plane of the planet's equator.

Micrometric measures of the diameters of such small objects are impossible; but from photometric observations, Professor E. C. Pickering, assuming that they have the same reflecting power as that of Mars itself, estimates the diameter of Phobos as about seven miles, and that of Deimos as five or six. According to this, Phobos, at the time of full moon, as seen from the planet's surface, would have an apparent diameter of about one-fifth that of our moon, and would probably give about one-fiftieth as much light. Deimos would be hardly more than a brilliant star, like Venus.

259. Habitability of Mars. — As to this question, we can only say that, while the conditions on Mars are certainly very different from those prevailing on the earth, the difference is less than in the case of any other heavenly body which we can see with our present means of observation; and if life, such as we know life upon the earth, can exist upon any of them, Mars is the place. But there is at present no scientific ground for belief one way or the other as to the habitability of "other worlds than ours," passionately as the doctrine has been affirmed and denied by men of opposite opinions.

CHAPTER IX.

THE PLANETS CONTINUED.

THE ASTEROIDS. — INTRA-MERCURIAN PLANETS AND THE ZODIACAL LIGHT. — THE MAJOR PLANETS, JUPITER, SATURN, URANUS, AND NEPTUNE.

THE ASTEROIDS, OR MINOR PLANETS.

260. THE asteroids[1] are a multitude of small planets circling around the sun in the space between Mars and Jupiter. It was early noticed that between Mars and Jupiter there is a gap in the series of planetary distances, and when Bode's Law (Art. 219) was published in 1772, the impression became very strong that there must be a missing planet in the space, — an impression greatly strengthened when Uranus was discovered in 1781, at a distance precisely corresponding to that law.

The first member of the group was found by the Sicilian astronomer, Piazzi, on the very first night of the present century (Jan. 1, 1801). He named it *Ceres*, after the tutelary divinity of Sicily. The next year *Pallas* was discovered by Olbers. *Juno* was found in 1804 by Harding, and in 1807 Olbers, who had broached the theory of an exploded planet, discovered the fourth, *Vesta*, the only one which is bright enough ever to be easily seen by the naked eye. The search was kept up for some years longer, but without success, because the searchers

---

[1] They were first called "*asteroids*" (*i.e.*, "star-like" bodies) by Sir William Herschel early in the century, because, though really planets, the telescope shows them only as stars, without a sensible disc.

did not look for small enough objects. The fifth asteroid (Astræa) was found in 1845 by Hencke, an amateur who had resumed the subject by studying the fainter stars. In 1847 three more were discovered, and every year since then has added from one to thirty. They are usually designated by their "numbers," but all the older ones also have names : thus, Ceres is ①, Thule is ㉚, etc. In May, 1895, the list included 401 duly "numbered," besides about half a dozen more to which no number could be assigned, because they had not then been sufficiently observed to make it certain that they were not old ones rediscovered : for of the older ones eight or nine are now "adrift," *i.e.*, they have not been observed for many years, and we do not know exactly their present position. Since 1891 the catalogue has been growing with rather inconvenient rapidity on account of the substitution of photography for the old-fashioned method of planet-hunting. A large camera is strapped on the back of a telescope driven by clock-work, and a negative, covering from 5° to 10° square of the heavens, is taken with an exposure of several hours. The thousands of stars that appear upon the plate all show neat, round discs, if the observer has kept his telescope steadily pointed ; but if there is a planet anywhere in the field it will move quite perceptibly during the long exposure, and its image upon the plate will be, not a dot, but a *streak*, which can be recognized at a glance. Sometimes three or four planets thus "show up" upon a single plate, — old ones as well as new of course ; but a few nights' observation will usually furnish data from which the orbits can be computed with sufficient accuracy to decide all doubtful questions. Wolf of Heidelberg, who first introduced the method, and Charlois of Nice, have been especially successful in this kind of asteroid-hunting. Among the old-fashioned planet-hunters Palisa of Vienna took the lead as the discoverer of 71 ; the late Dr. Peters of Clinton, N. Y., stood second with 48.

261. Their Orbits. — The *mean distances* of the different asteroids from the sun differ pretty widely, and the *periods*, of course, correspond. Medusa, (149), is probably the nearest to the sun of those at present known, its distance being 2.13 (astronomical units), or 198,000000 miles, with a period of 3 years and 40 days. Brucia, (323), may, however, have a period shorter by a day or two. Thule, (279), is the most remote, with a mean distance of 4.30 (400,000000 miles) and a period only 10 days less than 9 years.

The *inclinations* of the orbits to the ecliptic average nearly 8°. The orbit of Pallas, (2), is inclined at an angle of 35°, and seven others exceed 25°. The *eccentricity* of the orbits is very large in many cases. Aethra, (132) (one of the eight or nine that are adrift) has the largest eccentricity (0.38), and ten others have an eccentricity exceeding 0.30. It should be noted that the orbits of these planets are subject to very great "disturbances" from the attraction of Jupiter, and this makes the calculation of their motions much more laborious than that of the larger planets. Very few of them, therefore, are followed up closely ; only those that for some reason or other possess a special interest at some given time.

262. The Bodies Themselves. — The four first discovered, and one or two others, when examined with a powerful telescope, show discs that are perceptible, but too small for satisfactory measurement with ordinary telescopes. By photometric observations, assuming — what is by no means certain — that their albedo is about the same as that of Mars, it has been estimated that Vesta, the brightest, has a diameter of about 320 miles, and that the other three of the first four may be two-thirds as large. In 1895, however, Mr. Barnard of the Lick Observatory measured the diameters of Ceres, Pallas and Vesta, micrometrically, and obtained results that differ from these very widely, and should probably be preferred. He finds Ceres to be the largest, with a diameter of 488 miles. For Pallas, Vesta and Juno he gets diameters of

304, 248 and 118 miles respectively. As this work was done with a " double-image " micrometer, which avoids irradiation, the results are more reliable than his measures of Jupiter and Saturn, quoted later. None of the rest can well exceed 100 miles in diameter; and the more newly discovered ones, which are just fairly visible in a telescope with an aperture of 10 or 12 inches, cannot be many times larger than the moons of Mars, — say from 10 to 20 miles in diameter.

As to the individual masses and densities, we have no certain knowledge.

Assuming the correctness of Barnard's measures, and that the density of Ceres is about the same as that of the rocks which compose the earth's crust, her mass may be as great as $\frac{1}{4000}$ that of the earth. If so, *gravity on her surface* would be about $\frac{1}{18}$ of gravity here, so that a body would fall about twelve inches in the first second. Of course, on the smaller asteroids it would be much less.

From the perturbations of Mars, Leverrier has estimated that the *aggregate mass* of the whole swarm cannot exceed *one-fourth the mass of the earth,* — something more than double that of Mars.

The united mass of those at present known would make only a small fraction of such a body, — hardly a thousandth of it; probably, however, those still undiscovered are very numerous.

263. Origin. — As to this we can only speculate. It is hardly possible to doubt, however, that this swarm of little rocks in some way represents a single planet of the " terrestrial " group. A commonly accepted view is that the material, which, according to the nebular hypothesis, once formed a ring (like one of the rings of Saturn), and ought to have collected to make a single planet, has failed to be so united; and the failure is ascribed to the perturbations produced by the next neighbor, the giant Jupiter, whose powerful attraction is supposed to have torn the ring to pieces, and thus prevented its normal development into a planet.

Another view is that the asteroids may be fragments of an exploded planet. If so, there must have been not one, but many, explosions, first of the original body, and then of the separate pieces; for it is demonstrable that *no single* explosion could account for the present tangle of orbits.

INTRA-MERCURIAN PLANETS AND THE ZODIACAL LIGHT.

264. Intra-Mercurian Planets. — It is very possible, indeed not improbable, that there is a considerable quantity of matter circulating around the sun inside the orbit of Mercury. This is suggested by an otherwise unexplained perturbation of its orbit. It has been somewhat persistently supposed that this intra-Mercurian matter is concentrated into one, or possibly two, planets of considerable size, and such a planet has several times been reported as discovered, and has even been named *Vulcan*. The supposed discoveries have never been confirmed, however, and the careful observations of total solar eclipses during the past ten years make it practically certain that there is no "Vulcan." Possibly, however, there is a family of intra-Mercurian asteroids; but they must be very minute, or some of them would certainly have been found either during eclipses or crossing the sun's disc; a planet as much as 200 miles in diameter could hardly have escaped discovery.

265. The Zodiacal Light. — This is a faint beam of light extending from the sun both ways along the ecliptic. In the evening it is best seen in the early spring, and in our latitude then extends about 90° eastward from the sun; in the tropics, it is said that it can be followed quite across the sky. The region near the sun is fairly bright and even conspicuous, but the more distant portions are extremely faint and can be observed only in places where there is no illumination of the air by artificial lights. Its spectrum is a simple, continuous spectrum, without markings of any kind, so far as can be observed.

We emphasize this, because of late it has been mistakenly reported that the bright line which characterizes the spectrum of the Aurora Borealis appears in the spectrum of the zodiacal light.

The cause of the phenomenon is not certainly known. Some imagine that the zodiacal light is only an extension of the solar corona (whatever that may be), which is not perhaps unlikely; but on the whole the more prevalent opinion seems to be that it is due to sunlight reflected from myriads of small meteoric bodies circling around the sun, nearly in the plane of the ecliptic, thus forming a thin, flat sheet (something like one of Saturn's rings), which extends far beyond the orbit of the earth.

THE MAJOR PLANETS. — JUPITER.

266. Jupiter, the nearest of the major planets, stands next to Venus in the order of brilliance among the heavenly bodies, being fully five or six times as bright as Sirius, and decidedly superior to Mars, even when Mars is nearest. It is not, like Venus, confined to the twilight sky, but at the time of opposition dominates the heavens all night long.

Its orbit presents no marked peculiarities. The *mean distance* of the planet from the sun is a little more than five astronomical units (483,000000 miles), and the *eccentricity* of the orbit is not quite $\frac{1}{20}$, so that the actual distance ranges about 21,000000 miles each side of the mean. At an average opposition, the planet's distance from the earth is about 390,000000 miles, while at conjunction it is distant about 580,000000.

The *inclination* of its orbit to the ecliptic is only 1° 19'. Its *sidereal* period is 11.86 years, and the *synodic* is 399 days (a figure easily remembered), — a little more than a year and a month; *i.e.*, each year Jupiter comes to opposition a month and four days later than in the preceding year.

267. Dimensions, Mass, Density, etc. — The planet's apparent diameter varies from 50″ to 32″, according to its distance from the earth. The disc, however, is distinctly oval, so that while the equatorial diameter is 88,200 miles, the polar diameter is only 83,000. The *mean* diameter (see Art. 112) is 86,500 miles, or very nearly eleven times that of the earth.[1]

Its *surface*, therefore, is 119, and its *volume* or bulk 1300 times that of the earth. It is by far the largest of all the planets, — larger, in fact, than all the rest united.

Its *mass* is very accurately known, both by means of its satellites and from the perturbations it produces upon certain asteroids. It is $\frac{1}{1048}$ of the sun's mass, or about 316 times that of the earth.

Comparing this with its volume, we find its *mean density* to be 0.24; *i.e.*, less than one-fourth the density of the earth, and almost precisely the same as that of the sun. Its *surface gravity* is about $2\frac{2}{3}$ times that of the earth, but varies nearly 20 per cent between the equator and poles of the planet on account of the rapid rotation.

268. General Telescopic Aspect, Albedo, etc. — In a small telescope the planet is a fine object; for a magnifying power of only 60 makes its apparent diameter, even when remotest, equal to that of the moon. With a large instrument and a magnifying power of 200 or 300, it is magnificent, the disc being covered with an infinite variety of detail, interesting in outline and rich in color, changing continually as the planet turns on its axis. For the most part the markings are

---

[1] A series of measures made by Barnard in 1893 with the Lick telescope, gives an equatorial diameter of 90,190 miles, and a polar diameter of 84,560 ; but these figures probably need a correction for irradiation.

arranged in "belts" parallel to the planet's equator, as shown in Fig. 55.

The left-hand one of the two larger figures is from a drawing by Trouvelot (1870), and the other from one by Vogel (1880). The smaller figure below represents the planet's ordinary appearance in a three-inch telescope.

Near the limb the light *is less brilliant than in the centre of the disc*, and the belts there fade out. The planet shows no

FIG. 55. — Telescopic Views of Jupiter.

perceptible phases, but the edge which is turned away from the sun is usually sensibly darker than the other. According to Zöllner, the mean *albedo* of the planet is 0.62, which is extremely high, that of white paper being only 0.78. The question has been raised whether Jupiter is not to some extent *self-luminous*, but there is no proof and little probability that such is the case.

269. Atmosphere and Spectrum. — The planet's atmosphere must be very extensive. The forms which we see with the telescope are all evidently *atmospheric*. In fact, the low mean density of the planet makes it very doubtful whether there is anything solid about it anywhere, — whether it is anything more than a ball of fluid overlaid by cloud and vapor.

The spectrum of the planet differs less from that of mere reflected sunlight than might have been expected, showing that the light is not obliged to penetrate the atmosphere to any great depth before it encounters the reflecting envelope of cloud. There are, however, certain unexplained dark shadings in the red and orange parts of the spectrum that are probably due to the planet's atmosphere, and seem to be identical in position with certain bands which, in the spectra of Uranus and Neptune, are much more intense.

270. Rotation. — Jupiter rotates on its axis more swiftly than any other of the planets. Its sidereal day has a length of *about* 9 hours, 55 minutes, but the time can be given only approximately, because different results are obtained from different spots, according to their nature and their distance from the equator, — the differences amounting to six or seven minutes. Speaking generally, spots near the equator indicate a shorter period of rotation than those near the poles, just as is the case with the sun. *White* spots also make the circuit quicker than dark spots near them.

In consequence of the swift rotation, the planet's *oblateness* or "polar compression" is quite noticeable, — about $\frac{1}{17}$. The inclination of the planet's equator to its orbit is only 3°, so that there can be no well-marked seasons on the planet due to such causes as our own seasons.

271. Physical Condition. — This is obviously very different from that of the earth or Mars. No permanent markings are found upon the disc, though occasionally there are some which may be called "sub-permanent" as, for instance, the *great red spot* shown in Fig. 55. This was first noticed in 1878, became extremely conspicuous for several years, and still (1895) remains visible as a faded ghost of itself. Were it not that during the years of its visibility it has changed the length of its apparent rotation by about six seconds (from 9 hours, 55 minutes, 34.9 seconds to 9 hours, 55 minutes, 40.2 seconds), we might suppose it permanently attached to the planet's surface, and evidence of a coherent mass underneath. As it is, opinion is divided on this point; the phenomenon is as puzzling as the canals of Mars.

Many things in the planet's appearance indicate *a high temperature,* as, for instance, the abundance of clouds, and the swiftness of their transformations; and since on Jupiter the solar light and heat are only $\frac{1}{27}$ as intense as here, we are forced to conclude that it gets very little of its heat from the sun, but is probably hot on its own account, and for the same reason that the sun is hot; viz., as the result of a process of condensation. In short, it appears very probable that the planet is a sort of *semi-sun,* — hot, though not so hot as to be sensibly self-luminous.

272. Satellites. — Jupiter has five satellites, four of them large and easily seen with a very small telescope, while the fifth, discovered by Barnard in 1892, is extremely small and visible only in the largest instruments.

The four large satellites were the first heavenly bodies ever *discovered.* Galileo found them in January, 1610, within a few weeks after the invention of his telescope.

They are now usually known as the first, second, etc., in the order of their distance from the planet. The distances range from 262,000 to 1,169,000 miles, being respectively 6, 9, 15, and 26 radii of the planet (nearly). Their sidereal periods range from 42 hours to $16\frac{2}{3}$ days. Their orbits are sensibly circular, and lie very nearly in the plane of the equator. The third satellite is much the largest, having a diameter of about 3600 miles, while the others are between 2000 and 3000.

For some reason, the fourth satellite is a very dark-complexioned body, so that when it crosses the planet's disc it looks like a black spot hardly distinguishable from its own shadow: the others, under similar circumstances, appear bright, dark, or invisible, according to the brightness of the part of the planet which happens to form the background. In Fig. 55 a satellite and its shadow are visible together near the eastern limb of the planet. In the case of the fourth satellite, a certain regularity in its changes of brightness suggests that it probably follows the example of our moon in always keeping the same face towards the planet. Prof. W. Pickering reports that they show certain curious and regularly recurring changes of form, which indicate that they are not solid masses, but whirling clouds or swarms of minute particles; his observations, however, have not yet received satisfactory confirmation.

The fifth satellite is at a distance of about 112,500 miles, and its period of revolution is $11^{\text{h}}\ 57^{\text{m}}\ 22.^{\text{s}}6$. Its diameter is probably less than 100 miles.

273. Eclipses and Transits. — The orbits of the satellites are so nearly in the plane of the planet's orbit that with the exception of the fourth, which sometimes escapes, they are eclipsed at every revolution. When the planet is either at opposition or conjunction, the shadow, of course, is directly behind it, and we cannot see the eclipse at all. At other times we ordinarily see only the beginning or the end; but when the planet is very near quadrature the shadow projects so far to one side that the whole eclipse of every satellite, except the first, takes place clear of the disc, and both the disappearance and reappearance can be seen.

Two important uses have been made of these eclipses: they have been employed for the determination of longitude, and they *furnish the means of ascertaining the time required by light to traverse the space between the earth and the sun.* (See Appendix, Arts. 431–434.)

SATURN.

274. This is the most remote of the planets known to the ancients. It appears as a star of the first magnitude (outshining all of them, indeed, except Sirius), with a steady, yellowish light, not varying much in appearance from month to month, though in the course of 15 years it alternately gains and loses nearly 50 per cent of its brightness with the changing phases of its rings; for it is unique among the heavenly bodies, a great globe attended by eight satellites and surrounded by a system of rings, which has no counterpart elsewhere in the universe so far as known.

Its *mean distance* from the sun is about $9\frac{1}{2}$ astronomical units, or 886,000000 miles; but the distance varies over 100,000000 miles on account of the considerable *eccentricity* of the orbit (0.056). Its least distance from the earth is about 774,000000 miles, the greatest, about 1028,000000. The *inclination* of the orbit to the ecliptic is $2\frac{1}{2}°$. The *sidereal period* is about $29\frac{1}{2}$ years, the *synodic* period being 378 days, or a year and a fortnight nearly.

275. Dimensions, Mass, etc. — The apparent mean diameter of the planet varies according to the distance from 14″ to 20″. The planet is more flattened at the poles than any other (nearly $\frac{1}{10}$), so that while the equatorial diameter is about 75,000 miles, the polar is only 68,000: the mean diameter (Art. 112) being not quite 73,000,[1]— a little more than *nine* times that of the earth. Its *surface* is about 84 times that of the earth, and its *volume* 770 times. Its *mass* is found (by

---

[1] Barnard's recent measures give a diameter about 1000 miles larger.

Fig. 56. — Saturn and his Rings.

means of its satellites) to be 95 times that of the earth, so that its mean *density* comes out only *one-eighth* that of the earth, — actually *less than that of water!* It is by far the least dense of all the planetary family.

Its mean *superficial gravity* is about 1.2 times as great as gravity upon the earth, varying, however, nearly 25 per cent between the equator and the pole, so that at the planet's equator it is practically the same as upon the earth. It rotates on its axis in about 10 hours, 14 minutes, but different spots give various results, as in the case of Jupiter.

The equator of the planet is inclined about 27° to the plane of its orbit — about 28° to the ecliptic.

276. Surface, Albedo, Spectrum. — The disc of the planet, like that of Jupiter, is shaded at the edge, and like Jupiter it shows a number of belts arranged parallel to the equator. The equatorial belt is very bright, and is often of a delicate pinkish tinge. The belts in higher latitudes are comparatively faint and narrow, while just at the pole there is usually a cap of olive green (see Fig. 56).

Zöllner makes the mean *albedo* of the planet 0.52, about the same as that of Venus.

The planet's spectrum is substantially like that of Jupiter, but the dark bands are more pronounced. These bands, however, do not appear in the spectrum of the *ring*, which probably has very little atmosphere. As to its physical condition and constitution, the planet is probably much like Jupiter, though it does not seem to be " boiling " quite so vigorously.

277. The Rings. — The most remarkable peculiarity of the planet is its *ring system*. The globe is surrounded by three thin, flat, concentric rings, like circular discs of paper pierced through the centre. They are generally referred to as *A. B*, and *C*, *A* being the exterior one.

Galileo *half* discovered them in 1610; *i.e.*, he saw with his little
telescope two appendages, one on each side of the planet; but he could
make nothing of them, and after a while he lost them. The problem
remained unsolved for nearly fifty years, until Huyghens explained
the mystery in 1655. Twenty years later D. Cassini discovered that
the ring is *double*; *i.e.*, composed of two concentric rings, with a dark
line of separation between them; and in 1850, Bond of Cambridge,
U.S., discovered the third "dusky" or "gauze" ring between the
principal ring and the planet. (It was discovered a fortnight later,
independently, by Dawes, in England.)

The outer ring, *A*, has a diameter of about 170,000 miles,
and a width of about 11,000. Cassini's division is about 2000
miles wide ; the ring *B*, which is much the broadest of the

Fig. 57. – The Phases of Saturn's Rings.

three, is about 17,000. The semi-transparent ring, *C*, has a
width of about 10,000 miles, leaving a clear space of from
8000 to 9000 miles in width between the planet's equator and
its inner edge. The thickness of the rings is extremely small,
probably not over 100 miles, as proved by the appearance
presented, when once in 15 years we view them edgewise.

The recent researches of H. Struve show that the *mass* of
the rings, and their mean density are also surprisingly small,
so small that the rings exert hardly more influence on

the motion of the satellites than if they were composed of
"immaterial light," to use his own expression.

278. Phases of the Rings. — The plane of the rings coincides with the plane of the planet's equator, and is inclined
about 28° to the ecliptic. It, of course, remains parallel
to itself at all times. Twice in a revolution of the planet,
therefore, this plane sweeps across the orbit of the earth (too
small to be shown in the figure — Fig. 57), occupying nearly a
year in so doing; and whenever the plane passes between the
earth and the sun the dark side of the ring is towards us, and
the edge alone is visible, as when the planet is at 1 or 2;
when it is at the intermediate points 3 and 4 the rings present
their widest opening.

When the ring is exactly edgewise towards us only the largest telescopes can see it, like a fine needle of light piercing the planet's ball,
as in the uppermost engraving of Fig. 56. It becomes obvious at
such times that the thickness of the rings is not uniform, since considerable irregularities appear upon the line of light at different
points. The last period of disappearance was in 1892 ; the next will
be in 1907.

279. Structure of the Rings. — It is now universally admitted that they are not continuous sheets, either solid or
liquid, but mere *swarms of separate particles*, each particle pursuing its own independent orbit around the planet, though all
moving nearly in a common plane.

The idea was first suggested by J. Cassini, in 1715, but was lost
sight of until again brought into notice by Bond, in 1850. A little
later Pierce proved, from mechanical considerations, that the rings
could not be *solid ;* and not long after Maxwell showed that they
could not be "continuous sheets" of any kind, either solid or liquid,
but might be composed of separate particles moving independently.
More recently, Muller and Seeliger have shown from photometric
observations that the variations in the brightness of the ring correspond to this "meteoric theory" ; and still more recently (in 1895),

Keeler has demonstrated, by a most beautiful and delicate spectroscopic observation, that the outer edge of the ring in its rotation
really moves more slowly than the inner, just as the theory requires.

It remains uncertain whether the rings constitute a system that is
permanently stable, or whether they are liable ultimately to be
broken up and disappear.

280. Satellites. — Saturn has eight of these attendants, the
largest of which was discovered by Huyghens in 1655. It
looks like a star of the ninth magnitude, and is easily seen
with a three-inch telescope. The smallest one, the seventh in
order of distance from the planet, was discovered by Bond, at
Cambridge (U.S.) in 1848.

Since the order of discovery does not agree with that of distance, it
has been found convenient to designate them by the names assigned
by Sir John Herschel, as follows, beginning with the most remote
viz. : —

   Iapĕtus (Hyperion), Titan, Rhea, Dione, Tethys;
   Encelădus, Mimas.

(The name, Hyperion, was not given by Herschel, but interpolated after its discovery by Bond.)

The range of the system is enormous. Iapetus has a distance of
2,225,000 miles, with a period of 79 days, nearly as long as that of
Mercury. On the western side of the planet, this satellite is always
much brighter than upon the eastern, showing that, like our own
moon, it keeps the same face towards its primary.

Titan, as its name suggests, is by far the largest. Its distance is
about 770,000 miles, and its period a little less than 16 days. It is
probably 3000 or 4000 miles in diameter, and, according to Stone, its
mass is $\frac{1}{4600}$ of Saturn's, or about double that of our moon. The orbit
of Iapetus is inclined nearly 10° to the plane of the rings, but all of
the other satellites move almost exactly in their plane, and all the five
inner ones in orbits nearly circular.

URANUS.

281. Urănus (not U-rā'nus) was the first *planet* ever "discovered," and the discovery created great excitement and brought the highest honors to the astronomer. It was found accidentally by the elder Herschel on March 13, 1781, while "sweeping" for interesting objects with a seven-inch reflector of his own construction.. He recognized it at once by its disc as something different from a star, but supposed it to be a peculiar sort of a comet, and its planetary character was not demonstrated until nearly a year had passed. It is easily visible to a good eye as a star of the sixth magnitude.

Its *mean distance* from the sun is about 19 times that of the earth, or about 1800,000000 miles, and the *eccentricity* of its orbit is about the same as that of Jupiter's. The *inclination* of the orbit to the ecliptic is very slight — only 46'. The *sidereal period* is 84 years, and the *synodic*, 369½ days.

In the telescope it shows a greenish disc about 4'' in diameter, which corresponds to a *real diameter* of about 32,000 miles. This makes its *bulk* about 66 times that of the earth. The planet's *mass* is found from its satellites to be about 14.6 times that of the earth; its *density*, therefore, is 0.22 — about the same as that of Jupiter and the sun.

The *albedo* of the planet, according to Zöllner, is very high, 0.64, — even a little above that of Jupiter. The *spectrum* exhibits intense dark bands in the red, due to some unidentified substance in the planet's atmosphere. These bands explain the marked greenish tint of the planet's light. The atmosphere is probably dense.

The disc is obviously oval, with an ellipticity of about $\frac{1}{14}$. There are no clear markings upon it, but there seem to be faint traces of something like belts. No spots are visible from which to determine the planet's diurnal rotation.

282. Satellites. — The planet has four satellites, — Ariel, Umbriel, Titania, and Oberon, — Ariel being the nearest to the planet.

The two brightest, Oberon and Titania, were discovered by Sir William Herschel a few years after his discovery of the planet; Ariel and Umbriel, by Lassell, in 1851.

They are among the smallest bodies in the solar system, and the most difficult to see.   •

Their orbits are sensibly circular, and all lie in one plane, which *ought* to be, and probably is, coincident with the plane of the planet's equator.

They are very *close packed* also, Oberon having a distance of only 375,000 miles, and a period of 13 days, 11 hours, while Ariel has a period of 2 days, 12 hours, at a distance of 120,000 miles. Titania, the largest and brightest of them, has a distance of 280,000 miles, somewhat greater than that of the moon from the earth, with a period of 8 days, 17 hours.

The most remarkable thing about this system remains to be mentioned. The plane of their orbits is inclined 82°.2, or almost perpendicularly, to the plane of the ecliptic; and in that plane they revolve *backwards*.

NEPTUNE.

283. Discovery. — The discovery of this planet is reckoned the greatest triumph of mathematical astronomy. Uranus failed to move precisely in the path computed for it, and was misguided by some unknown influence to an extent which could almost be seen with the naked eye. The difference between the actual and computed places in 1845 was the "intolerable quantity" of nearly two minutes of arc.

This is a little more than half the distance between the two principal components of the double-double star, Epsilon Lyræ, the northern one of the two little stars which form the small equilateral triangle with Vega (Arts. 67 and 375). A very sharp eye detects the duplicity of Epsilon without the aid of a telescope.

One might think that such a minute discrepancy between observation and theory was hardly worth minding, and that to consider it "intolerable" was putting the case very strongly. But just these minute discrepancies supplied the data which were found sufficient for calculating the position of a great unknown world, and bringing it to light. As the result of a most skilful and laborious investigation, Leverrier (born 1811, died 1877) wrote to Galle in substance : —

"Direct your telescope to a point on the ecliptic in the constellation of Aquarius, in longitude 326°, and you will find within a degree of that place a new planet, looking like a star of about the ninth magnitude, and having a perceptible disc."

The planet was found at Berlin on the night of Sept. 23, 1846, in exact accordance with this prediction, within half an hour after the astronomers began looking for it, and within 52' of the precise point that Leverrier had indicated.

We cannot here take the space for a historical statement, further than to say that the English Adams fairly divides with Leverrier the credit for the mathematical discovery of the planet, having solved the problem and deduced the planet's approximate place even earlier than his competitor. The planet was being searched for in England at the time it was found in Germany. In fact, it had already been observed, and the discovery would necessarily have followed in a few weeks, upon the reduction of the observations.

284. Error of the Computed Orbit. — Both Adams and Leverrier, besides calculating the planet's position in the sky, had deduced elements of its orbit and a value for its mass, which turned out to be seriously wrong, and certain high authorities have therefore characterized the discovery as a "happy accident." This is not so, however. While the data and methods employed were not sufficient to determine the planet's *orbit* with accuracy, they were adequate to ascertain the planet's *direction* from the earth. The computers informed the observers *where to point their telescopes,* and this was all that was necessary for finding the planet.

285. The Planet and its Orbit. — The planet's *mean distance* from the sun is a little more than 2800,000000 miles (800,-000000 miles nearer the sun than it should be according to Bode's Law). The orbit is very nearly circular, its *eccentricity* being only 0.009. The *inclination* of the orbit is about $1\frac{3}{4}°$. The *period* of the planet is about 164 years (instead of 217, as as it should have been according to Leverrier's computed orbit), and the orbital velocity is about $3\frac{1}{8}$ miles per second.

Neptune appears in the telescope as a small star of between the eighth and ninth magnitudes, absolutely invisible to the naked eye, though easily seen with a good opera-glass. Like Uranus, it shows a greenish disc, having an apparent diameter of about $2''.6$. The real diameter of the planet is about 35,000 [1] miles, and the *volume* a little more than 90 times that of the earth.

Its *mass*, as determined by means of its satellite, is about 18 times that of the earth, and its *density* 0.20.

The planet's *albedo*, according to Zöllner, is 0.46, a trifle less than that of Saturn and Venus.

There are no visible markings upon its surface, and nothing certain is known as to its rotation.

The spectrum of the planet appears to be like that of Uranus, but of course is rather faint.

It will be noticed that Uranus and Neptune form a "pair of twins," very much as the earth and Venus do, being almost alike in magnitude, density, and many other characteristics.

286. Satellite. — Neptune has one satellite, discovered by Lassell within a month after the discovery of the planet itself. Its distance is about 223,000 miles, and its period $5^d 21^h$. Its orbit is inclined to the ecliptic at an angle of 34° 48', and it moves *backward* in it from east to west, like the satellites of Uranus. From its brightness, as compared with that of Neptune itself, its diameter is estimated as about the same as that of our own moon.

---

[1] 33,000 according to Barnard.

287. The Solar System as seen from Neptune. — At Neptune's distance the sun itself has an apparent diameter of only a little more than one minute of arc, — about the diameter of Venus when nearest us, and too small to be seen as a disc by the naked eye, if there are eyes on Neptune. The solar light and heat are there only $\frac{1}{900}$ of what we get at the earth.

Still, we must not imagine that the Neptunian sunlight is feeble as compared with starlight, or even moonlight. Even at the distance of Neptune the sun gives a light nearly equal to 700 full moons. This is about 80 times the light of a standard candle at one metre's distance, and is abundant for all visual purposes. In fact, as seen from Neptune, the sun would look very like a large electric arc lamp, at a distance of a few yards.

288. Ultra-Neptunian Planets. — Perhaps the breaking down of Bode's Law at Neptune may be regarded as an indication that the solar system terminates there, and that there is no remoter planet ; but of course it does not make it certain. If such a planet exists, it is sure to be found sooner or later, either by means of the disturbances it produces in the motion of Uranus and Neptune, or else by the methods of the asteroid hunters, although its slow motion will render its discovery in that way difficult. Quite possibly such a discovery may come within a few years as a result of the photographic star-charting operations now in progress.

288*. Stability of the Solar System. — It is an interesting and important question, once long and warmly discussed, whether the so-called "perturbations" which result from the mutual attractions of the planets can ever seriously derange the system. It is now nearly a century since Laplace and Lagrange first demonstrated that they cannot. The system is stable in itself, all the planetary disturbances *due to gravitation* being either of such a character, or so limited in extent, that they can never do any harm.

It does not follow, however, that because the mutual attractions of the planets are thus harmless, there may not be other causes which would act disastrously. Many such are conceivable, — such, for instance, as the retardation of the speed of the planets which would be caused by the presence ot a resisting medium in space, or by the encounter of the system with a sufficiently dense and extended cloud of meteors.

But so far as we can now judge, the ultimate cooling of the sun (Art. 193) is likely to extinguish life upon the planets long before the mechanical destruction of the system can occur from any such external causes.

CHAPTER X.

COMETS AND METEORS.

THE NUMBER, DESIGNATION, AND ORBITS OF COMETS. —
THEIR CONSTITUENT PARTS AND APPEARANCE. — THEIR
SPECTRA AND PHYSICAL CONSTITUTION. — THEIR PROB-
ABLE ORIGIN. — REMARKABLE COMETS. — AEROLITES.
THEIR FALL AND CHARACTERISTICS. — SHOOTING STARS.
METEORIC SHOWERS. — CONNECTION BETWEEN COMETS
AND METEORS.

289. Comets — their Appearance and Number. — The word
"comet" (derived from the Greek *kōmé*) means a "hairy star."
The appearance is that of a rounded cloud of luminous fog
with a star shining through it, often accompanied by a long
fan-shaped train or "tail" of hazy light. They present themselves from time to time in the heavens, mostly when unex-
pected, move across the constellations in a path longer or
shorter according to circumstances, and remain visible for some
weeks or months, until they fade out and vanish in the distance. The large ones are magnificent objects, sometimes as
bright as Venus, and visible by day; with a head as large as
the moon, and having a train which extends from the horizon
to the zenith, and is really long enough to reach from the earth
to the sun. Such comets are rare, however. The majority are
faint wisps of light, visible only with the telescope. Fig. 58
is a representation of Donati's comet of 1858, which was one of
the finest ever seen.

FIG. 58.—Naked-eye View of Donati's Comet, Oct. 4, 1858.   (Bond.)

In ancient times, comets were always regarded with terror, as at least presaging evil, if not actively malignant, and the notion still survives in certain quarters, though the most careful research goes to prove that they really do not exert upon the earth the slightest perceptible influence of any kind.

Thus far our lists contain about 675, about 400 of which were observed before the invention of the telescope, and therefore must have been fairly bright. Of the 275 observed since then, only a small proportion have been conspicuous to the naked eye, perhaps one in five. The total number that visit the solar system must be enormous; for there is seldom a time when one at least is not in sight, and even with the telescope we see only the few which come near the earth and are favorably situated for observation.

290. Designation of Comets. — A remarkable comet generally bears the name of its discoverer or of some one who has "acquired its ownership," so to speak, by some important research concerning it. Thus we have Halley's, Encke's, and Donati's comets. The ordinary telescopic comets are designated only by the year of discovery with a letter indicating the order of discovery in that year, as comet "1890 a "; or still again, with the year and a Roman numeral denoting the order of perihelion passage, as 1890 I, the latter method being the most used. In some cases a comet bears a double name, as the Lexell-Brooks comet (1889 V), which was investigated by Lexell in 1770, and discovered by Brooks on its recent return in 1889.

291. Duration of Visibility and Brightness. — The great comet of 1811 was observed for seventeen months, and the little comet, known as 1889 I, for more than two years, the longest period of visibility on record. On the other hand, the whole appearance sometimes lasts only a week or two. The average is probably not far from three months.

As to brightness, comets differ widely. About one in five reaches the naked-eye limit, and a very few, say four or five in a century, are bright enough to be seen in the day-time. The great comet of 1882 was the last one so visible.

292. Their Orbits. — A large majority of the comets move in orbits that are sensibly *parabolas* (see Appendix, Arts. 439–440). A comet moving in such an orbit approaches the sun from an enormous distance, far beyond the limits of the solar system, sweeps once around the sun, and goes off, never to come back. The parabola does not return into itself and form a closed curve, like the circle and ellipse, but recedes to infinity. Of the 280 orbits that have been computed, more than 200 appear to be of this kind. About 70 orbits are more or less distinctly elliptical, and about half a dozen are perhaps hyperbolas (see Appendix, Art. 440); but the hyperbolas differ so slightly from parabolas that the hyperbolic character is not really certain in a single case.

Comets which have elliptical orbits of course return at regular intervals. Of the 70 elliptical orbits, there are about a dozen to which computation assigns periods near to or exceeding 1000 years. These orbits approach parabolas so closely that their real character is still rather doubtful. About 55 comets, however, have orbits which are distinctly and certainly elliptical, and 30 have periods of less than one hundred years. Fifteen of these 30 have been actually observed at two or more returns at perihelion. As to the rest of them, some are now due within a few years, and some have probably been lost to observation, either like Biela's comet (Art. 312), or by having their orbits transformed by perturbations.

293. The first comet ascertained to move in an elliptical orbit was that known as Halley's, with a period of about seventy-six years, its periodicity having been discovered by Halley in 1681. It has since been observed in 1759 and 1835, and is expected again about 1911. The second of the periodic comets (in the order of discovery) is Encke's, with the shortest period known, — only three and one-half years. Its periodicity was discovered in 1819, though the comet itself had been observed several times before.   Fig. 59 shows the orbits of a number of short period comets (it would cause confusion to insert more of them), and also a part of the orbit of Halley's comet.   These comets all have periods ranging from three and one-half to eight years, and it will be noticed that they all pass very near to the orbit of Jupiter.   Moreover, each comet's orbit crosses that of Jupiter *near one of its nodes*, the node being marked by a short cross-line on the

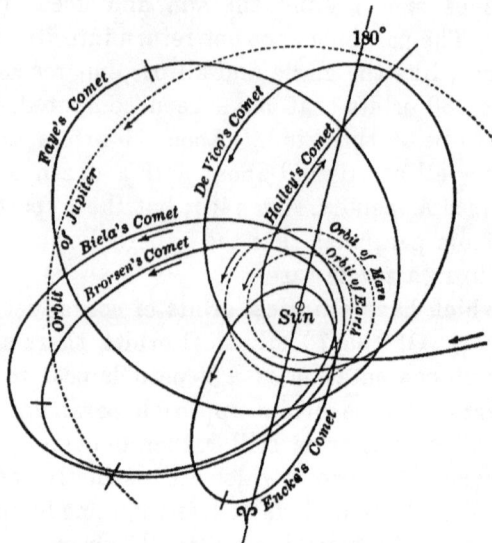

FIG. 59. — Orbits of Short-period Comets.

comet's orbit.   The fact is very significant, showing that these comets at times come very near to Jupiter, and it points to an almost certain connection between that planet and these bodies.

294. Comet Groups.— There are several instances in which a number of comets, certainly distinct, chase each other along almost exactly the same path, at an interval usually of a few months or years, though they sometimes appear simultaneously.   The most remarkable of these "comet groups" is that composed of the great comets of 1668, 1843, 1880, 1882, and 1887.   It is of course nearly certain that the comets of such a "group" have a common origin.

295. Perihelion Distance, etc. — Eight of the 280 cometary orbits, thus far determined, approach the sun within less than 6,000000 miles, and four have a perihelion distance exceeding 200,000000. A single comet (that of 1729) had a perihelion distance of more than four 'astronomical units,' or 375,000000 miles. It must have been an enormous one, to be visible at all under the circumstances. There may, of course, be any number of comets with still greater perihelion distances, because as a rule we are only able to see such as come reasonably near the earth, and this is probably only a small percentage of the total number that visit the sun.

The inclinations of cometary orbits range all the way from zero to 90°. As regards the direction of motion, all the elliptical comets having periods of less than 100 years move *direct*, *i.e.*, from west to east, except Halley's comet and Tempel's comet of 1886. Other comets show no decided preponderance either way.

296. Parabolic Comets are Visitors. — The fact that the orbits of most comets are sensibly parabolic, and that their planes have no evident relation to the ecliptic, indicates (though it does not absolutely prove) that these bodies do not in any proper sense belong to the solar system, but are only visitors. Such comets come to us precisely as if they simply dropped towards the sun from an enormous distance among the stars; and they leave the system with a velocity which, if no force but the sun's attraction acts upon them, will carry them away to an infinite distance, or until they encounter the attraction of some other sun. Their motions are just what might be expected of ponderable masses moving among the stars under the law of gravitation.

A slightly different view is advocated by some high authorities, and is perhaps tenable, — that these comets come from a great distance indeed, but not from among the stars. It may be that our solar system, in its journey through space (Art. 342), is accompanied by outlying clouds of nebulous matter, and that these are the source of the comets. It is argued that if this were not the case the number of *hyperbolic* orbits would be much greater, because we should *meet* so many more comets than could overtake us.

297. Origin of Periodic Comets. — But while the parabolic comets are thus probably strangers and visitors, there is a question as to the periodic comets, which move in elliptical orbits. Are we to regard them as native citizens, or only as naturalized foreigners, so to speak? It is evident that, somehow or other, many of them stand in peculiar relations to Jupiter, Saturn, and other planets, as already indicated in Art. 293. All short period comets (those which have periods ranging from three to eight years) pass very close to the orbit of Jupiter, and are now recognized and spoken of as Jupiter's "family of comets"; more than twenty are known at present. Similarly, Saturn is credited with two comets, and Uranus with three, one of them being Tempel's comet, which is closely connected with the November meteors, and is due on its next return in 1900. Finally, Neptune has a family of six, among them Halley's comet, and two others which have returned a second time to perihelion since 1880.

298. The Capture Theory. — The generally accepted theory as to the origin of these "comet-families" is one first suggested by Laplace nearly 100 years ago, — that the comets which compose them have been *captured* by the planet to which they stand related. A comet entering the system in a parabolic orbit, and passing near a planet, will be disturbed, — either accelerated or retarded. If it is accelerated, it is easy to prove that the original parabolic orbit will be changed to an hyperbola, and the comet will never be seen again, but will pass out of the system forever; but if it is *retarded*, the orbit becomes *elliptical*, and the comet will return at each successive revolution to the place where it was first disturbed.

But this is not the end. After a certain number of revolutions, the planet and the comet will come together a second
time at or near the place where they met before. The result
may then be an acceleration, which will send the comet out
of the system finally; but it is an even chance at least, that
it may be a second retardation, and that the orbit and period
may thus be further diminished : and this may happen over and
over again, until the comet's orbit falls so far inside that of the
planet that there is no further disturbance to speak of. Given
time enough and comets enough, and the result would inevitably be such a comet-family as really exists. We may add
that certain researches of Professor Newton of New Haven
and others, upon the position and distribution of cometary
orbits, decidedly favor the idea that these bodies do not
originate in the solar system, but come to us from interstellar
space.

The late R. A. Proctor declined to accept this capture theory, and
maintained with much vigor and ability the theory that comets and
meteor-swarms have been ejected from the great planets by eruptions
of some sort. We cannot take space here to discuss the theory, which
is really not quite so wild as at first it seems; but the objections to it
are serious, and we think fatal.

299. The Lexell-Brooks Comet. — The "capture" theory
has recently received an interesting illustration in the case
of a little comet, 1889 V, discovered by Mr. Brooks of Geneva,
N.Y., in July, 1889. It was soon found to be moving with a
period of about seven years, in an elliptical orbit which passes
very near to that of Jupiter. (We remark in passing that
this comet in August divided into four fragments; see Art.
314.) On investigating the orbit more carefully, Dr. S. C.
Chandler of Cambridge (U.S.) discovered that, in 1886, the
comet and the planet had been close together for some months,
and that as a consequence the comet's orbit must have been
greatly changed, the previous orbit having been a much larger
one with a probable period of nearly twenty-seven years.

Now, in 1770, a famous comet appeared, which is known as Lexell's, because Lexell computed its orbit. It was bright, and came very near the earth, and, according to Lexell's calculations, was then moving in an orbit with a period of only five and a half years, — the first instance of a short-period comet on record; but it was never seen again. Its failure to reappear, in 1776, was easily accounted for by the fact that its orbit did not then bring it anywhere near the earth. But it should have reappeared in 1781, and for a long time its disappearance was very mysterious, until Laplace, some years later, showed that, in 1779, the comet must have come very close to Jupiter, perhaps as near as some of its satellites, and that in consequence the attraction of that planet had probably sent it into a new orbit, not observable from the earth.

More recent investigations by Leverrier, some thirty years ago, show that while the data are insufficient to determine the comet's subsequent orbit with certainty, one of the *possible* orbits would have had a period a little less than twenty-seven years. This would bring it back, in 1886, after four revolutions, to the same place which it had occupied in 1779; now nine of Jupiter's periods are 106¾ years, so that he, also, would have returned to the same place.

To make a long story short, Mr. Chandler showed it to be extremely probable that Brooks's comet, 1889 V, is identical with Lexell's comet of 1770. Jupiter first transformed its orbit from a parabola to an ellipse, with a period of five and a half years; then removed it from our sphere of observation; and, again, after a century or more, has brought it back. What will happen at the next encounter of the comet with the planet it is not yet possible to predict.

The still more recent calculations of Dr. Poor of Baltimore, based in part on later and more accurate observations than those available to Chandler, appear, however, to make it rather more likely that Brooks's comet is not identical with Lexell's, but only a member of the same comet group (Art.

294). The question will have to await another return of the comet for final decision. Dr. Poor finds that the comet in 1886 passed between Jupiter and the orbit of its first satellite, within about 200,000 miles of the planet's surface.

PHYSICAL CONSTITUTION OF COMETS.

300. Constituent Parts of a Comet. — (a) The *essential* part of a comet, that which is always present and gives the comet its name, is the *coma*, or nebulosity, a hazy cloud of faintly luminous transparent matter.

(b) Next we have the *nucleus*, which, however, is wanting in many comets, and makes its appearance only as the comet comes near the sun. It is a bright, more or less star-like point near the centre of the comet. In some cases it is double, or even multiple.

(c) The *tail* or *train* is a stream of light which commonly accompanies a bright comet, and is sometimes present even with a telescopic one. As the comet approaches the sun, the tail follows it; but as the comet moves away from the sun, it precedes. It is always, speaking broadly, directed away from the sun, though its precise form and position are determined partly by the comet's motion. It is practically certain that it consists of extremely rarefied matter which is thrown off by the comet and powerfully *repelled* by the sun.

It certainly is not — like the smoke of a locomotive or train of a meteor — simply *left behind* by the comet, because as the comet is receding from the sun the tail goes before it, as has been said.

(d) *Jets and Envelopes.* — The head of a comet is often veined by short jets of light, which appear to be spurted out from the nucleus; and sometimes the nucleus throws off a series of concentric envelopes, like hollow shells, one within the other. These phenomena, however, are seldom observed in telescopic comets.

301. Dimensions of Comets. — The *volume* or bulk of a comet is often enormous, almost inconceivably so, if the tail is included in the estimate. The head, as a rule, is from 40,000 to 50,000 miles in diameter (comets less than 10,000 miles in diameter would stand little chance of discovery). Comets exceeding 150,000 miles are rather rare, though there are several on record.

The comet of 1811 at one time had a diameter of fully 1,200000 miles, 40 per cent larger than that of the sun. The head of the comet of 1680 was 600,000 miles in diameter, and that of Donati's comet, of 1858, about 250,000. Holmes' comet (1892) exceeded 800,000.

The diameter of the head changes continually and capriciously; on the whole, while the comet is approaching the sun the head usually *contracts*, expanding again as it recedes.

No entirely satisfactory explanation is known for this behavior, but Sir John Herschel has suggested that the change is merely optical; that near the sun a part of the nebulous matter is evaporated by the solar heat and so *becomes invisible*, condensing and reappearing again when the comet gets to cooler regions.

The *nucleus* ordinarily has a diameter ranging from 100 miles up to 5000 or 6000, or even more. Like the comet's head it also varies greatly in diameter, even from day to day, so that it is probably not a solid body. Its changes, however, do not seem to depend in any regular way upon the comet's distance from the sun, but rather upon its activity in throwing off jets and envelopes.

The *tail* of a comet, as regards simple magnitude, is by far the most imposing feature. Its length is seldom less than from 5,000000 to 10,000000 miles. It frequently attains 50,000000, and there are several cases where it has exceeded 100,000000; while its diameter at the end remote from the comet varies from 1,000000 to 15,000000.

302. Mass of Comets. — While the bulk of comets is thus enormous, their *masses* are apparently insignificant, in no case at all comparable with that of our little earth, even. The evidence on this point, however, is purely *negative;* it does not enable us in any case to say just what the mass really is, but only to say *how great it is not;* *i.e.,* it only proves that a comet's mass is *less* than $\frac{1}{100000}$ of the earth's,[1] — how much less we cannot yet find out. The evidence is derived from the fact that no sensible perturbations are produced in the motions of a planet when a comet comes even very near it, although in such a case the comet itself is fairly "sent kiting," thus showing that gravitation has its full effect between the two bodies.

Lexell's comet, in 1770, and Biela's comet on several occasions, have come so near the earth that the length of the comet's period was changed by several weeks, while the year was not altered by so much as a single second. It would have been changed by many seconds if the comet's mass were as much as $\frac{1}{100000}$ that of the earth.

303. Density of Comets. —This is, of course, almost inconceivably small, the mass of comets being so minute and their volumes so enormous. If the head of a comet, 50,000 miles in diameter, has a mass $\frac{1}{100000}$ that of the earth, its mean density must be about $\frac{1}{8000}$ that of the air at the sea-level, — far below that of the best air-pump vacuum. As for the tail, the density must be almost infinitely lower yet. It is nearer to an "airy nothing" than anything else we know of.

The extremely low density of comets is shown also by their transparency. Small stars can be seen through the head of a comet 100,000 miles in diameter, even very near its nucleus, and with hardly a perceptible diminution of their lustre.

We must bear in mind, however, that the low *mean* density of a comet does not necessarily imply a low density of its constituent particles. A comet may be to a considerable extent composed of small

---

[1] One one-hundred-thousandth of the earth's mass is about ten times the mass of the earth's whole atmosphere, and is equivalent to the mass of an iron ball about 150 miles in diameter.

heavy bodies, and still have a low *mean* density, provided they are far enough apart. There is much reason, as we shall see, for suppcsing that such is really the case, — that the comet is largely composed of small meteoric stones, carrying with them a certain quantity of enveloping gas.

Another point should be referred to. Students often find it impossible to conceive how such impalpable "dust clouds" can move in orbits like solid masses, and with such enormous velocities. They forget that *in a vacuum* a feather falls as

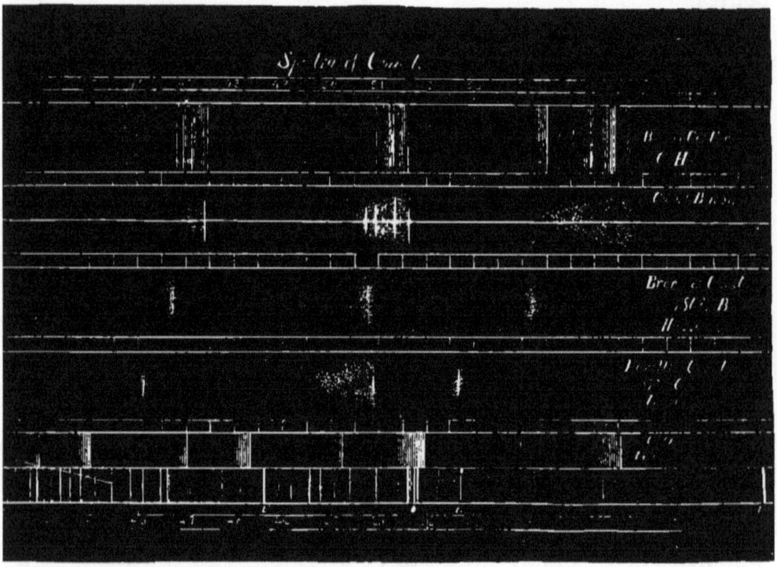

FIG. 60.—Comet Spectra.

(For convenience in engraving, the *dark* lines of the solar spectrum in the lowest strip of the figure are represented as *bright*.)

swiftly as a stone. Interplanetary space is a vacuum far more perfect than anything we can produce by air-pumps, and in it the lightest bodies move as freely and swiftly as the densest, since there is nothing to resist their motion. If all the earth were suddenly annihilated except a single feather, the feather would keep right on and continue the same orbit, with unchanged speed.

304. The Light of Comets. — To some extent their light may be mere reflected sunlight; but in the main it is light emitted by the comet itself under the stimulus of solar action. That the light depends in some way on the sun is shown by the fact that its brightness usually varies with its distance from the sun, according to the same law as that of a planet.

But the brightness frequently varies rapidly and capriciously without any apparent reason; and that the comet is

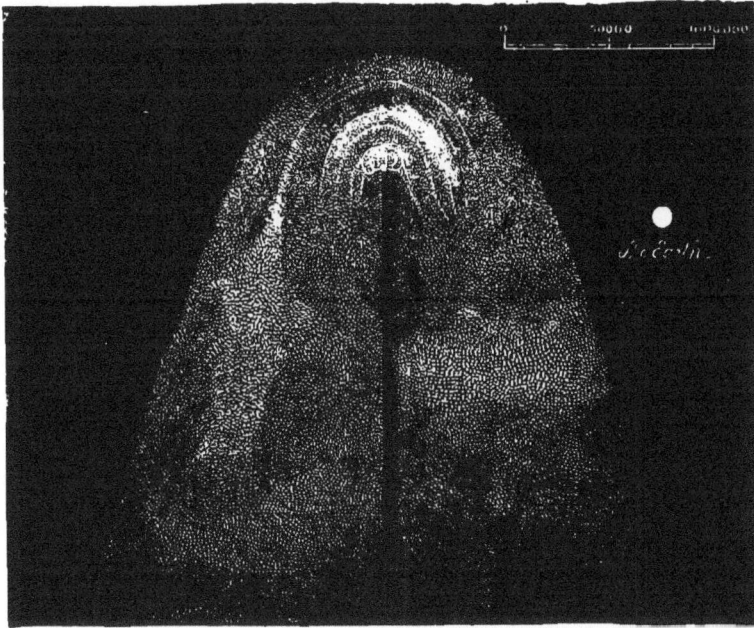

Fig. 61. — Head of Donati's Comet, Oct. 5, 1858. (Bond.)

self-luminous when near the sun is proved by its *spectrum*, which is not at all like the spectrum of reflected sunlight, but is a spectrum of *bright bands*, three of which are usually seen, and have been identified repeatedly and certainly with the spectrum of gaseous hydrocarbons. (All the different hydrocarbon gases give the same spectrum at the temperature of a Bunsen burner.) This spectrum is absolutely identical with that given by the blue base of a candle flame, or, better, by a Bunsen burner consuming ordinary coal gas.

Occasionally a fourth band is seen in the violet, and when the comet approaches unusually near the sun, the bright lines of sodium, and other metals (probably *iron*), sometimes appear. There seem to be cases, also, when different bands replace the ordinary ones, and Holmes's comet in 1892 showed a purely continuous spectrum. Fig. 60 represents the ordinary comet spectrum, compared with the solar spectrum and with that of a candle flame. The spectrum makes it almost certain that hydrocarbon gases are present in considerable quantity, and that these gases are somehow rendered luminous; not probably by any general *heating*, however, for there is no reason to think that the general temperature of a comet is very high. Nor must we infer that the hydrocarbon gas, because it is so conspicuous in the spectrum, necessarily constitutes most of the comet's mass: more likely it is only a very small fraction of the whole.

305. Phenomena that accompany a Comet's Approach to the Sun. — When the comet is first discovered, it is usually a mere round, hazy cloud of faint nebulosity, a little brighter near the middle. As it approaches the sun it brightens rapidly, and the nucleus appears. Then on the sunward side the nucleus begins to emit luminous jets, or else to throw off more or less symmetrical envelopes, which follow each other at intervals of some hours, expanding or growing fainter, until they are lost in the nebulosity of the head.

Fig. 61 shows the envelopes as they appeared in the head of Donati's comet of 1858. At one time seven of them were visible together: very few comets, however, exhibit this phenomenon with such symmetry. More frequently the emissions from the nucleus take the form of jets and streamers.

↑ To Sun

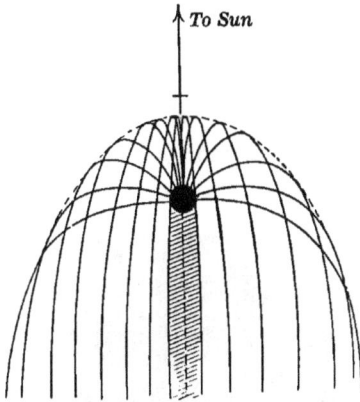

FIG. 62. — Formation of a Comet's Tail by
Matter expelled from the Head.

306. Formation of Tail. — The tail appears to be formed of material which is first projected from the nucleus of the comet towards the sun, and then afterwards repelled by the sun, as illustrated by Fig. 62. At least this theory has the great advantage over all others which have been proposed that it not only accounts for the phenomenon in a general way, but admits of being worked out in detail and verified mathematically, by comparing the actual size and form of the planet's tail, at different points in the orbit, with that indicated by the theory; and the accordance is generally very satisfactory.

The repelled particles are still subject to the sun's gravitational attraction, and the *effective* force acting upon them is therefore the difference between the gravitational attraction and the electrical (?) repulsion. This *difference* may or may not be in favor of the attraction, but in any case, the sun's attracting force is, at least, lessened. The consequence is that those repelled particles, as soon as they get a little away from the comet, begin to move around the sun in *hyperbolic* orbits (see Art. 439), which lie in the plane of the comet's orbit, or nearly so, and are perfectly amenable to calculation.

In the case of a great comet the tail is usually a sort of curved, hollow cone, including the head of the comet at its smaller extremity; in smaller comets the tail is generally a comparatively narrow streamer where it issues from the head of the comet, brushing out as it recedes, and often showing, *in photographs,* peculiar knots and condensations, which are not visible with the telescope.

The tail is *curved* because the repelled particles, after leaving the comet's head, retain their original motion, so that they are arranged not along a straight line drawn from the sun to

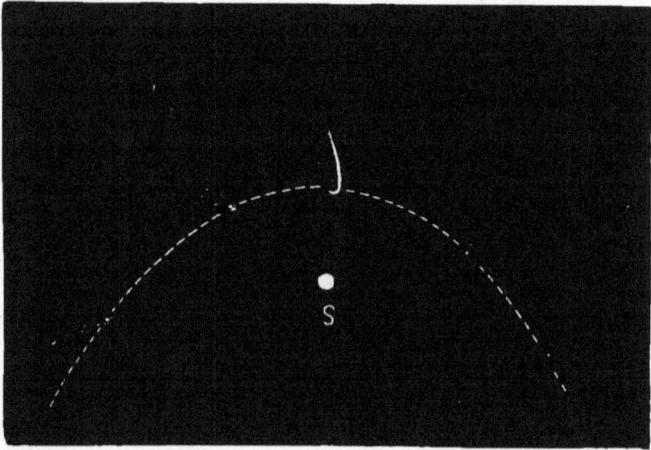

FIG. 63. — A Comet's Tail at Different Points in its Orbit near Perihelion.

the comet, but on a curve convex to the comet's motion, as shown in Fig. 63; but the stronger the repulsion the less the curvature, and the straighter the tail. The nature of the force which repels the particles of a comet is, of course, only a matter of speculation; but the idea that it is *electrical* generally prevails, though the detailed explanation is not easy. There is no reason to suppose that the matter driven off to form the tail is ever recovered by the comet.

307. Types of Comets' Tails. — Bredichin, of Pulkowa, has found that the trains of comets may be classified under three different types, as indicated by Fig. 64.

First, the *long, straight rays,* composed of matter upon which the solar repulsion is from ten to fifteen times as great as the attraction of gravity, so that the particles leave the comet with a velocity of four or five miles a second, which is afterwards increased until it becomes enormous. The nearly straight rays, shown in Fig. 58, belong to this type. For plausible reasons, which, however, we cannot stop to explain, Bredichin supposes these straight rays to be composed of *hydrogen*.

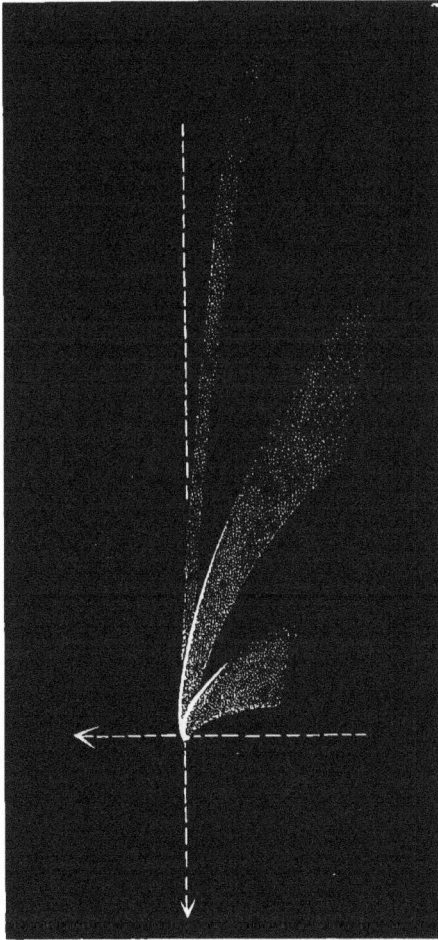

The *second* type is the ordinary, curved, plume-like train, like the principal tail of Donati's comet. In trains of this type, supposed to be due to hydrocarbon vapors, the repulsive force varies from 2.2 times the attraction of gravity for particles on the convex edge of the train, to half that amount for those on the inner edge. The spectrum is the same as that of the comet's head.

Third, a few comets show tails of still a third type, — short, stubby brushes, violently curved, and due to matter on which the repulsive force is feeble as compared with gravity. These are assigned by Bredichin to metallic vapors of considerable density, with an admixture of sodium, etc.

Fig. 64.

Bredichin's Three Types of Cometary Tails.

308. Unexplained and Anomalous Phenomena. — A curious

phenomenon, not yet explained, is the dark stripe which, in a

large comet approaching the sun, runs down the centre of the
tail, looking very much as if it were a shadow of the comet's
head. It is certainly not a shadow, however, because it usually
makes more or less of an angle
with the sun's direction. It is
well shown in Fig. 61. When the
comet is at a greater distance from
the sun, this central stripe is usually bright, as in Fig. 65 ; and in
the case of small comets, generally all the tail they show is
such a narrow streamer.

FIG. 65.

Bright-centred Tail of Coggia's Comet,
June, 1874.

Not unfrequently, moreover, comets possess anomalous tails,—
tails directed sometimes straight toward the sun, and sometimes at
right angles to that direction. Then sometimes there are luminous
sheaths, which seem to envelop the head of the comet and project
towards the sun (Fig. 66), or little clouds of cometary matter which
leave the main comet like puffs of smoke from a bursting bomb, and
travel off at an angle until they fade away (see Fig. 66). None of
these appearances are contradictory to the theory above stated,
though they are not yet clearly included in it.

309. The Nature of Comets. — All things considered, the
most probable hypothesis as to the constitution of a comet, so
far as we can judge at present, is that its head is a swarm of
small meteoric particles, widely separated (say pin-heads, many
yards apart), each carrying with it an envelope of rarefied gas
and vapor, in which light is produced either by electric dis-
charges between the solid particles, or by some action due to
the rays of the sun. As to the size of the constituent particles, opinions differ widely. Some maintain that they are
large rocks : Professor Newton calls a comet a " gravel bank " :
others say that it is a mere " dust-cloud." The unquestion-
able close connection between meteors and comets (Art. 327)
almost compels some " meteoric hypothesis."

310. Danger from Comets. — In all probability there is none. It has been supposed that comets might do us harm in two ways, — either by actually striking the earth, or by falling into the sun and thus producing such an increase of solar heat as to burn us up.

As regards the possibility of a collision between a comet and the earth, the event is certainly possible. In fact, if the earth lasts long enough, it is practically sure to happen, for there are several comet's orbits which pass nearer to the earth's orbit than the semi-diameter of the comet's head. As to the consequence of such a collision, it is impossible to speak with absolute confidence for want of certain knowledge as to the constitution of a comet. If the solid "particles" of which the main portion of the comet is probably composed are no larger than pin-heads, the result would be only a fine meteoric shower ; if on the other hand they weigh tons, the bombardment would be a very serious matter. It is possible too that the mixture of the comet's gases with our atmosphere might be a source of danger.

The encounters, however, will be very rare. If we accept the estimate of Babinet, they will occur on the average once in about 15,000000 years.

If a comet actually strikes the sun, which would necessarily be a very rare phenomenon, it is not likely that the least harm will be done. The collision might generate about as much heat as the sun radiates in eight or nine hours; but the cometary particles would pierce the photosphere, and their heat would be liberated mostly below the solar surface, simply expanding by some slight amount the diameter of the sun, but making no particular difference in the amount of its radiation for the time being. There might be, and very likely would be, a flash of some kind at the solar surface when the shower of meteors struck it; but probably nothing that the astronomer would not take delight in observing.

311. Remarkable Comets. — Our space does not permit us to give full accounts of any considerable number. We limit ourselves to three, which for various reasons are of special interest.

Biela's comet is, or rather *was*, a small comet some 40,000 miles in diameter, at times barely visible to the naked eye, and sometimes showing a short tail. It had a period of 6.6 years, and was the second comet of short period known, having been discovered by Biela, an Austrian officer, in 1826 (the periodicity of Encke's comet had been discovered seven years earlier). Its orbit comes within a few thousand miles of the earth's orbit, the distance varying somewhat, of course, on account of perturbations; but the approach is sometimes so close that, if the comet and the earth should happen to arrive at the same time, there would be a collision. At its return, in 1846, it *split into two*. When first seen on Nov. 28th, it was one and single. On Dec. 19th it was distinctly pear-shaped, and ten days later it was divided.

The twin comets travelled along for four months, at an almost unchanging distance of about 165,000 miles, without any apparent effect upon each other's *motions*, but both very active from the physical point of view, showing remarkable variations and alterations of brightness entirely unexplained. In August, 1852, the twins were again observed, then separated by a distance of about 1,500000 miles; but it was impossible to tell which was which. Neither of them has ever been seen again, though they must have returned many times, and more than once in a very favorable position.

312. There remains, however, another remarkable chapter in the story of this comet. In 1872, on Nov. 27th, just as the earth was crossing the track of the lost comet, but some millions of miles behind where the comet ought to be, she encountered a wonderful meteoric shower. As Miss Clerke expresses it, perhaps a little too positively, " it became evident that Biela's comet was shedding over us the pulverized products of its disintegration." A similar meteoric shower occurred again in 1885, when the earth once more crossed the track of the comet ; and still again in 1892.

It is not certain whether the meteor swarms thus encountered were the *remains of the comet itself,* or whether they were other small bodies merely *following in its path.* The comet must have been several millions of miles ahead of the place where these meteor swarms were met, unless it has been set back in its orbit since 1852 by some unexplained and improbable perturbations. But the comet cannot be found, and if it still exists and occupies the place it ought to, it must have somehow lost the power of shining.

313. The Great Comet of 1882. — This is the most recent of the brilliant comets that have been observed, and will long be remembered not only for its magnificent beauty, but for the great number of unusual phenomena which it presented. It was first seen in the southern hemisphere about Sept. 3d, but not in the northern until the 17th, the day on which it arrived at perihelion. On that day it was independently discovered within 2° or 3° of the sun, near noon, by several observers, who had not before heard of its existence. It was visible to the naked eye *in full sunshine* for more than a week after its perihelion passage. It then became a splendid object in the morning sky, and continued to be observed for six months.

That portion of the orbit visible from the earth coincides almost exactly with the orbits of four other comets, — those of 1668, 1843, 1880, and 1887, with which it forms a "comet group," as already mentioned (Art. 294). The perihelion distances of the comets of this group are all less than 750,000 miles, so that they pass within 300,000 miles of the sun's surface; *i.e.*, right through the corona, and with a velocity exceeding 300 miles a second; and yet this passage through the corona does not disturb their motion perceptibly.

The orbit of the comet of 1882 turns out to be a very elongated ellipse with a period of about 800 years. The period of the comet of 1880 appears to be only seventeen years, while the orbits of the other three are sensibly parabolic.

314. Early in October the comet presented the ordinary features. The nucleus was round, a number of well-marked envelopes were visible in the head, and the dark stripe down

the centre of the tail was sharply defined. Two weeks later
the nucleus had been broken up and transformed into a
crooked stream, some 50,000 miles in length, of five or six
bright points: the envelopes had vanished from the head, and
the dark stripe was replaced by a bright central spine.

At the time of perihelion the comet's spectrum was filled
with countless bright lines. Those of sodium were easily
recognizable, and continued visible for weeks; the other lines

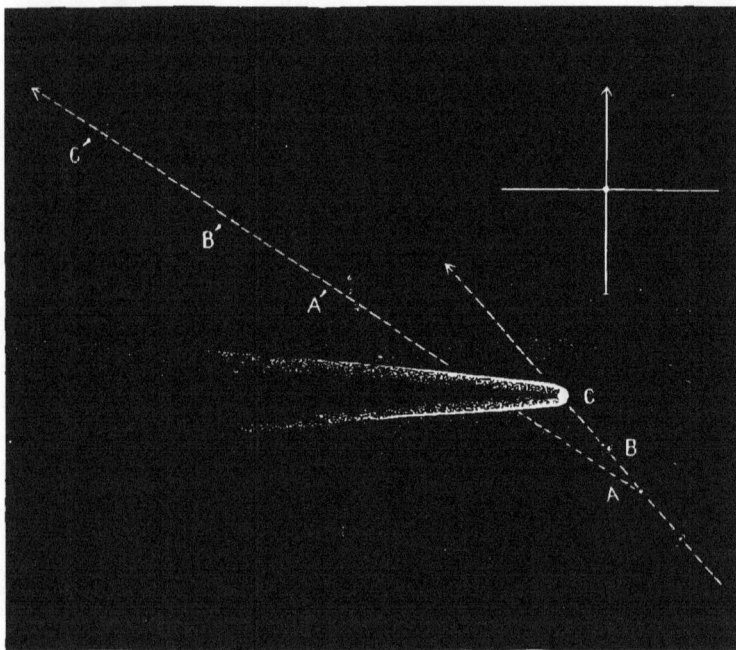

Fig. 66. — The "Sheath," and the Attendants of the Comet of 1882.

continued only a few days and were not certainly identified,
although the general aspect of the spectrum indicated that
iron, manganese. and calcium were probably present. By the
middle of October it had become simply the normal comet
spectrum, with the ordinary hydrocarbon bands.

The comet was so situated that the tail was directed nearly away from the earth, and so was not seen to good advantage, never having an apparent length exceeding 35°. The actual length, however, at one time was more than 100,000000 miles.

A unique, and so far unexplained, phenomenon was a faint, straight-edged "sheath" of light, which enveloped the portions of the comet near the head, and projected 3° or 4° in front of it, as shown in Fig. 66. Moreover, there were certain shreds of cometary matter accompanying the comet, at a distance of 3° or 4° when first seen, but gradually receding and growing fainter. This also was something new in cometary history, though the Lexell-Brooks comet, 1889 V, has since shown the same thing.

Holmes's comet of 1892–3 was in many ways remarkable. When first discovered it was already visible to the naked eye, and was apparently almost stationary, fast increasing in size as if swiftly approaching. For a time a popular impression prevailed that it was Biela's lost comet, and might strike the earth, which led to something like a "newspaper panic" in certain quarters. It was, however, really receding, and never came nearer than 150,000000 miles. It was never conspicuous, and had no nucleus or notable train ; but its bulk was enormous : at one time its diameter exceeded 800,000 miles. It experienced many capricious changes of apparent size and brightness, and its spectrum was purely *continuous*, — a thing unprecedented in comets. It moves in an orbit like that of an asteroid, with its perihelion just outside the orbit of Mars, and its aphelion close to that of Jupiter, its period being a few days less than seven years.

314.* Photography of Comets. — It is now possible to photograph comets, and the photographs bring out numerous peculiarities and details which are not visible to the eye even with telescopic aid. This is especially the case in the comet's tail. Fig. 66* is from a photograph of Rordame's comet of 1893, for which we are indebted to the kindness of Professor Holden, director of the Lick Observatory. As the camera was kept pointed at the head of the comet (which was moving pretty rapidly) the star-images during the hour's

COMET RORDAME, JULY 13, 1893.

Photographed by W. J. HUSSEY, at the LICK Observatory.

exposure are drawn out into parallel streaks, the little irregularities being due to faults of the clock-work and vibrations
of the telescope. The knots and streamers which characterize
the comet's tail were none of them visible in the telescope,
and are not the same shown upon plates taken the day before
and the day after. Other plates, made the same evening a
few hours earlier and later, indicate that the "knots" were
swiftly receding from the comet's head at a rate exceeding
150,000 miles an hour.

In 1892 Barnard *discovered* a small comet, by the streak it left
upon one of his star-plates.

METEORS AND SHOOTING-STARS.

315. Meteorites. — Occasionally bodies fall upon the earth
out of the sky. Such a body during its flight through the air
is called a "Meteorite" or "Bolide," and the pieces which fall
to the earth are called "Meteorites," "Aerolites," "Urano-
liths," or simply "meteoric stones."

If the fall occurs at night, a ball of fire is seen, which moves
with an apparent velocity depending upon the distance of the
meteor and the direction of its motion. The fire-ball is generally followed by a luminous train, which sometimes remains
visible for many minutes after the meteor itself has disappeared. The motion is usually somewhat irregular, and here
and there along its path the meteor throws off sparks and fragments, and changes its course more or less abruptly. Sometimes it vanishes by simply fading out in the air, sometimes
by bursting like a rocket. If the observer is near enough, the
flight is accompanied by a heavy, continuous roar, emphasized
now and then by violent detonations.

The observer must not expect to hear the explosion at the moment
when he sees it, since sound travels only about twelve miles a minute.

If the fall occurs by day, the luminous appearances are mainly wanting, though sometimes a white cloud is seen, and the train may be visible. In a few cases aerolites have fallen almost silently, and without warning.

316. The Aerolites themselves. — The mass that falls is sometimes a single piece, but more usually there are many fragments, sometimes numbering thousands; so that, as the old writers say, "it rains stones." The pieces weigh from 500 pounds to a few grains, the aggregate mass sometimes amounting to a number of tons. By far the greater number of aerolites are *stones*, but a few, perhaps three or four per cent of the whole number, are masses of nearly pure *iron* more or less alloyed with *nickel*.

The total number of meteorites which have fallen and been gathered into cabinets since 1800 is about 250, — only 10 of which are *iron* masses. Nearly all, however, contain a large percentage of iron, either in the metallic form or as sulphide. Between 25 and 30 of the 250 fell within the United States, the most remarkable being those of Weston, Conn., in 1807; New Concord, Ohio, 1860; Amana, Iowa, 1875; Emmett County, Iowa, 1879 (mainly iron); and Johnson County, Ark., 1886 (iron).

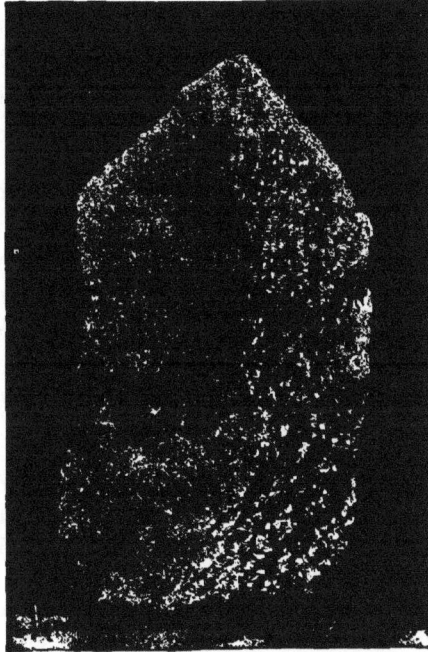

FIG. 67.

Fragment of one of the Amana Meteoric Stones.

Twenty-five of the chemical elements have been found in these bodies, including *helium* (Art. 181); but not one new element, though a large number of new *minerals* appear in them, and seem to be characteristic of aerolites.

The most distinctive external feature of a meteorite is the thin, black, varnish-like crust that covers it. It is formed by the melting of the surface during the meteor's swift flight through the air, and in some cases penetrates the mass in cracks and veins. The surface is generally somewhat uneven, having "thumb-marks" upon it, — hollows, probably formed by the fusion of some of the softer minerals. Fig. 67 is from a photograph given in Langley's "New Astronomy," where the body is designated, perhaps a little too positively, as "part of a comet."

317. Path and Motion. — When a meteor has been observed from a number of different stations, its path can be computed. It usually is first seen at an altitude of between 80 and 100 miles, and disappears at an altitude of between 5 and 10. The length of the path may be anywhere from 50 to 500 miles. In the earlier part of its course, the velocity ranges from 10 to 40 miles a second, but this is greatly reduced before the meteor disappears.

In observing these bodies, the object should be to obtain as accurate an estimate as possible of the *altitude* and *azimuth* of the meteor, at moments which can be identified, and also of the *time* occupied in traversing definite portions of the path. The altitude and azimuth will enable us to determine the height and position of the meteor, while the observations of the time are necessary in computing its velocity. By night the stars furnish the best reference points from which to determine its position. By day, one must take advantage of natural objects or buildings to define the meteor's place, the observer marking the precise spot where he stood. By taking the proper instrument to the place afterwards, it is then easy to ascertain the bearings and altitude. As to the time of flight, it is usual for the observer to begin to repeat rapidly some familiar verse of doggerel when the meteor is first seen, reiterating it until the meteor disappears. Then by rehearsing the same before a clock, the number of seconds can be pretty accurately determined.

318. The Light and Heat of Meteors. — These are due simply to the destruction of the meteor's velocity by the fric-. tion and resistance of the air. When a body moving with a high velocity is stopped by the resistance of the air, by far the greater part of its energy is transformed into heat. Lord Kelvin has demonstrated that the heating effect in the case of a body moving through the air with a velocity exceeding ten miles a second, is the same as if it were "immersed in a flame having a temperature at least as high as the oxyhydrogen blow-pipe"; and, moreover, this temperature is independent of the *density* of the air, — depending only on the velocity of the meteor. Where the air is dense, the total *quantity* of heat (*i.e.*, the number of calories developed in a given time) is of course greater than where the air is rarified; but the virtual *temperature* of the air itself where it rubs against the surface is the same in either case. During the meteor's flight, its surface, therefore, is raised to a white heat and melted, and the liquefied portions are swept off by the rush of air, condensing as they cool to form the train. In some cases this train remains visible for many minutes, — a fact not easily explained. It seems probable that the material must be phosphorescent.

319. Origin of Meteors. — They cannot be, as some have maintained, the *immediate* product of eruptions from volcanoes, either terrestrial or lunar, since they reach our atmosphere with a velocity which makes it certain that they come to us from the depths of space. There is no proof that they have originated in any way different from the larger heavenly bodies. At the same time many of them resemble each other so closely as almost to compel the surmise that these, at least, had a common source. It is not perhaps impossible that such may be fragments which long ago were shot out from now extinct *lunar* volcanoes, with a velocity which made planets of them for the time being. If so, they have since been travelling in independent orbits until they encountered the earth at the point where her orbit crosses theirs. Nor is it impossible that some of them were thrown out by *terrestrial* eruptions when the earth was young; or that they have been ejected from the planets, or even from the stars. It is only certain that during the period immediately preceding their arrival upon the earth, they have been travelling in long ellipses, or parabolas, around the sun.

SHOOTING-STARS.

320. Their Nature and Appearance. — These are the evanescent, swiftly moving, star-like points of light which may be seen every few minutes on any clear moonless night. They make no sound, nor has anything been known to reach the earth's surface from them.

For this reason it is probably best to retain, provisionally, at least, the old distinction between them and the great meteors from which aerolites fall. It is quite possible that the distinction has no real ground, — that shooting-stars, as is maintained by many, are just like other meteors, except that being so small they are entirely consumed in the air; but then, on the other hand, there are some things which rather favor the idea that the two classes differ in about the same way as asteroids do from comets. We know that an *aerolitic meteor* is a compact mass of rock. It is possible, or even likely, that a *shooting-star*, on the contrary, is a little *dust-cloud*, — like a puff of smoke.

321. Number of Shooting-stars. — Their number is enormous. A single observer averages from four to eight an hour; but if the observers are sufficiently numerous, and so placed as to be sure of noting all that are visible from a given station, about eight times as many are counted. From this it has been estimated that the total number which enter our atmosphere daily must be between 10,000000 and 20,000000, the average distance between them being some 200 miles.

Besides those which are visible to the naked eye, there is a still larger number of meteors which are so small as to be observable only with the telescope.

The average hourly number about 6 o'clock in the morning is double the hourly number in the evening; the reason being that in the morning we are in *front* of the earth, as regards its orbital motion, while in the evening we are in the rear. In the evening we see only such as overtake us; in the morning we see all that we either meet or overtake.

322. Elevation, Path, and Velocity. — By observations made at stations 30 or 40 miles apart, it is easy to determine these data with some accuracy. It is found that on the average the shooting-stars appear at a height of about 74 miles, and disappear at an elevation of about 50 miles, after traversing a course 40 or 50 miles long, with a velocity from 10 to 50 miles a second, — about 25 on the average. They do not begin to be visible at so great a height as the aerolitic meteors; and they are more quickly consumed, and therefore do not penetrate the atmosphere so deeply.

323. Brightness, Material, and Mass. — Now and then a shooting-star rivals Jupiter, or even Venus, in brightness. A considerable number are like first-magnitude stars; but the great majority are faint. The bright ones generally leave trains. Occasionally it has been possible to get a "snap shot," so to speak, at the *spectrum* of a meteor, and in it the bright lines of sodium and magnesium (probably) are fairly conspicuous among many others which cannot be identified by such a hasty glance.

Since these bodies are consumed in the air, all that we can hope to get of their material is their "ashes."

In most places its collection and identification is, of course, hopeless; but the Swedish naturalist Nordenskiold thought that it might be found in the polar snows. In Spitzbergen he therefore melted several tons of snow, and on filtering the water he actually detected in it a sediment containing minute globules of oxide and sulphide of iron. Similar globules have also been found in the products of deep-sea dredging. They may be meteoric, but what we now know of the distance to which smoke and fine volcanic dust is carried by the wind make it not improbable that they may be of purely terrestrial origin.

We have no way of determining the exact mass of a shooting-star, but from the *light* it emits as seen from a known distance, an approximate estimate can be formed, since we know roughly how much energy corresponds to the production of a given amount of light. It is likely, on the whole, that an ordinary meteor and a good electric incandescent lamp do not differ widely in what is called their 'luminous efficiency'; *i.e.*, the percentage of their total energy which is converted into visible light. Calculations on this basis indicate that ordinary shooting-stars are very minute, weighing only a small fraction of an ounce, — from less than a grain up to 50 or 100 grains for a very large one.

324. Meteoric Showers. — There are occasions when these bodies, instead of showing themselves here and there in the sky at intervals of several minutes, appear in showers of thousands; and at such times they do not move at random, but all their paths diverge or radiate from a single point in the sky known as *the radiant; i.e.*, their paths produced backward all pass through this point, though they do not usually start there. Meteors which appear near the radiant are apparently stationary, or describe paths which are very short, while those in the more distant regions of the sky pursue long courses. The radiant keeps its place among the stars sensibly unchanged during the whole continuance of the shower; it may be for hours and even days, and the shower is named accordingly from the place of the radiant. Thus we have the

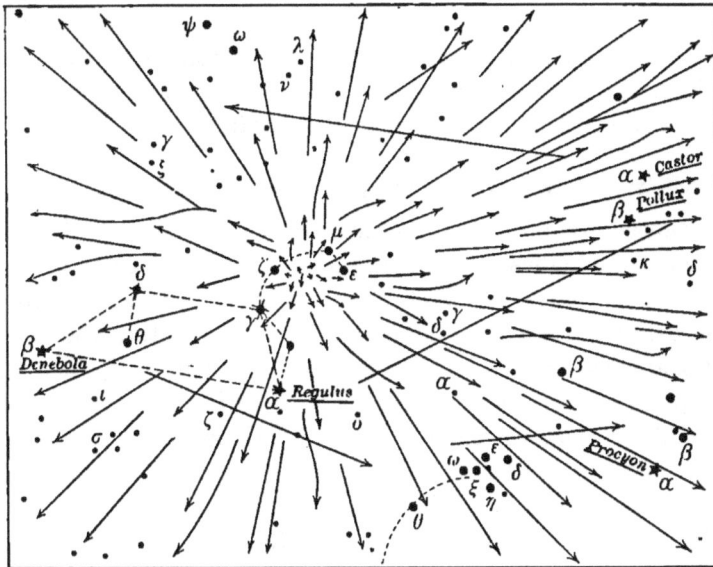

Fig. 68. — The Meteoric Radiant in Leo, Nov. 13, 1866.

"Leonids," or meteors whose radiant is the constellation of Leo; the "Andromedes" (or Bielids); the "Perseids"; the "Lyrids," etc.

Fig. 68 represents the tracks of a large number of the Leonids of 1866, showing the positions of the radiant near Zeta Leonis.

The radiant is explained as a mere effect of perspective. The meteors are all moving in lines nearly parallel with each other when encountered by the earth, and the radiant is simply the perspective "vanishing-point" of this system of parallels. Its position depends entirely on the *direction* of the motion of the meteors with respect to the earth. For various reasons, however, the paths of the meteors, after they enter the air, are not exactly parallel, and in consequence the radiant is not a mathematical point, but a "spot" in the sky, often covering an area of 3° or 4° square.

Probably the most remarkable of all the meteoric showers that ever occurred was that of the Leonids on Nov. 12th, 1833. The number of meteors at some stations was estimated as high as 100,000 an hour, for five or six hours. "The sky was as full of them as it ever is of snow-flakes in a storm."

325. Dates of Meteoric Showers. — Such meteoric showers are caused by the earth's encounter with a swarm of little meteors ; and since this swarm pursues a regular orbit around the sun, the earth can meet it only when she is at the point where her orbit cuts this path. The encounter, therefore, must always happen on or near the same day of the year, except as in time the meteoric orbits shift their positions on account of perturbations. The Leonid showers, therefore, always appear on the 13th of November, within a day or two, and the Andromedes on the 27th or 28th[1] of the same month.

In some cases the meteors are distributed along their whole orbit, forming a sort of elliptical ring, and are rather widely scattered. In that case the shower recurs every year, and may continue for several weeks, as is the case with the Perseids, which appear in early August. On the other hand, the flock may be concentrated, and then the shower will occur only when the earth and the meteor swarm both arrive at the orbit-crossing together. This is the case with both the Leonids and the Andromedes. The showers then occur not every year, but only at intervals of several years, though always on or near the same day of the month. For the Leonids, the interval is about thirty-three years, and for the Bielids about thirteen years.

326. The meteors which belong to the same group have a marked family resemblance. The Perseids are yellow, and

---

[1] In 1892 the shower occurred on Nov. 23d.

move with medium velocity. The Leonids are very swift (we meet them), and they are of a bluish green tint, with vivid trains. The Bielids are sluggish (they overtake the earth), are reddish, being less intensely heated than the others, and they usually have only feeble trains. During these showers no sound is heard, no sensible heat perceived, nor do any masses of matter reach the ground: with one exception, however, that on Nov. 27th, 1885, a piece of meteoric iron fell at Mazapil, in Northern Mexico, during the shower of Andromedes, which occurred that evening. The coincidence may be accidental, but is certainly interesting. Many high authorities speak confidently of this piece of iron as a piece of Biela's comet itself; and this brings us to one of the most important astronomical discoveries of the last half-century.

327. The Connection between Comets and Meteors. — At the time of the great meteoric shower of 1833, Professors Olmsted and Twining of New Haven were the first to recognize the "radiant," and to point out its significance as indicating that the meteors must be members of a swarm of bodies revolving around the sun in a permanent orbit. In 1864 Professor Newton of New Haven, taking up the subject anew, showed by an examination of the old records that there had been a number of great meteoric showers about the middle of November at intervals of thirty-three or thirty-four years; and he predicted confidently the repetition of the shower on Nov. 13th or 14th, 1866. It occurred as predicted, and was observed in Europe; and it was followed by another, in 1867, which was visible in America, the meteoric swarm being extended in so long a procession as to require more than two years to cross the earth's orbit. The researches of Newton and Adams showed that the flock was moving in a long ellipse with a period of thirty-three years.

328. Identification of Meteoric and Cometary Orbits. — Within a few weeks after the shower of 1866 it was found

that the orbit pursued by these meteors was identical with
that of a comet, known as Tempel's, which had been visible
about a year before; and about the same time Schiaparelli
showed that the Perseids, or August meteors, move in an orbit
identical with that of the bright comet of 1862. Now a single
coincidence might be accidental, but hardly two. Five years
later came the shower of Andromedes, following in the track
of Biela's comet; and among the more than a hundred distinct

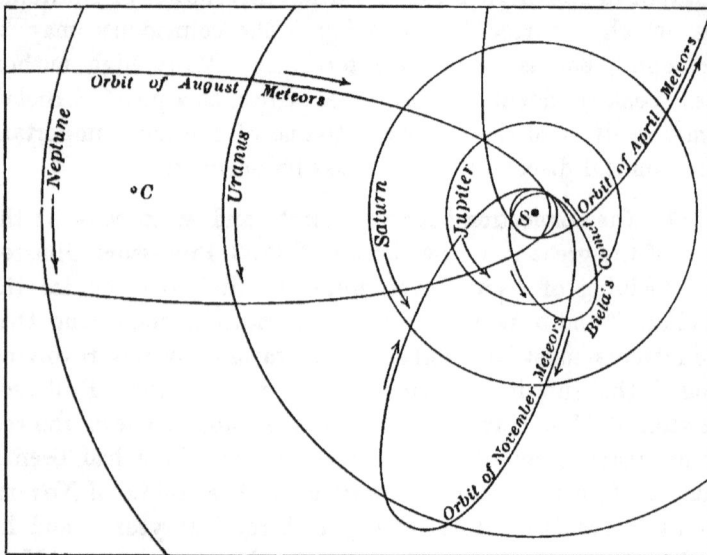

Fig. 69. — Orbits of Meteoric Swarms.

meteor swarms now recognized, Professor Alexander Herschel
finds five others which are similarly related, each to its special
comet. It is no longer possible to doubt that there is a real
and close connection between these comets and their attendant meteors. Fig. 69 represents four of the orbits of these
cometo-meteoric bodies.

329. Nature of the Connection. — This cannot be said to be
ascertained. In the case of the Leonids and Andromedes, the

meteoric swarm *follows* the comet, but this does not seem to be so in the case of the Perseids, which scatter along more or less abundantly every year. The prevailing belief is that the comet itself is only the thickest part of the meteoric swarm, and that the clouds of meteors scattered along its path are the result of its *disintegration;* but this is by no means certain.

It is easy to show that if the comet really is such a swarm, it must at each return to perihelion gradually break up more and more, and disperse its constituent particles along its path, until the compact swarm has become a diffuse ring. The longer the comet has been moving

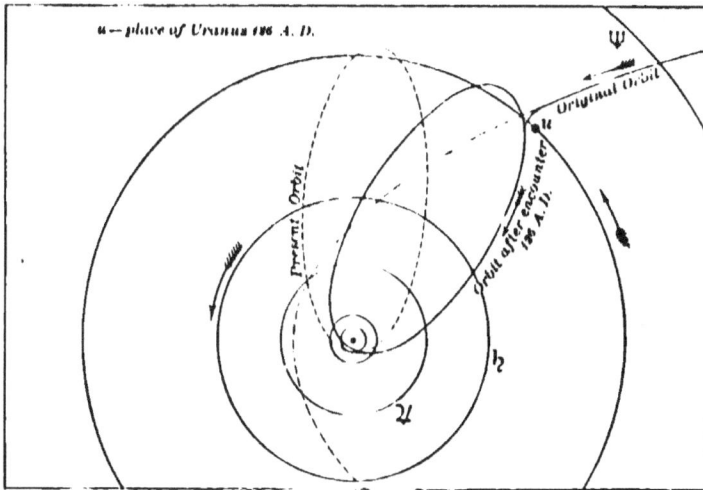

FIG. 70. — Origin of the Leonids.

around the sun, the more uniformly the particles will be distributed. The Perseids, therefore, are supposed to have been in the system for a long time, while the Leonids and Andromedes are believed to be comparatively new comers. Leverrier, indeed, has gone so far as to indicate the year 126 A.D. as the time at which Uranus captured Tempel's comet, and brought it into the system, as illustrated by Fig. 70. But the theory that meteoric swarms are the product of cometary disintegration assumes the premise that comets enter the system as compact clouds, which, to say the least, is not yet certain.

330. Mr. Lockyer's Meteoritic Hypothesis. — Within the last few years Mr. Lockyer has been enlarging the astronomical importance of meteors. The probable meteoric constitution of the zodiacal light, as well as of Saturn's rings, and of the comets, has long been recognized; but he goes much farther, and maintains that all the heavenly bodies are either meteoric swarms, more or less condensed, or the final products of such condensation; and upon this hypothesis he attempts to explain the evolution of the planetary system, the phenomena of variable and colored stars, the various classes of stellar spectra, and the forms and structure of the nebulæ, — indeed pretty much everything in the heavens from the Aurora Borealis to the sun. As a "working hypothesis," his theory is unquestionably important, and has attracted much attention, but it does not bear criticism in all its details.

CHAPTER XI.

THE STARS.

THEIR NATURE, NUMBER, AND DESIGNATION. — STAR
CATALOGUES AND CHARTS. — PROPER MOTIONS AND
THE MOTION OF THE SUN IN SPACE. — STELLAR PAR-
ALLAX. — STAR MAGNITUDES. — VARIABLE STARS. —
STELLAR SPECTRA.

331. THE solar system is surrounded by an immense void
peopled only by stray meteors. The nearest star, as far as
our present knowledge goes, is one whose distance is more
than 200,000 times as great as our distance from the sun, —
so remote that from it the sun would look no brighter than
the Pole-star, and no telescope yet constructed would be able
to show a single one of all the planets. As to the nature of
the stars, their spectra indicate that they are bodies resembling our sun, — that is, incandescent, and each shining with
its own peculiar light. Some are larger and hotter than the
sun, others smaller and cooler; some, perhaps, large but
hardly luminous at all. They differ enormously among themselves, not being, as once thought, as much alike as individuals
of the same race, but differing as widely as animalcules from
elephants.

332. Number of Stars. — Those which are visible to the eye,
though numerous, are by no means countless. If we take a
limited region, for instance, the bowl of the Dipper, we shall
find that the number we can see within it is not very large, —

hardly a dozen. In the whole celestial sphere, the number of stars bright enough to be distinctly seen by an average eye is only between 6000 and 7000, even in a perfectly clear and moonless sky; a little haze or moonlight will cut down the number fully one-half. At any one time not more than 2000 or 2500 are fairly visible, since near the horizon the small stars (which are vastly the more numerous) all disappear. The total number which could be seen by the ancient astronomers well enough to be observable with their instruments is not quite 1100. With even the smallest telescope, however, the number is enormously increased. A common opera-glass brings out at least 100,000, and with a 2½ inch telescope Argelander made his "Durchmusterung" of the stars north of the equator, more than 300,000 in number. The Lick telescope, 36 inches in diameter, probably makes visible at least 100,000000.

333. Constellations. — The stars are grouped in so-called "constellations," many of which are extremely ancient. All of those of the zodiac and most of those near the north pole antedate history. Their names are, for the most part, drawn from the Greek and Roman mythology, many of them being connected in some way or other with the Argonautic Expedition. In some cases the eye, with the help of a lively imagination, can trace in the arrangement of the stars a vague resemblance to the object which gives the name to the constellation; but generally no reason is obvious for either name or boundaries.

We have already, in Chapter II., given a brief description of those constellations which are visible in the United States, with maps and directions for tracing them.

334. Designation of the Stars. — In Art. 24 we have already indicated the different methods by which the brighter stars are designated, — by proper names, position in the constellation, or by letters of the Greek and Roman alphabets. But these methods do not apply to the telescopic stars, at least to any considerable extent. Such stars we identify by their catalogue number; that is, we refer to them as No. so-and-so in some one's star-catalogue. Thus Ll., 21,185 is read "Lalande, 21,185," and means the star that is so numbered in Lalande's catalogue. At present not far from 800,000 stars are catalogued, so that, except in the Milky Way, every star visible in a three-inch telescope can be found and identified. Of course all the bright stars which have names, have letters also, and are sure to be found in every catalogue which covers their part of the heavens. A conspicuous star, therefore, has usually many "aliases," and sometimes great care is necessary to avoid mistakes on this account.

335. Star-catalogues are carefully arranged lists of stars, giving their positions (*i.e.*, their right ascensions and declinations, or latitudes and longitudes) for a given date, and usually also indicating their so-called magnitudes or brightness. The earliest of these star-catalogues was made about 125 B.C. by Hipparchus of Bithynia, the first of the world's great astronomers, and gives the latitudes and longitudes of 1080 stars. This catalogue was republished by Ptolemy 250 years later, the longitudes being corrected for precession; and during the Middle Ages several other catalogues were made by Arabic astronomers and those that followed them. The modern catalogues are numerous; some, like Argelander's "Durchmusterung," give the places of a great number of stars rather roughly, merely as a means of ready identification. Others are "catalogues of precision," like the Pulkowa and Greenwich catalogues, which give the places of only a few hundred so-called "fundamental" stars, determined as accurately as possible, each star by itself. Finally, we have the so-called "zones," which give the place of many thousands of stars, determined accurately but not independently; that is, their positions are determined by reference to the fundamental stars in the same region of the sky.

336. Mean and Apparent Places of the Stars. — The modern star-catalogue contains the *mean* right ascension and declination of its stars at the beginning of some designated year; *i.e.*, the place the star would occupy if there were no nutation, or aberration (Art. 126, and Appendix, 435). To get the actual (apparent) right ascension and declination of a star for some given date, which is what we always want in practice, the catalogue place must be "reduced" to that date; *i.e.*, it must be corrected for precession, etc. The operation is an easy one with modern tables and formulæ, but tedious when many stars are in question.

337. Star-charts and Stellar Photography. — For some purposes, accurate star-charts are even more useful than catalogues. The old-fashioned and laborious way of making such charts was by "plotting" the results of zone observations, but at present it is being done by means of photography, vastly better and more rapidly. A co-operative international campaign is now in progress, the object of which is to secure a photographic chart of all the stars down to the 14th magnitude. The work is well advanced, but its completion can hardly be expected much before the end of the century. One of the most remarkable things about the photographic method is that there appears to be no limit to the faintness of the stars that can be photographed with a good instrument. By increasing the time of exposure, smaller and smaller stars are continually reached. With the ordinary plates and exposure-times not exceeding twenty minutes, it is now possible to get distinct photographs of stars that the eye cannot possibly see with the same telescope.

Fig. 71 represents the photographic telescope (fourteen inches diameter, and eleven feet focus, of the Paris observatory). The other instruments engaged in the star-chart campaign are substantially like it, though differing more or less in minor details.

FIG. 71. — Photographic Telescope of the Paris Observatory.

Professor Pickering of Cambridge, U.S., is also planning an independent work of the same kind, with an instrument having a four-lens object-glass of twenty-four inches diameter and eleven feet focus. It will take much larger plates and require much shorter exposures than the Paris instrument, and so will do the work much more rapidly. It is intended to map with it first the northern circumpolar region, and then to transfer it to the southern hemisphere.

STAR MOTIONS.

338. The stars are ordinarily called "fixed," in distinction from the planets, or "wanderers," because they keep their positions and configurations sensibly unchanged with respect to each other for long periods of time. Delicate observations, however, demonstrate that the fixity is not absolute, but that the stars are really in motion. Moreover, by the spectroscope their rate of motion towards or from the earth can in some cases be approximately measured. In fact, it appears that the velocities of the stars are of the same order as those of the planets. The stars are flying through space far more swiftly than cannon-balls, and it is only because of their enormous distance from us that they appear to change their positions so slowly.

339. Proper Motion. — If we compare a star's position (right ascension and declination) as determined to-day with that observed 100 years ago, it will always be found to have changed considerably. The difference is due in the *main* to precession (Art. 125) ; but after allowing for all such merely apparent motions of a star, it generally turns out that within a century the star has really altered its place more or less with reference to others near it, and this real shifting of its place is called its "proper motion." Of two stars side by side in the same telescopic field of view, the proper motions may be directly opposite, while, of course, the apparent motions (due to precession, etc.) will be sensibly the same.

Even the largest of these proper motions is very small. The largest at present known, that of the so-called "run-away star," 1830 Groombridge, is only 7″ a year. (This star is not visible to the naked eye.) About a dozen stars are known to have an annual proper motion exceeding 3″, and about 150, so far as known at present, exceed 1″. The proper motions of the bright stars average higher than those of the faint, as might be expected, since on the average the bright ones are probably nearer. For the first-magnitude stars, the average is about $\frac{1}{4}$″ annually; and for the sixth-magnitude stars, the smallest visible to the naked eye, it appears to be about $\frac{1}{25}$″.

Motions of this kind were first detected in 1718 by Halley, who found that since the time of Hipparchus the star Arcturus had moved towards the south nearly a whole degree, and Sirius about half as much.

340. Velocity of Star Motions. — The proper motion of a star gives us very little knowledge as to the star's real motion in miles per second. The proper motion is derived from the comparison of star-catalogues of different dates, and is only the value in seconds of arc of that part of its motion which is perpendicular to the line of sight. A star moving straight towards us or from us has no proper motion at all; *i.e.*, no change of apparent place which can be detected by comparing observations of its position.

We can, however, in some cases fix a minor limit to the velocity of a star. We know, for instance, that the distance of the star, 1830 Groombridge, is certainly not less than 2,000000 'astronomical units,' and, therefore, since its yearly path subtends an angle of 7″ at the earth, the length of the path must at least equal sixty-nine astronomical units a year, which corresponds to a velocity of over 200 miles a second. The real velocity must be more than this, but how much greater we cannot determine until we know how much the star's distance exceeds 2,000000 units, and also how fast it is moving towards or from us.

In many cases a number of stars in the same region of the sky have a motion practically identical, making it almost certain that they are real neighbors and in some way connected, — probably by community of origin. In fact, it seems to be the rule rather than the exception that stars which are apparently near each other are real comrades; they show, as Miss Clerke expresses it, a distinctly "gregarious" tendency.

341. Motion in the Line of Sight. — Within the last thirty years a method[1] has been developed by which any swift motion of a star, directly towards or from us, may be detected by means of the spectroscope.

If a star is approaching us, the lines of its spectrum will apparently be shifted towards the violet, according to Doppler's principle (Art. 179), and *vice versa*, if it is receding from us. Visual observations of this sort, first made by Huggins in 1868, and since then by many others, have succeeded in demonstrating the reality of these motions in the line of sight and in roughly measuring some of them. Recently Vogel of Potsdam has taken up the investigation photographically, and has obtained results that are far more satisfactory than any before reached. He photographs the spectrum of the star and the spectrum of hydrogen gas, or some other substance whose lines appear in the star spectrum, together upon the same plate, the light from both being admitted through the same slit. If the star is not moving towards or from us, its lines will coincide precisely with those of the comparison spectrum; otherwise, they will deviate one way or the other.

---

[1] It is not, as students sometimes think, by changes in the apparent size and brightness of a star. Theoretically, of course, a star which is approaching us must grow brighter, but even the nearest star of all, Alpha Centauri (Art. 343) is so far away that if it were coming directly towards us at the rate of 100 miles a second, it would require more than 8000 years to make the journey; so that in a century its brightness would only change about two per cent, — far too little to be observed.

Fig. 72 is from one of his negatives of the spectrum of Beta Orionis (Rigel), in which one of its dark lines is compared with the corresponding bright lines in the spectrum of hydrogen. The dark line of the stellar spectrum (bright in the negative) is shifted towards the red by an amount which indicates that at the time the star was rapidly receding.

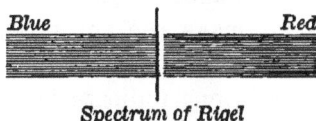

Blue   Red

Spectrum of Rigel

FIG. 72. — Displacement of $H\gamma$ Line in the Spectrum of β Orionis.

For the most part, these motions of the stars, so far as ascertained, seem to range between zero and fifty miles a second, with still higher speeds in a few exceptional cases.

342. The "Sun's Way." — The proper motions of the stars are due partly to their own real motions, but partly also to the motion of the sun, which, like the other stars, is travelling through space, taking with it its planets. Sir William Herschel was the first to investigate and determine the direction of this motion a century ago. The principle involved is this : On the whole the stars appear to drift bodily in a direction opposite to the sun's real motion. Those in that quarter of the sky which we are approaching open out from each other, and those in the rear close up behind us. The motions of the individual stars lie in all possible directions, but when we deal with them by thousands, the individual is lost in the general, and the prevailing drift appears.

About twenty different determinations of the point, towards which the sun's motion is directed, have been made by various astronomers. There is a reasonable and almost surprising accordance of results, and they all show that the sun is moving towards a point in the constellation of Hercules, having a right ascension of about 267° (17ʰ 48ᵐ), and a declination of about 32° north. This point is called the "apex of the sun's way." As to the velocity of this motion of the sun, it comes out as about 0″.05 annually, seen from the average distance of the standard sixth-magnitude star. It is assumed by high authorities, on grounds that we cannot stop to discuss, that this distance is about 20,000000 astronomical units, which would make the sun's velocity about sixteen miles a second, as determined from the "proper motions." The spectroscopic observations indicate that it is about eleven miles, and this result is probably more trustworthy.

THE PARALLAX AND DISTANCE OF STARS.

343. When we speak of the "parallax" of the sun, of the moon, or of a planet, we always mean the "diurnal" or "geocentric" parallax (Art. 139); *i.e.*, the apparent semi-diameter of the earth as seen from the body. In the case of a *star*, this kind of parallax is practically nothing, never reaching $\frac{1}{20000}$ of a second of arc. The expression, "parallax of a star," always refers, on the contrary, to its "annual" or "helio-

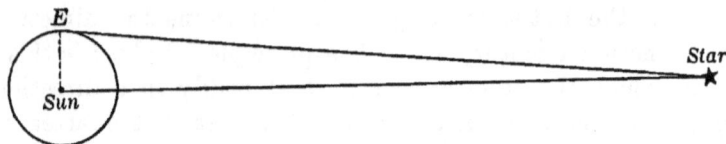

FIG. 73.— The Annual Parallax of a Star.

centric" parallax; *i.e.*, the apparent semi-diameter, not of the earth, but of the *earth's orbit*, as seen from the star. In Fig. 73 the angle at the star is its parallax.

Even this heliocentric parallax, in the case of most stars, is far too small to be detected by our present instruments : it never reaches a single second of arc. But in a few instances it has been actually measured. Alpha Centauri, which is our nearest neighbor, so far as yet known, has a parallax of about 0".9, according to the earlier observers, or only 0".75, according to the latest authorities. There are but four or five other stars at present known which have a parallax more than half as great as this. (For the method of determining stellar parallax, see Appendix, Arts. 441–443.)

344. Unit of Stellar Distance; the Light-year. — The distances of the stars are so enormous that even the radius of the earth's orbit, the "astronomical unit" hitherto employed, is far too small for convenience. It is better, and now usual, to take as the unit of stellar distance the so-called *light-year;* *i.e.*, the distance which light travels in a year. This is about 63,000 times the distance of the earth from the sun.

This number is found by dividing the number of seconds in a year by 499, the number of seconds required by light to make the journey from the sun to the earth (Appendix, Art. 432).

A star with a parallax of $1''$ is at a distance of 3.26 light-years, and in general the distance in light-years equals $\dfrac{3.26}{p''}$, where p'' is the parallax of the star expressed in seconds.

So far as can be judged from the scanty data, it appears that few if any stars are nearer than four light-years from the solar system; that the naked-eye stars are probably, for the most part, within 200 or 300 years; and that many of the remoter stars must be thousands, or even tens of thousands, of light-years away.

For the parallaxes of a number of stars, see Table V., Appendix.

THE LIGHT OF THE STARS.

345. Star Magnitudes. — As has already been mentioned (Art. 23), Hipparchus and Ptolemy arbitrarily divided the stars into six "magnitudes" according to their brightness, the stars of the sixth magnitude being those which are barely perceptible by an ordinary eye, while the first class comprise about twenty of the brightest. After the invention of the telescope the same system was extended to the smaller stars, though without any special plan, so that the "magnitudes" assigned to telescopic stars by different observers are very discordant.

Heis enumerates the stars clearly visible to the naked eye, north of the 35th parallel of south declination, as follows:—

First Magnitude, 14.		Fourth Magnitude, 313.	
Second	" 48.	Fifth	" 854.
Third	" 152.	Sixth	" 2010.

Total, 3391.

It will be noticed how rapidly the numbers increase for the smaller magnitudes. Nearly the same holds good also for the telescopic stars, though below the tenth magnitude the rate of increase falls off.

346. Light-ratio and "Absolute Scale" of Star Magnitudes. — The scale of magnitudes ought to be such that the "light-ratio," or number of times by which the brightness of any star exceeds that of a star which is one magnitude smaller, should be the same throughout the whole extent of the scale. This relation was roughly, but not accurately, observed by the older astronomers, and very recently Professor Pickering of Cambridge, U. S., and Professor Pritchard of Oxford, England, have made photometric measurements of the brightness of all the naked-eye stars visible in our latitude, and have re-classified them according to the so-called "absolute scale," which uses a light-ratio equal to the fifth root of 100, (2.51); i.e., upon this scale a star of the third magnitude is just 2.51 times brighter than one of the fourth.

This ratio is based upon an old determination of Sir John Herschel's, who found that the average first-magnitude star is just about a hundred times as bright as a star of the sixth magnitude, five magnitudes fainter.

On this scale, *Altair* (Alpha Aquilæ) and *Aldebaran* (Alpha Tauri) may be taken as standard first-magnitude stars, while the Pole-star and the two pointers are very nearly of the standard second magnitude.

Of course, in indicating the brightness of stars with precision, fractional numbers must be used; that is, we have stars of 2.4 magnitude, etc.

Stars that are brighter than Aldebaran or Altair have their bright-ness denoted by a *fraction*, or even by a *negative* number; thus the absolute magnitude of Vega is 0.2, and of Sirius −1.4. The necessity of these negative and fractional magnitudes for bright stars is rather unfortunate, but not really of much importance.

347. Magnitudes and Telescopic Power. — If a good telescope just shows a star of a certain magnitude, we must have a telescope with its aperture larger in the ratio of 1.58 : 1, in order to show stars one magnitude smaller; (1.58 = $\sqrt{2.51}$). A tenfold increase in the diameter of an object-glass theoretically carries the power of vision just five magnitudes lower. ·

It is usually estimated that the 12th magnitude is the limit of vision for a 4-inch glass. It would require, therefore, a 40-inch glass to reach the 17th magnitude of the absolute scale.

Our space does not permit any extended discussion of the methods by which the brightness of stars is measured, a subject which has of late attracted much attention (see General Astronomy, Arts. 823–829).

348. Starlight compared with Sunlight. — Zöllner and others have endeavored to determine the amount of light [1] re-ceived by us from certain stars, as compared with the light of the sun. According to him, Sirius gives us about $\frac{1}{7000\,000000}$ as much light as the sun does, and Capella and Vega about $\frac{1}{50000\,000000}$. At this rate, the standard first-magnitude star, like Altair, should give us about $\frac{1}{90000\,000000}$, and it would take, therefore, about nine million million stars of the sixth magnitude to equal the sun. These numbers, however, are very uncertain. The various determinations for Vega vary more than fifty per cent.

Assuming what is roughly true, that Argelander's magnitudes agree with the absolute scale, it appears that the 324,000 stars of his " Durch-

---

[1] Undoubtedly, the stars send us *heat* also, and attempts have been made to measure it ; but there is no reason for supposing that the propor-tion of stellar heat to solar differs much from the proportion of star*light* to sunlight ; and if so, the heat of a star must be far below the possibility of measurement by any apparatus yet at our command.

musterung," all of them north of the celestial equator, give a light
about equivalent to 240 or 250 first-magnitude stars. How much
light is given by stars smaller than the $9\frac{1}{2}$ magnitude (which was his
limit) is not certain. It must greatly exceed that given by the larger
stars. As a rough guess we may estimate that the total starlight of
both the northern and southern hemispheres is equivalent to about
3000 stars like Vega, or 1500 at any one time. According to this, the
starlight on a clear night is about $\frac{1}{70}$ of the light of a full moon, or
about $\frac{1}{33\,000\,000}$ that of sunlight. More than 95 per cent of it comes
from stars which are entirely invisible to the naked eye.

349. Amount of Light emitted by Certain Stars. — When
we know the distance of a star in astronomical units, it is easy
to compute the amount of light it really emits as compared
with that given off by the sun. It is only necessary to multiply the light we now get from it (expressed as a fraction of
sunlight) by the square of the star's distance in astronomical
units. Thus, the distance of Sirius is about 550,000 units, and
the light we receive from it is $\frac{1}{7\,000\,000\,000}$ of sunlight. Multiplying this fraction by the square of 550,000, we find that
Sirius is really radiating more than forty times as much light
as the sun. As for several other stars, whose distance and
light have been measured, some turn out brighter, and some
darker than the sun. The range of variation is very wide,
and in brilliance the sun holds apparently about a medium
rank among its kindred.

350. Why the Stars differ in Brightness. — The apparent
brightness of a star, as seen from the earth, depends both on
its distance and on the quantity of light it emits, and the
latter depends on the extent of its luminous surface and upon
the brightness of that surface. As Bessel long ago suggested,
"there may be as many *dark* stars as bright ones." Taken as
a class, the bright stars undoubtedly *average* nearer to us than
the fainter ones, and just as undoubtedly they also average
larger in diameter and more intensely luminous; but when we

compare any particular bright star with another fainter one, we can seldom say to which of these different causes it owes its superiority. We cannot assert that the faint star is smaller or darker or more distant than that particular bright star, unless we know something more about it than the simple fact that it is fainter.

351. Dimensions of the Stars. — The stars are so far away that their apparent diameters are· altogether too small to be measured by any known form of micrometer. The sun at the distance of the nearest star would measure[1] not quite $0''.01$ across. Micrometers, therefore, do not help us in the matter, and until very recently we were absolutely without any positive knowledge as to the real size of a single one of the stars. But in 1889, by a spectroscopic method, more fully explained in Art. 360, Vogel succeeded in showing that the bright variable star, Algol (Beta Persei) (Art. 358), must have a diameter of about 1,160,000 miles, while its invisible companion is about 840,000 miles in diameter, or just about the size of the sun.

VARIABLE STARS.

352. Classes of Variables. — Many stars are found to change their brightness more or less, and are known as "variable." They may be classed as follows : —

 I. Stars which change their brightness slowly and continuously.

 II. Those that fluctuate irregularly.

 III. Temporary stars which blaze out suddenly and then disappear.

 IV. Periodic stars of the type of "Omicron Ceti," usually having a period of several months.

---

[1] This does not refer, of course, to the "spurious disc" of the star (Appendix, Art. 408), which is many times larger.

V. Periodic stars of the type of "Beta Lyræ," usually having short periods.

VI. Periodic stars of the "Algol" type, in which the period
is usually short, and the variation is like what might
be produced if the star were periodically "eclipsed"
by some intervening object.

353. Gradual Changes. — The number of stars which are
certainly known to be gradually changing in brightness is surprisingly small. On the whole, the stars present not only in
position, but in brightness also, sensibly the same relations as
in the catalogues of Hipparchus and Ptolemy.

There are, however, a few instances in which it can hardly be
doubted that considerable alteration has occurred even within the last
two or three centuries. Thus, in 1610 Bayer lettered Castor as Alpha
Geminorum, while Pollux, which he called Beta Geminorum, is now
considerably brighter. There are about a dozen other similar cases
known, and a much larger number is suspected.

It is commonly believed that a considerable number of stars have
disappeared since the first catalogues were made, and that many new
ones have come into existence. While it is unsafe to deny absolutely
that such things may have happened, it can be said, on the other
hand, that *not a single case of the kind is certainly known.* The discrepancies between the older and newer catalogues are all accounted
for by some error or other that has already been discovered.

354. Irregular Fluctuations. — The most conspicuous star
of the second class is Eta Argûs (not visible in the United
States). It varies all the way from above the first magnitude
(in 1843 it stood next to Sirius) down to the seventh magnitude (invisible to the eye), which has been its status ever
since 1865, though recently it is reported as slightly brighten-
ing up again. Alpha Orionis, Alpha Herculis, and Alpha
Cassiopeiæ behave in a similar way, except that their variation is quite small, never reaching an entire magnitude.

355. Temporary Stars. — There are eleven well-authenti-
cated instances of stars which have blazed up suddenly, and
then gradually faded away (see General Astronomy, Arts.
842–845). The most remarkable of these is that known as
Tycho's, which appeared in the constellation of Cassiopeia
(Art. 28) in November, 1572, was for some days as bright as
Venus at her best, and then gradually faded away, until at
the end of sixteen months it became invisible. (There were
no telescopes then.) It is not certain whether it still exists as
a telescopic star: so far as we can judge it may be either of
half a dozen which are near the place determined by Tycho.

There has been a curious and utterly unfounded notion that this
star was the "Star of Bethlehem" and would reappear to herald the
second advent.

A temporary star which appeared in the constellation Corona
Borealis, in May, 1866, is interesting as having been spectroscopically examined when near its brightest (second magnitude). It then showed the same bright lines of hydrogen
which are conspicuous in the solar prominences. Before its
outburst it was an eighth-magnitude star of Argelander's catalogue, and within a few months it returned to its former low
estate, which it still retains.

A more recent instance is that of a sixth-magnitude star
which in August, 1885, suddenly appeared in the midst of the
great nebula of Andromeda (Art. 377). In a few months it
totally disappeared, even to the largest telescopes. Still
more recently (in 1892) a star of the $4\frac{1}{2}$ magnitude appeared
in the constellation of Auriga. At first its spectrum was very
complicated, showing lines both dark and bright, the bright
lines of Hydrogen and Helium being especially conspicuous.
The lines were so displaced as to indicate, in the luminous
gases, velocities of more than 500 miles a second. In April
the star became invisible, but brightened up again in the
autumn, and then showed an entirely different spectrum,
closely resembling that of a nebula (Art. 380). The phe-

nomena of this star have called out a great deal of discussion, and cannot be considered to have reached a satisfactory explanation.

356. Variables of the "Omicron Ceti" Type. — These objects behave almost exactly like a temporary star in remaining most of the time faint, suddenly blazing out, and then

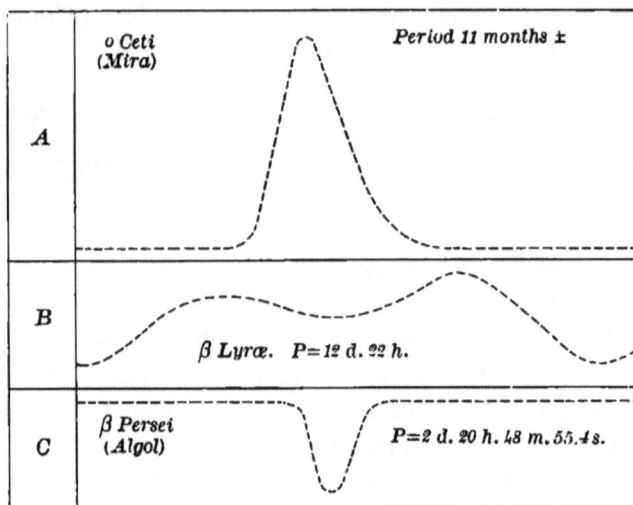

FIG. 74. — Light-curves of Variable Stars.

gradually fading away, — *but they do it periodically.* Omicron Ceti, or Mira (*i.e.,* "the wonderful") is the type. It was discovered in 1596, and was the first variable star known. During most of the time it is of the ninth magnitude, but at intervals of about eleven months it runs up to the fourth, third, or even second magnitude, and then back again, the whole change occupying about 100 days, and the rise being much more rapid than the fall. It remains at its maximum about a week or ten days. The maximum brightness varies very considerably, and its period, while always about eleven months, varies to the extent of two or three weeks. The spectrum of the star when brightest is very beautiful, showing a large number of intensely bright lines, some of which are due to hydrogen and helium. Its light-curve is A in Fig. 74.[1]

Nearly half of all the known variables belong to this class, and a large proportion of them have periods which do not differ very widely from a year. Most of the periods, however, are more or less irregular. Some writers include the temporary stars in this class, maintaining that the only difference is in the length of their period.

357. Class V. — The variables of Class V. are mostly of short period, and are characterized by a continual rising and falling of brightness, running through the whole period. Sometimes there are two, or even three, maxima before the cycle is completed. The light-curve of Beta Lyræ, the type-star of this class (period about thirteen days) is B in Fig. 74.

358. The "Algol" Type. — In the stars of Class VI. the variation is precisely the reverse of that in Class IV. The star remains bright for most of the time, but apparently suffers a periodical eclipse. The periods are mostly very short, — only a few days, — and one little star in the constellation of Antlia has a period of less than eight hours.

Algol (Beta Persei) is the type-star. During most of the time it is·of the second magnitude, and it loses about five-sixths of its light at the time of obscuration. The fall of brightness occupies about $4\frac{1}{2}$ hours. The minimum lasts about 20 minutes, and the recovery of light takes about $3\frac{1}{2}$ hours. The period, a little less than three days, is known with great precision, to a single second indeed, and is given in connection with the light-curve of the star in Fig. 74. At present the period seems to be slowly shortening. Between fifteen and twenty variables are now known to belong to this class.

---

[1] The light-curve diagrams are not drawn to scale, and make no pretensions to exact accuracy.

359. Explanation of Variable Stars. — No single explanation will cover the whole ground. As to *progressive* changes, no explanation need be looked for. The wonder rather is that as the stars grow old, such changes are not more notable than they are.

As for irregular changes, no sure account can yet be given. Where the range of variation is small (as it is in most cases), one thinks of spots upon the surface of the star, more or less like sun spots; and if we suppose these spots to be much more extensive and numerous than are the sun spots, and also, like them, to have a regular period of frequency, and also that the star revolves upon its axis, we find in the combination a possible explanation of a large proportion of all the variable stars.

For the temporary stars, we may imagine either great eruptions of glowing matter, like solar prominences on an enormous scale ; or, with Mr. Lockyer, we may imagine that they, and most of the variable stars, are only swarms of meteors, rather compact but not yet having reached the condensed condition of our own sun. Outbursts of brightness are the result of collisions between such swarms. Stars of the Mira type, according to this theory, consist of two such swarms, the smaller revolving around the larger in a long oval, so that once in every revolution it brushes through the outer portions of the larger one. But the great irregularity in the periods of variables belonging to this class is hard to reconcile with a true orbital revolution, which usually keeps time accurately.

360. Explanation of the Algol Type. — The natural and most probable explanation of the behavior of these stars is that the periodical darkening is produced by the interposition of some opaque body between us and the star. This eclipse theory has lately received a striking confirmation from the spectroscopic work of Vogel, who has found by the method indicated in Art. 341 that about seventeen hours before the obscuration, Algol is receding from us at the rate of nearly twenty-seven

miles a second, while seventeen hours after the minimum it
approaches us at the same rate. This is just what it ought to
do, if it had a large, dark companion, and the two were revolving around their common centre of gravity in an orbit nearly
edgewise to the earth. When the dark star is rushing forward to interpose itself between us and Algol, Algol itself
must be moving backwards, and *vice versa* when the dark star
is receding after the eclipse. Vogel's conclusions are, that the
distance of the dark star from Algol is about 3,250000 miles;
that their diameters are respectively about 840,000 and
1,160000 miles; that their united mass is about two-thirds
that of the sun; and their density about one-fifth that of the
sun, — not much greater than that of cork. Still more
recently (1892), Mr. Chandler finds evidence from a slight
alternate shortening and lengthening of the star's period of
variation, that the pair are probably moving together around
a third (invisible) star in an orbit about as large as that of
Uranus, accomplishing the circuit in about 130 years. But
Tisserand suggests a different explanation.

**361.[1] Number and Designation of Variables, and their Range
cf Variation.** — Mr. Chandler's catalogue of known variables,
with its later supplements, includes 343 objects, and there is
also a considerable number of suspected variables.

About 200 of the 343 are distinctly periodic. The rest of
them are, some irregular, some temporary, and in respect to
many we have not yet certain knowledge whether the variation is or is not periodic.

Table IV., Appendix, contains a list of the naked-eye variables visible in the United States.

Such variable stars as had not names of their own before their
variability was discovered are at present generally indicated by the
letters *R*, *S*, *T*, etc.; *i.e.*, *R* Sagittarii is the first discovered variable in
the constellation of Sagittarius, *S* Sagittarii is the second, etc.

[1] See note on pages 301-2.

In a considerable number of the earlier discovered variables, the range of brightness is from two to eight magnitudes; that is, the maximum brightness exceeds the minimum from 6 to 1000 times. In the majority, however, the range is much less, — only a fraction of a magnitude.

It is worth noting that a large proportion of the variables, especially those of Classes IV. and V., are reddish in their color. This is not true of the Algol type.

STAR SPECTRA.

362. As early as 1824 Fraunhofer observed the spectra of a number of bright stars by looking at them with a small telescope with a prism in front of the object-glass. In 1864, as soon as the spectroscope had taken its place as a recognized instrument of research, it was applied to the stars by Huggins and Secchi. The former studied very few spectra, but very thoroughly, with reference to the identification of the chemical elements in certain stars. He found with certainty in their spectra the lines of *sodium, magnesium, calcium, iron,* and *hydrogen,* and more or less doubtfully a number of other metals. Secchi, on the other hand, examined great numbers of spectra, less in detail, but with reference to a classification of the stars from the spectroscopic point of view.

363. Secchi's Classes of Spectra. — He made four classes, as follows: —

I. Those which have a spectrum characterized by great intensity of the dark lines of hydrogen, all other lines being comparatively feeble or absent. This class comprises more than half of all the stars, — nearly all the white or bluish stars. Sirius and Vega are its types.

II. Those which show a spectrum resembling that of the sun; *i.e.,* marked with a great number of fine dark lines.

Capella (Alpha Aurigæ) and Pollux (Beta Geminorum) are conspicuous examples. The stars of this class are also numerous. The first and second classes together comprise fully seven-eighths of all the stars whose spectra are known.

Certain stars, like Procyon and Altair, seem to be intermediate between the first and second classes. The line of demarcation is by no means sharp.

III. Stars which show a spectrum characterized by dark bands, sharply defined at the upper or more refrangible edge, and shading out towards the red. Most of the red stars, and a large number of the variable stars, belong to this class. Some of them show, also, bright lines in their spectra.

IV. This class comprises only a few small stars, which, like the preceding, show dark bands, but shading in the opposite direction. Usually they also show a few bright lines. There are not a few anomalous stars that will not fall into any of these classes.

This classification is by no means entirely satisfactory, and various modifications have been proposed for it by Vogel, Lockyer, and others. On the whole, however, we give it as the best known and simplest, and sufficient for most purposes.

364. Photography of Stellar Spectra. — The observation of these spectra by the eye is very tedious and difficult, and photography has of late been brought in most effectively. Huggins, in England, and Henry Draper, in this country, were the pioneers, but incomparably the finest results in this line are those that have been obtained by Professor E. C. Pickering, of Cambridge, in connection with the Draper Memorial Fund. Pickering has recurred to the old method of Fraunhofer, using a prism, or prisms, in front of the object-glass of his photographic telescope, thus forming a "slitless spectroscope." The edges of the prism, or prisms, are placed east and west. If the clock-work of the instrument followed the star exactly, the spectrum formed on the sensitive plate would be a mere nar-row streak; but by allowing the clock to gain or lose slightly, the image of the star will move to the east or west by a very small quantity during the exposure, converting the streak into a band.

The slitless spectroscope has three great advantages : (1) it saves all the light which comes from the star, much of which in the usual form of the instrument is lost in the jaws of the slit ; (2) by taking advantage of the length of a large telescope, it produces a long spec-trum with even a single prism ; (3) and most important of all, it gives on the same plate, and with a single exposure, the spectra of all the many stars (sometimes more than a hundred) whose images fall upon the plate.

On the other hand, the giving up of the slit precludes the usual methods of identifying the lines and measuring their displacements, by actually confronting them with comparison spectra. For instance, it has not yet been found possible to use the slitless spectroscope for determining the absolute motions of the stars in the line of sight.

364.* With the eleven-inch telescope formerly belonging to Dr. Draper, and a battery of four enormous prisms placed in front of the object-glass, spectra are obtained with an ex-posure of thirty minutes, which before enlargement are fully three inches long from the F line to the ultra-violet extremity.

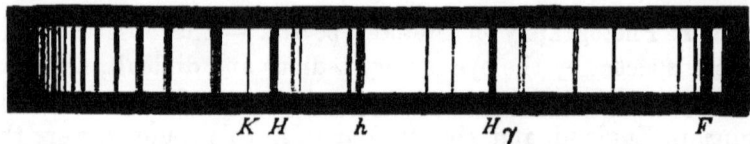

K H   h   H γ   F

FIG. 75. — Photographic Spectrum of Vega.

They easily bear tenfold enlargement, and show many hun-dreds of lines in the spectra of the stars which belong to Secchi's second class. Fig. 76 is from one of these photo-graphs of the spectrum of Vega. The photograph fails to show the lower portion of the spectrum, — i.e., the red, yellow, and green; but within a year or two the use of isochromatic plates has made it possible to deal with these colors also.

The spectra of all the naked-eye stars in the northern hemisphere have already been photographed and catalogued, and the work is well advanced in the southern hemisphere by parties sent out from Cambridge to South America. Many fainter stars have also been included, and the matter is to be followed up with the great Bruce telescope mentioned in Art. 337.

The admirable spectrographic work of Vogel has been already referred to in Art. 341.

365. Twinkling or Scintillation of the Stars. — This phenomenon is purely physical, and not in the least astronomical. It depends both upon the irregularities of refraction in the air traversed by the light on its way to the eye (due to winds and differences of temperature), and also on the fact that the star is optically a luminous *point* without apparent size, — a fact which, under the circumstances, gives rise to the optical phenomenon known as "*interference.*" Planets which have discs measurable with a micrometer do not sensibly twinkle.

The scintillation is of course greatest near the horizon, and on a good night it practically disappears at the zenith. When the image of a twinkling star is examined with the spectroscope, dark interference-bands are seen moving back and forth in its spectrum.

CHAPTER XII.

DOUBLE AND MULTIPLE STARS AND CLUSTERS. — NEBULÆ. — DISTRIBUTION OF STARS AND CONSTITUTION OF THE STELLAR UNIVERSE. — COSMOGONY AND THE NEBULAR HYPOTHESIS.

366. Double Stars. — The telescope shows numerous cases in which two stars lie so near each other that they can be separated only by a high magnifying power. These are "double stars," and at present more than 10,000 such couples are known. There is also a considerable number of triple stars, and a few which are quadruple. Fig. 76 represents a few of the best known objects of each class. The apparent distances generally range from 30″ downwards, very few telescopes being able to separate stars closer than a quarter of a second. In a large proportion of cases (perhaps a third of all) the two components are nearly equal in brightness, but in many they are very unequal; in that case (never when they are equal) they often present contrasts of color, and when they do, the smaller star (for some reason not known) always has a tint *higher in the spectrum* than that of the larger: if the larger is reddish or yellow, the small star will be green, blue, or purple.

Gamma Andromedæ and Beta Cygni are fine examples of colored doubles for a small telescope.

367. Stars Optically and Physically Double. — Stars may be double in two ways, — optically or physically. In the first case they are merely approximately in line with each other, as seen from the earth; in the second case, they are really near each other.   In the case of stars that are only optically double, it usually happens that after some years we can detect their mutual independence in the fact that their *relative motion is in*

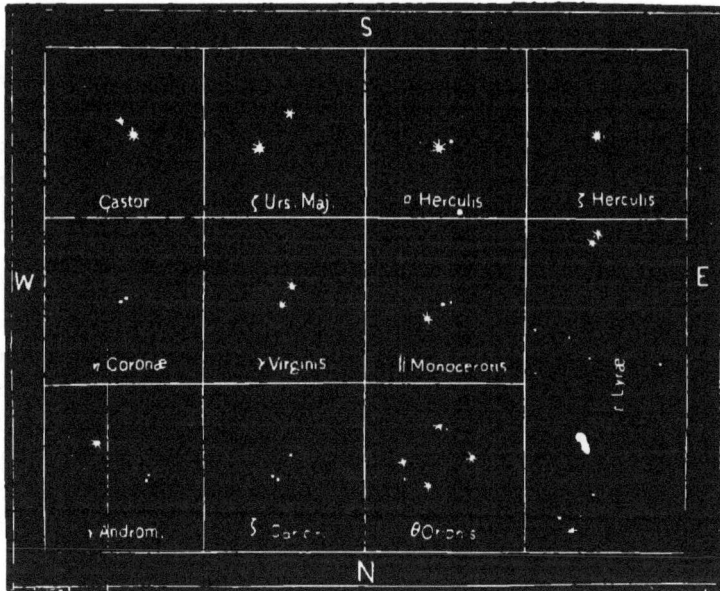

FIG. 76. — Double and Multiple Stars.

a straight line and *uniform; i.e.*, one of them drifts by the other in a line which is perfectly straight.   This is a simple consequence of the combination of their independent "proper motions."   If they are physically connected, we find on the contrary that the relative motion is in a *concave curve; i.e.*, taking either of them as a centre, the other one appears to move around it in a curve.

The doctrine of chances shows what direct observation confirms, that optical pairs must be comparatively rare, and that the great majority of double stars must be really physically connected, probably by the same attraction of gravitation which controls the solar system.

368. Binary Stars. — Stars thus physically connected are also known as "binary" stars. They revolve in elliptical orbits around their common centre of gravity, in periods which range from 14 years to 1500 (so far as at present known), while the apparent length of the ovals ranges from 40″ to 0″.5. The older Herschel, a little more than a century ago, first discovered this orbital motion of "binaries" in trying to ascertain the parallax of some of the few double stars which were known at his time. It was then supposed that they were simply optical pairs, and he expected to detect an annual displacement of one member of the pair with reference to the other, from which he could infer its annual parallax (Art. 343). He failed in this, but found instead a true orbital motion. The apparent orbit is always an ellipse; but this apparent orbit is the true orbit seen more or less obliquely; so that the larger star is not usually in the focus of the relative orbit pursued by the smaller one. If we assume what is probable, though certainly not *proved* as yet, that the orbital motion of the pair is under the law of gravitation, we know that the larger star must be in the focus of the *true* relative orbit of the smaller, and, moreover, that the latter must describe around it equal areas in equal times. By the help of these principles we can deduce from the apparent oval the true orbital ellipse; but the calculation is troublesome and delicate.

369. At present the number of pairs in which this kind of motion has been certainly detected exceeds 200, and it is continually increasing as our study of the double stars goes on. About fifty pairs have progressed so far, either having completed an entire revolution or a large part of one, that it is possible to determine their orbits with some accuracy.

The case of Sirius is peculiar. Nearly forty years ago it had been found from meridian-circle observations to be moving, for no assignable reason, in a small orbit, with a period of about fifty years. In 1862, Clark, the telescope-maker in Cambridge, U.S., found near it a minute companion, which explains everything; only we have to admit that this faint attendant, which does not give $\frac{1}{11000}$ as much light as Sirius itself, has a mass more than a quarter part as great. It seems to be one of Bessel's dark stars. Fig. 77 represents the apparent orbits of two of the best determined double-star systems, Gamma Virginis and Xi Ursæ Majoris.

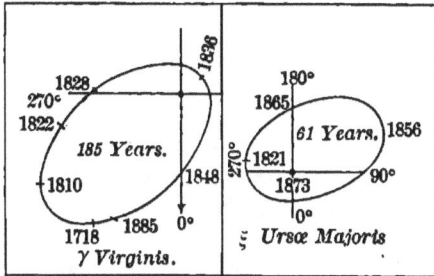

FIG. 77. — Orbits of Binary Stars.

370. Size and Form of the Orbits. — The dimensions of a double-star orbit can easily be obtained if we know its distance from us. Fortunately, a number of stars whose parallaxes have been ascertained are also binary; and assuming the best available data, we have the results, given in the little table which follows, the real semi-major axis of the orbit (in astronomical units) being always equal to the fraction $\frac{a''}{p''}$, in which a'' is the angular semi-major axis of the double star orbit in seconds of arc, and p'' the parallax of the star.

NAME.	Assumed Parallax.	Angular Semi-axis.	Real Semi-axis.	Period.	Mass. $\odot = 1.$
η Cassiopeiæ	0″.44	8″.64	19.6	195.r2	0.18
Sirius.	0.38	8.58	22.6	52.0?	4.26?
α Geminorum	0.20?	5.54	27.7?	266	0.30?
α Centauri	0.75	17.50	23.3	77.0	2.14
70 Ophiuchi	0.16	4.79	29.9	94.5	3.0
61 Cygni.	0.43	15.40?	35.8?	450.0?	0.23?

These double-star orbits are evidently comparable in magnitude with the larger orbits of the planetary system, none of those given being smaller than the orbit of Uranus, and none twice as large as that of Neptune.   In form they are much more eccentric than planetary orbits, and Professor See of Chicago has shown that this fact can be accounted for as a result of "tidal-evolution," operating upon a pair of nebulous masses formed by the separation of a parent nebula into two portions, which revolve around their common centre.

371. Masses of Binary Stars. — If we assume that the binary stars move under the law of gravitation, then when we know the semi-major axis of the orbit and the period of revolution, we can easily find the mass of the pair as compared with that of the sun; much more easily, indeed, than we can determine the mass of Mercury or the moon, strange as it may seem.   It is done simply by the following equation, which we give without demonstration (see General Astronomy, Arts. 530 and 878): —

$$(M + m) = S \left(\frac{a^3}{t^2} \right),$$

in which $(M + m)$ is the united mass of the two stars, S is the mass of the sun, a is the semi-major axis of the orbit of the double star *in astronomical units*, and t its period *in years*. The final column of the preceding table gives the masses of the star-pairs, resulting from such data as we now possess; but the reader must bear in mind that the margin of error is very considerable, because of the uncertainty of the orbits and *parallaxes* in question.   A very slight error in the parallax makes a very great error in the resulting mass.

372. Planetary Systems attending Stars. — It is a natural question whether some of the small companions that we see near large stars may not be the "Jupiters" of their planetary systems.   We can only say as to this that no telescope ever constructed could even come near to making visible a planet which bears to its primary any such relations of size, distance, and brightness, as Jupiter bears to the sun.

Viewed from our nearest neighbor among the stars, Jupiter would be
a little star of about the 21st magnitude, not quite 5″ distance from
the sun, which itself would look like a star of the second magnitude.
To render a star of the 21st magnitude barely visible (apart from all
the difficulties raised by the nearness of a larger star) would require
a telescope *more than twenty feet in diameter.* If any of the stars have
planetary systems accompanying them, we shall never be likely to see
them until our telescopes have attained a magnitude and power as yet
undreamed of.

373. Spectroscopic Binaries. — One of the most interesting
of recent astronomical results is the detection by the spectroscope of several pairs of double stars so close that no telescope
can separate them.   In 1889 the bright component of the well-
known double star Mizar (Zeta Ursæ Majoris, Fig. 76) was
found by Pickering to show the dark lines *double* in the photographs of its spectrum, at regular intervals of about fifty-two
days.   The obvious explanation is that this star is composed
of two, which revolve around their common centre of gravity
in an orbit which is turned nearly edgewise towards us.   (If it
was *exactly* edgewise, the star would be variable like Algol.)

When the stars are at right angles to the line from them to
us, one of the two will be moving towards us, while the other
is moving in an opposite direction ; and as a consequence,
the lines in their spectra will be shifted opposite ways, according to Doppler's principle (Art. 179).   Now since the two
stars are so close that their spectra overlie each other, the
result will be simply to make the lines in the compound spectrum look *double.*   From the distance apart of the lines, the
relative velocity of the stars can be found, and from this the
size of the orbit and the mass of the stars.   Thus it appears
that in the case of Mizar the relative velocity of the two
components is about 100 miles per second, the period about
104 days, and the distance between the two stars about the
same as the diameter of the orbit of Mars; from which it follows that their united mass is about forty times that of the
sun.

This makes Mizar really a *triple* star, the larger of the two that are seen with a small telescope being the one that is thus spectroscopically split.

374. The lines in the spectrum of Beta Aurigæ exhibit the same peculiarity, but the doubling occurs once in *four* days; the velocity being about 150 miles a second, and the diameter of the orbit about 8,000000 miles, while the united mass of the two stars is about two and a half times that of the sun.

These observations of Professor Pickering's were made by photographing the spectrum with the *slitless spectroscope* (Art. 364), and are possible only where the stars which compose the binary are both of them reasonably bright.

With his slit-spectroscope, Vogel (Art. 341), as has already been stated (Art. 360), has been able to detect a similar orbital motion in Algol, although the companion of the brighter star is itself invisible. More recently, in the case of the bright star Alpha Virginis (Spica), he has found a result of the same kind. At first the photographic observations of the spectrum of this star appeared very discordant. Some days they indicated that the star was moving *towards* us quite rapidly, and then again *from* us; but it is found that everything can be explained by the simple supposition that the star is double with a small companion, like that of Algol, not bright enough to show itself by its light, but heavy enough to make its partner swing around in an orbit about 6,000000 miles in diameter, once in four days, — the orbit not being quite edgewise to the earth, so that the dark companion does not eclipse Spica, as Algol is eclipsed by its attendant. Rigel (Beta Orionis) also shows traces of a similar periodic variation, though the observations have not yet been continued long enough to determine its period precisely. The variable stars, Delta Cephei and Beta Lyræ, behave in a similar manner, and probably are to be added to the list. These orbits, of course, are very much smaller than those of most of the telescopic binaries.

375. Multiple Stars (see Fig. 76). — In a considerable number of cases we find three or more stars connected in one system. Zeta Cancri consists of a close pair revolving in a nearly circular orbit, with a period somewhat less than sixty years, while a third star revolves in the same direction around them, at a much greater distance, and with a period not less than 500 years (not yet fully determined). Moreover, this third star is subject to a peculiar irregularity in its motion, which seems to indicate that it has an invisible companion very near the system, the system being really quadruple.

In Epsilon Lyræ we have a most beautiful quadruple system, composed of two pairs, each binary with a period of over 200 years. Moreover, since they have a common proper motion, it is probable that the two pairs revolve around each other in a period which can be reckoned only in thousands of years. In Theta Orionis, we have a remarkable object, in which the six components are not organized in pairs, but are at not very unequal distances from each other.

376. Clusters. — There are in the sky numerous groups of stars, containing from a hundred to many thousand members. A few of them are resolvable by the naked eye, as, for instance, the Pleiades (Fig. 78); some, like Præsepe in Cancer, break up under the power of even an opera-glass (Art. 52); but most of them require a large telescope to show the separate components. To the naked eye or small telescopes, if visible at all, they look like faint clouds of shining haze; but in a great telescope they are among the most magnificent objects the heavens afford. The cluster known as "13 Messier," not far from the "apex of the sun's way," is perhaps the finest.

The question at once arises whether the stars in such a cluster are comparable with our own sun in magnitude, and separated from each other by distances like that between the sun and Alpha Centauri, or whether they are really small (for stars) and closely packed, — whether the swarm is no more distant than the rest of the stars, or far beyond them.

Forty years ago the prevalent view was that these clusters were stellar universes, *galaxies*, like the group of stars to

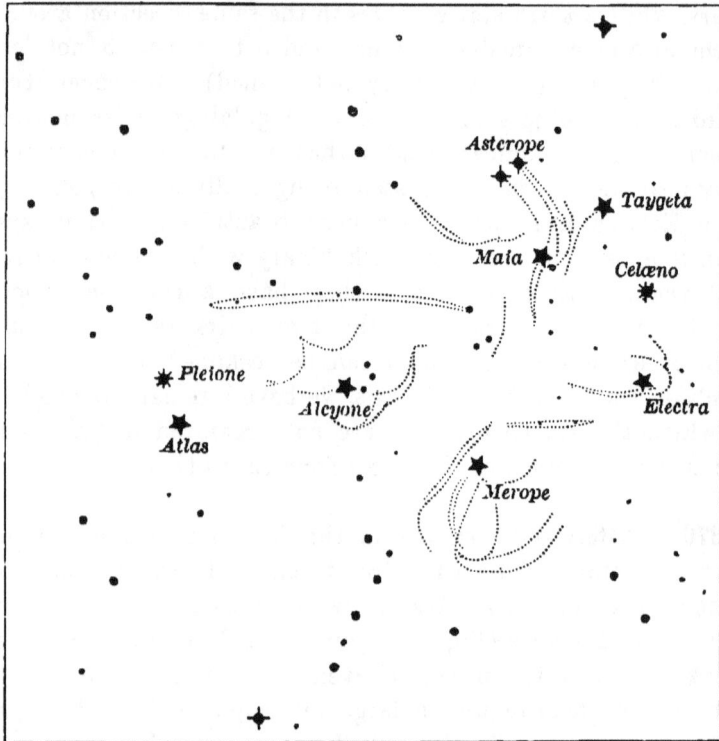

FIG. 78. — The Pleiades.

which it was supposed the sun belongs, — but so inconceivably remote that they dwindled to mere shreds of cloud.   It is now, however, quite certain that the opposite view is correct.   The star clusters are among *our* stars, and form a part of our own stellar universe.   Large and small stars are so associated in the same group as to leave no doubt on this point, although it has not yet been possible to determine the actual parallax and distance of any cluster.

NEBULÆ.

377. Besides the luminous clouds which, under the telescope, break up into separate stars, there are others which no
telescopic power resolves, and among them some which are
brighter than many of the clusters. These irresolvable objects, which now number
more than 8000, are "neb-
ulæ." Two or three of
them are visible to the
naked eye ; one, the brightest of all, and the one in
which the temporary star
of 1885 appeared, is in the
constellation of Androm-
eda (see Fig. 79). An-
other most conspicuous
and very beautiful nebula
is that in the sword of
Orion.

The larger and brighter
nebulæ are, for the most
part, irregular in form,
sending out sprays and
streams in all directions,

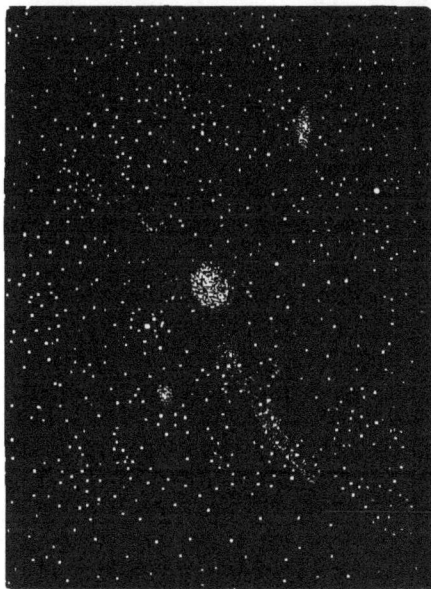

FIG. 79. — Telescopic View of the Great Nebula
in Andromeda.

and containing dark openings and "lanes." Some of them
are of enormous volume. The great nebula of Orion (which
includes within its boundary the multiple star, Theta Orionis)
covers several square degrees, and photographs show that
nearly the whole constellation is enveloped in a faint nebulosity, the wisps attaching themselves especially to the
brighter stars.

The nebula of Andromeda is not quite so extensive, but is
rather more regular in its form.

The smaller nebulæ are, for the most part, more or less nearly oval, and brighter in the centre. In the so-called "nebulous stars," the central nucleus is like a star shining through a fog. The "planetary nebulæ" are about circular and have nearly a uniform brightness throughout, while the rare "annular" or "ring nebulæ" are darker in the centre. Fig. 80 is a representation of the finest of these annular nebulæ, that in the constellation of Lyra. There are a number of nebulæ which exhibit a remarkable *spiral* structure in large telescopes. There are several *double* nebulæ,

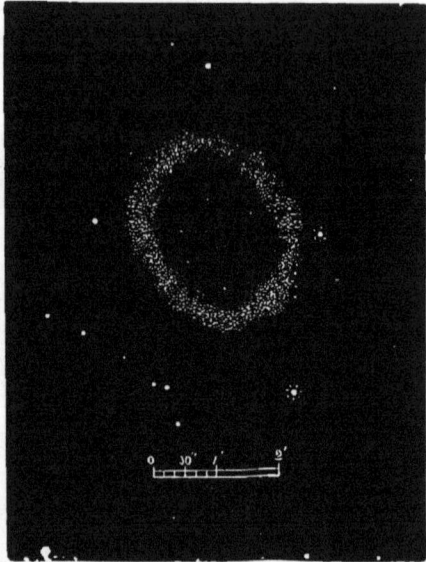

FIG. 80. — The Annular Nebula in Lyra.

and a few that are variable in brightness, though no regularity has yet been ascertained in their variation.

The great majority of the 8000 nebulæ are extremely faint, even in large telescopes, but the few that are reasonably bright are very interesting objects.

378. Drawings and Photographs of Nebulæ. — Until very lately the correct representation of a nebula was an extremely difficult task. More or less elaborate engravings exist of perhaps fifty of the more conspicuous of them, but photography has now taken possession of the field. The first success in this line was by Henry Draper of New York, in 1880, in photographing the nebula of Orion. Since his death, in 1882, great progress has been made both in Europe and in this

country, and at present the photographs are continually
bringing out new and before unsuspected features. Fig. 81.

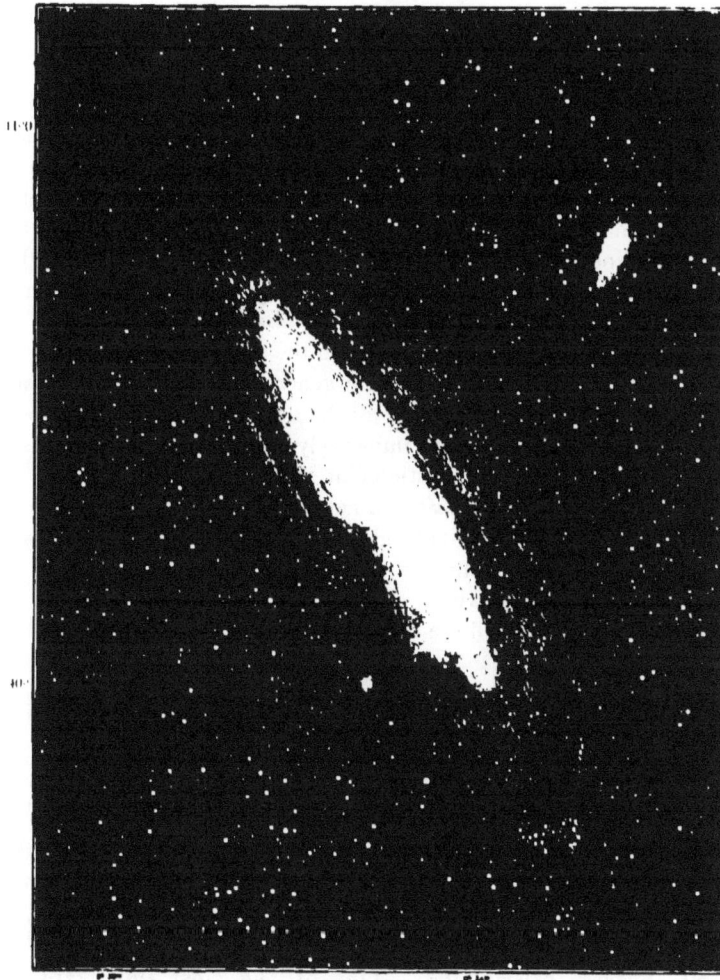

FIG. 81. — Mr. Roberts's Photograph of the Nebula of Andromeda.

for instance, is from a photograph of the nebula of Andromeda, taken by Mr. Roberts of Liverpool in 1888, and shows

that the so-called "dark lanes," which hitherto had been seen only as straight and wholly mysterious markings (Fig. 79), are really curved ovals, like the divisions in Saturn's rings. The photograph brings out clearly a distinct annular structure pervading the whole nebula, which as yet has never been made out satisfactorily by the eye with any telescope.

The photographs not only show new features in old nebulæ, but they reveal numbers of new nebulæ invisible to the eye with any telescope. Thus, in the Pleiades it has been found that almost all the larger stars have wisps of nebulosity attached to them, as indicated by the dotted lines in Fig. 78; and in a small territory, in and near the constellation of Orion, Pickering, with an eight-inch telescope, found upon his star-plates nearly as large a number of *new* nebulæ as of those that were previously known within the same boundary.

The photographs of nebulæ require generally an exposure of from one to two hours. The images of all the brighter stars that fall upon the plate, are, therefore, always immensely over-exposed, and seriously injure the picture from an artistic point of view.

The photographic brightness of a nebula, to use such an expression, is many times greater than its brightness to the eye, owing to the fact that its light consists mainly in rays which belong to the upper or blue portion of the spectrum. It has very little red or yellow in it. At least, this is so with all the nebulæ whose spectra are characterized by bright lines.

379. Changes in Nebulæ. — It cannot be stated with certainty that sensible changes have occurred in any of the nebulæ since they first began to be observed, — the early instruments were so inferior to modern ones that the older drawings cannot be trusted; but some of the differences between the older and more recent representations make it extremely likely that real changes are going on. Probably after a reasonable interval of time photography will settle the question.

380. Spectra of Nebulæ. — One of the most important of the early achievements of the spectroscope was the proof that the light of many nebulæ, if not all, proceeds from glowing gas of low density, and not from aggregations of stars.

Huggins, in 1864, first made the decisive observation by *finding bright lines in their spectra*. Thus far the spectra of all the nebulæ that show lines at all appear to be substantially the same. Four lines are usually easily observed, two of which are due to hydrogen; but the other two, which are brighter than the hydrogen lines, are not yet identified.

At one time the brightest of the four lines was thought to be due to nitrogen, and even yet the statement that such is the case is found in many books ; but it is now certain that, whatever it may be, nitrogen is not the substance. Very recently Mr. Lockyer has ascribed this line to *magnesium*, in connection with his meteoric hypothesis. But recent elaborate observations of Huggins and others show that this identification also is probably incorrect.

Fig. 82 shows the position of the principal lines so far as observed. In the brighter nebulæ a number of others are also sometimes seen, and photographs show many more, between thirty and forty in all;

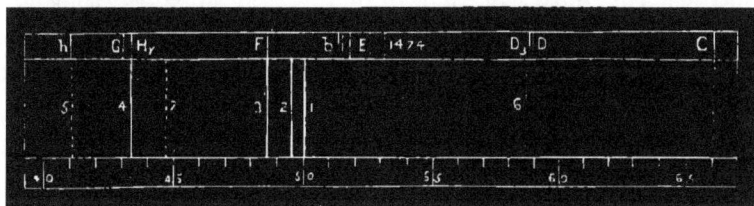

FIG. 82. — Spectrum of the Gaseous Nebulæ.

among them are several of the lines of *Helium*. Certain stars also show the nebular lines in their spectra ; and Mr. Campbell has found one or two which show bright hydrogen lines, extending out on each side of the star-spectrum in such a way as to indicate an immense envelope of the gas surrounding the star itself. Keeler has succeeded in measuring the motion of several of the brightest nebulæ by the displacement of their spectrum-lines.

381. Not all nebulæ show the bright-line spectrum. Those which do (about half the whole number) are of a greenish tint, at once recognizable in a large telescope. The *white* nebulæ, with the nebula of Andromeda, the brightest of all, at their head, present only a plain continuous spectrum, unmarked by lines of any kind. This, however, does not necessarily indicate that the luminous matter is not gaseous, for a gas *under pressure* gives a continous spectrum, like an incandescent solid or liquid. The telescopic evidence as to the non-stellar constitution of nebulæ is the same for all; no nebula resists all attempts at resolution (*i.e.*, breaking up into stars) more stubbornly than that of Andromeda.[1]

As to the real constitution of those bodies, we can only speculate. The fact that the luminous matter in them is mainly gaseous does not at all make it certain that they do not also contain dark matter, either liquid or solid. What proportion of it there may be, we have at present no means of knowing.

382. Distance and Distribution of Nebulæ. — As to the distance, we can only say that, like the star clusters, they are within the stellar universe and not beyond its boundaries. This is clearly shown by the nebulous stars, first pointed out and discussed by the older Herschel. We find all gradations, from a star with a little faint nebulosity around it, to nebulæ which show only the faintest spot of light in the centre. It is confirmed also by such peculiar associations of the stars and nebulæ as we find in the Pleiades. Moreover, in certain curious luminous masses, known as the "Nubeculæ," near the south pole, we have stars, star clusters, and nebulæ promiscuously intermingling.

Taking the sky generally, however, the distribution of the nebulæ is in contrast with that of the stars. The stars, as we shall see, crowd together near the Milky Way. The nebulæ, on the other hand, are most numerous just where the stars are fewest, as if the stars had somehow used up the substance of which the nebulæ are made.

---

[1] Some years ago it was stated that Lord Rosse's telescope had partially resolved the nebula of Andromeda and the nebula of Orion. This turned out to be a mistake.

THE SIDEREAL HEAVENS.

383. The Galaxy, or Milky Way. — This is a luminous belt of irregular width and outline, which surrounds the heavens nearly in a great circle. It is very different in brightness in different parts, and is marked here and there by dark bars and patches, which at night look like overlying clouds. For about a third of its length (between Cygnus and Scorpio) it is divided into two roughly parallel streams. The telescope shows it. to be made up almost entirely of small stars from the eighth magnitude down; it contains, also, numerous star clusters, but very few true nebulæ.

The galaxy intersects the ecliptic at two opposite points not far from the solstices, and at an angle of nearly 60°, the north "galactic pole" being, according to Herschel, in the constellation of Coma Berenices. As Herschel remarks, —

"The 'galactic plane' is to the sidereal universe much what the plane of the ecliptic is to the solar system, — a plane of ultimate reference, and the ground plan of the stellar system."

384. Distribution of Stars in the Heavens. — It is obvious that the distribution of the stars is not even approximately uniform. They gather everywhere into groups and streams; but, besides this, the examination of any of the great star-catalogues shows that the average number to a square degree increases rapidly and pretty regularly from the galactic pole to the galaxy itself, where they are most thickly packed. This is best shown by the "star-gauges" of the older Herschel, each of which consists merely in an enumeration of the stars visible in a single field of view. He made 3400 of these gauges, and his son followed up the work at the Cape of Good Hope with 2300 more in the south circumpolar regions. From these data it appears that near the pole of the galaxy, the average number of stars in a single field of view is only about 4; at 45° from the galaxy, a little over 10; while on the galactic circle itself it is 122.

Herschel, starting from the unsound assumption that the stars are all of about the same size and brightness and separated by approximately equal distances, drew from his observations numerous untenable conclusions as to the form and structure of the "galactic cluster" to which the sun was supposed to belong, — theories for a time widely accepted, and even yet more or less current in popular text-books, though in many points certainly incorrect.

But although the apparent brightness of the stars does not depend entirely, or even mainly, upon their distance, it is certain that *as a class* the faint stars are really more remote, as well as smaller and darker than the brighter ones. We may, therefore, safely draw a few inferences, which, so far as they go, in the main agree with those of Herschel.

385. Structure of the Stellar Universe. — I. The great majority of the stars we see are included within a space having, roughly, the form of a rather thin, flat disc, like a watch, with a diameter eight or ten times as great as its thickness, our sun being not very far from its centre.

II. Within this space the naked-eye stars are distributed with some uniformity, but not without a tendency to cluster, as shown in the Pleiades. The smaller stars, on the other hand, are strongly "gregarious," and are largely gathered into groups and streams which have comparatively vacant spaces between them.

III. At right angles to the galactic plane the stars are scattered more evenly and thinly than in it, and we find on the sides of the disc the comparatively starless region of the nebulæ.

IV. As to the Milky Way itself, it is not certain whether the stars which compose it form a sort of thin, flat, continuous *sheet*, or whether they are arranged in a sort of *ring* with a comparatively empty space in the middle, where the sun is situated, not far from its centre.

As to the size of the disc-like space which contains most of the stars, very little can be said positively. Its diameter must be as great as 20,000 or 30,000 *light-years*, — how much greater it may be we cannot even guess; and as to the "beyond," we are still more ignorant. If, however, there are other stellar systems of the same order as our own, these systems are neither the nebulæ, nor the clusters which the telescope reveals, but are far beyond the reach of any instrument at present existing.

386. Do the Stars form a System? — It is probable (though not certain) that gravitation operates between the stars, as indicated by the motion of the binaries. The stars are certainly moving very swiftly in various directions, and the question is whether these motions are governed by gravitation, and are "orbital" in the ordinary sense of the word.

There has been a very persistent belief that somewhere there is an enormous *central sun,* around which the stars are all circulating in the same way as the planets of the solar system move about our own sun. This belief has been abundantly proved to be unfounded. It is now certain that there is no such great body dominating the stellar universe.

387. Maedler's Hypothesis. — Another less improbable doctrine is that there is a general revolution of the mass of stars around the *centre of gravity* of the whole, — a revolution nearly in the plane of the Milky Way. Some years ago, Maedler, in his speculations, concluded (though without sufficient reason) that this centre of gravity of the stellar system was not far from Alcyone, the brightest of the Pleiades, and, therefore, that this star was in a sense the 'central sun'; and the idea is frequently met with in popular writings. It has no basis of reason, however, nor is there yet proof or probability of any such general revolution.

388. On the whole, the most reasonable view seems to be that the stars are moving much as bees do in a swarm, each star mainly under the control of the attraction of its nearest neighbors, though influenced more or less, of course, by that of the general mass. If so, the paths of the stars are not "orbits" in the strict sense; that is, they are not paths which return into themselves, the forces which at any moment act upon a given star being so nearly balanced that its motion must be sensibly in a straight line for thousands of years at a time.

The *solar* system is an absolute despotism, the sun supreme. Among the stars, on the other hand, there is no central power, but the system is a pure democracy, in which the individuals are controlled by the influence of their neighbors, and by the authority of the whole community to which they themselves belong.

COSMOGONY.

389. One of the most interesting topics of speculation re-lates to the process by which the present state of things has come about. In a forest, to use an old comparison of Herschel's, we see around us trees in all stages of their life-history, from the sprouting seedlings to the prostrate and decaying trunks of the dead. Is the analogy applicable to the heavens, and can we hope by a study of the present condition and behavior of the bodies around us to come to an understanding of their past history and probable future? Possibly to some extent. But human life is so short that the processes of change are hardly perceptible, and our telescopes and spectro-scopes reveal but little of the "true inwardness" of things, so that speculation is continually baffled, and its results can seldom be accepted as secure. Still, some general conclusions seem to have been reached, which are *likely* to be true; but the pupil is warned that they are not to be regarded as established in any such sense as the law of gravitation and the theory of planetary motion.

In a general way we may say that the shrinkage of clouds of rarefied matter into more compact masses under the force of gravitation, the production of heat by this shrinkage, the effect of this heat upon the mass itself and upon neighboring bodies, — these principles cover nearly all the explanations that can thus far be given for the present condition of the heavenly bodies.

390. Genesis of the Planetary System. — Our planetary system is clearly no accidental aggregation of bodies. Masses of matter coming haphazard to the sun would move (as comets actually do move) in orbits which, though necessarily conic sections, would have every degree of inclination and eccentricity. In the planetary system this is not so. Numerous relations exist for which gravitation does not at all account, and for which the mind demands an explanation.

We note the following as the principal : —

1. The orbits of the planets are all *nearly circular* (*i.e.*, never very eccentric).

2. They are all nearly *in one plane* (excepting those of some of the asteroids).

3. The revolution of all, without exception, is *in the same direction*.

4. There is a curious and regular progression of distances (expressed by Bode's Law; which, however, breaks down with Neptune).

As regards the planets themselves : —

5. The plane of every planet's rotation nearly coincides with that of its orbit (probably excepting Uranus).

6. The direction of rotation is the same as that of the orbital revolution (excepting, probably, Uranus and Neptune).

7. The plane of orbital revolution of the planet's *satellites* coincides nearly with that of the planet's rotation, wherever this has been ascertained.

8. The direction of the satellites' revolution also coincides with that of the planet's revolution (with the same limitation).

9. The largest planets rotate most swiftly.

391. Now this arrangement is certainly an admirable one for a planetary system, and therefore some have argued that the Deity constructed the system in that way, perfect from the first. But to one who considers the way in which other perfect works usually attain their perfection, — their processes of growth and development, — this explanation seems improbable. It appears far more likely that the planetary system was formed by growth than that it was built outright. The theory which, in its main features, is now generally accepted, as supplying an intelligible explanation of the facts, is that known as the "nebular hypothesis." In a more or less crude and unscientific form, it was first suggested by Swedenborg and Kant, and afterwards, about the beginning of the present century, was worked out in mechanical detail by Laplace. On the whole, we may say that while, in its main outlines, the theory is probably true, it also probably needs serious modifications in its details.

392. Laplace's Nebular Hypothesis. — He maintained (a) that at some time in the past[1] the matter which is now gathered into the sun and planets was in the form of a "nebula."

(b) This nebula, according to him, was a cloud of *intensely heated gas* (questionable).

(c) Under the action of its own gravitation, the nebula assumed *a form approximately globular, with a motion of rotation*, the whirling motion depending upon the accidental differences in the original velocities and densities of the different

---

[1] As to the origin of the nebula itself, he did not speculate. There was no assumption on his part, as is often supposed, that the matter was first *created* in the nebulous condition. He assumed only that as the egg may be taken as the starting-point in the life-history of an animal, so the nebula is to be regarded as the starting-point of the life history of the planetary system. He did not raise the question whether the egg is older than the hen or not.

parts of the nebula. As the contraction proceeded, the swiftness of the rotation would necessarily increase for mechanical reasons.

(d) In consequence of its whirling motion, the globe would necessarily become flattened at the poles, and ultimately, as the contraction went on, the centrifugal force at the equator would there become equal to gravity, and rings of nebulous matter would be detached from the central mass, like the rings of Saturn. In fact, Saturn's rings suggested this feature of the theory.

(e) The ring thus formed would for a time revolve as a whole, but would ultimately break, *and the material would collect into a globe revolving around the central nebula as a planet.*[1]

Laplace supposed that the ring would revolve as if it were solid, the particles at the outer edge moving more swiftly than those at the inner (questionable). If this were always so, the planet formed would necessarily *rotate* in the same direction in which the ring had *revolved.*

(f) The planet thus formed would throw off rings of its own, and so form for itself a system of satellites.

393. This theory obviously explains most of the facts of the solar system, which were enumerated in the preceding article, though some of the exceptional facts (such as the short periods of the satellites of Mars, and the retrograde motions of those of Uranus and Neptune) cannot be explained by it *alone* in its original form. But even these exceptions do not *contradict* it, as is sometimes supposed.

As to the modifications required by the theory, while they alter the mechanism of the development in some respects, they do not touch the main results. It is rather more likely, for instance, that the original nebula was a cloud of ice-cold dust

---

[1] It has been suggested by Huggins and others that the two small nebulæ near the great nebula of Andromeda (Fig. 81) may be planets in process of formation.

than incandescent gas and "fire-mist," to use a favorite expression; and it is likely that planets and satellites were often separated from the mother-orb otherwise than in the form of rings.

Nor is it possible that a thin, wide ring could revolve in the same way as a solid mass; the particles near the inner edge must make their revolution in periods much shorter than those upon the circumference, or the ring would tear to pieces. But this very fact makes it possible to account for the peculiar backward motion of the satellites of Uranus and Neptune, thus removing one of the main objections to the theory in its original form.

Many things, also, make it questionable whether the outer planets are so much older than the inner ones, as Laplace's theory would indicate. It is not impossible that they may even be younger.

Our limits do not permit us to enter into a discussion of Darwin's "tidal theory" of satellite formation, which may be regarded as in a sense supplementary to the nebular hypothesis; nor can we more than mention Faye's proposed modification of it. According to him, the inner planets are the oldest.

394. Lockyer's Meteoritic Hypothesis. — Within the last two years Mr. Lockyer has vigorously revived a theory which has been from time to time suggested before; viz., that all the heavenly bodies in their present state are mere *clouds of meteors*, or have been formed by the condensation of such clouds; and it is an interesting fact, as Professor G. H. Darwin has recently shown, that a large swarm of meteors, in which the individuals move swiftly in all directions, would, in the long run and as a whole, behave almost exactly, from a mechanical point of view, in the same way as one of Laplace's hypothetical gaseous nebulæ.[1]

----

[1] This is not very strange, after all. According to the modern "kinetic theory of gases" (Rolfe's "Physics," page 157), a meteor cloud is mechani-

The spectroscopic observations upon which Mr. Lockyer rests his attempted demonstration are many of them very doubtful; but that does not really discredit the main idea, except so far as the question of the origin and nature of the light of the heavenly bodies is concerned. He makes the light in all cases depend upon the collisions between the meteors, and finds in the spectra of the heavenly bodies evidence of the presence of materials with which we are familiar in the meteorites which fall upon the earth's surface. These identifications are in many cases questionable, — in some certainly incorrect, — and it seems much more likely that the luminosity depends to a great degree upon other than mere mechanical actions.

395. Stars, Star-clusters, and Nebulæ. — It is obvious that the nebular hypothesis in all its forms applies to the explanation of the relations of these different classes of bodies to each other. In fact, Herschel, appealing only to the "law of continuity," had concluded, before Laplace published his theory, that the nebulæ develop sometimes into clusters, sometimes into double or multiple stars, and sometimes into single stars. He showed the existence in the sky of all the intermediate forms between the nebula and the finished star. For a time, about forty years ago, while it was generally believed that all the nebulæ were only star-clusters, too remote to be resolved by existing telescopes, his views fell rather into abeyance; but they regained acceptance in their essential features when the spectroscope demonstrated the substantial difference between gaseous nebulæ and the star-clusters.

396. Conclusions from the Theory of Heat. — Kant and Laplace, as Newcomb says, seem to have reached their results by reasoning *forwards*. Modern science comes to very similar

---

cally just the same thing as a mass of gas *magnified.* The kinetic theory asserts that gas is only a swarm of minute molecules, the peculiar gaseous properties depending upon the collisions of these molecules with each other and with the walls of the enclosing vessel. Magnify sufficiently the molecules and the distances between them, and you have a meteoric cloud.

conclusions by working *backwards* from the present state of things.

Many circumstances go to show that the *earth* was once much hotter than it now is. As we penetrate below the surface, the temperature rises nearly a degree (Fahrenheit) for every sixty feet, indicating a white heat at the depth of a few, miles; the earth at present, as Sir William Thomson says, " is in the condition of a stone that has been in the fire and has cooled at the surface."

The *moon* bears apparently on its surface the marks of the most intense igneous action, but seems now to be entirely chilled.

The *planets*, so far as we can make out with the telescope, exhibit nothing at variance with the view that they were once intensely heated, while many things go to establish it. Jupiter and Saturn, Uranus and Neptune, do not seem yet to have cooled off to anything like the earth's condition.

As to the *sun*, we have in it a body continuously pouring forth an absolutely inconceivable quantity of heat without any visible source of supply. As has been explained already (Art. 192), the only rational explanation of the facts, thus far presented, is that which makes it a huge, cloud-mantled ball of elastic substance, slowly shrinking under its own central gravity, and thus generating heat.[1] A shrinkage of about 300 feet a year in the sun's diameter will account for the whole annual output of radiant heat and light.

397. Age of the System. — Looking *backward*, then, and trying to imagine the course of time and of events *reversed*, we see the sun growing larger and larger, until at last it has

---

[1] So far we have no decisive evidence whether the sun has passed its maximum of temperature or not. Mr. Lockyer thinks its spectrum (resembling as it does that of Capella and the stars of the second class) proves that it is now on the *downward grade* and growing cooler; but others do not consider the evidence conclusive.

expanded to a huge globe that fills the largest orbit of our
system. How long ago this may have been, we cannot state
with certainty. If we could assume that the amount of heat
yearly radiated by the solar surface had remained constantly
the same through all those ages, and, moreover, that all the
radiated heat came solely from the slow contraction of the
sun's mass, apart from any considerable original capital in
the form of a high initial temperature, and without any re-
enforcement of energy from outside sources, — if we could
assume these premises, it is easy to show that the sun's past
history must cover about 15,000000 or 20,000000 years. But
such assumptions are at least doubtful; and if we discard
them, all that can be said is that the sun's age must be
greater, and probably many times greater, than the limit we
have named.

398. Future Duration of the System. — Looking *forward*,
on the other hand, from the present towards the future, it is
easy to conclude with certainty that if the sun continues its
present rate of radiation and contraction, and receives no sub-
sidies of energy from without, it must, within 5,000000 or
10,000000 years, become so dense that its constitution will be
radically changed. Its temperature will fall and its function
as a sun will end. Life on the earth, as we know life, will be
no longer possible when the sun has become a dark, rigid,
frozen globe. At least this is the inevitable consequence of
what now seems to be the true account of the sun's condition
and activity.

399. The System not Eternal. — One conclusion seems to
be clear: That the present system of stars and worlds is
not an *eternal* one. We have before us everywhere evidence
of continuous, irreversible progress from a definite beginning
towards a definite end. Scattered particles and masses are
gathering together and condensing, so that the great grow con-

tinually larger by capturing and absorbing the smaller. At the same time the hot bodies are losing their heat and distributing it to the colder ones, so that there is an unremitting tendency towards a uniform, and therefore *useless*, temperature throughout our whole universe: for heat is available as energy (*i.e.*, *it can do work*) only when it can pass from a warmer body to a colder one. The continual warming up of cooler bodies at the expense of hotter ones always means a loss, therefore, not of energy, — for that is indestructible, — but of *available* energy. To use the ordinary technical term, energy is continually "*dissipated*" by the processes which constitute and maintain life on the universe. This dissipation of energy can have but one ultimate result, that of absolute stagnation when the temperature has become everywhere the same.

If we carry our imagination backwards, we reach "a beginning of things," which has no intelligible antecedent; if forwards, we come to an end of things in dead stagnation. That in some way this end of things will result in a "new heavens and a new earth" is, of course, probable, but science as yet can present no explanation of the method.

Note to Article 361. New Variables and Variable-star Clusters.

Since the summer of 1895 there has been a rapid increase in the number of known variables, largely as the result of the examination of photographs of different portions of the sky made at Cambridge, U. S., and at Arequipa. About thirty variables have thus been brought to light, among them one star of the Algol type.

Mr. Chandler, from visual observations, has detected a variable of the Beta Lyræ type, designated as U, Pegasi, which is interesting as having a period of only 5^h 31^m, by far the shortest known.

All these new variables are too small to be visible to the naked eye, and all except the Algol variables show peculiar banded spectra, generally with bright lines in them.

In 1893 and 1895 three small "temporary" stars were found upon the Arequipa plates, indicating that the appearance of such stars is not very unusual, though only a few become bright enough to be seen without a telescope.

But the most remarkable discovery is that of variable-star clusters. Several have been found, but the most remarkable thus far are the two known as Messier 3 and Messier 5. In the first no less than 96 variables have been detected, and in the second nearly 60. In Messier 5 the changes are so rapid that the comparison of photographs taken only two hours apart brings them out very strikingly.

APPENDIX.

———•o¦o¦o¦oo———

CHAPTER XIII.

ASTRONOMICAL INSTRUMENTS.

THE CELESTIAL GLOBE.— THE TELESCOPE: SIMPLE, ACHRO-
MATIC, AND REFLECTING. — THE EQUATORIAL. — THE
FILAR MICROMETER. — THE TRANSIT INSTRUMENT. —
THE CLOCK AND CHRONOGRAPH. — THE MERIDIAN CIR-
CLE. — THE SEXTANT.

400. The Celestial Globe. — The celestial globe is a ball,
usually of papier-mâché, upon which are drawn the circles of
the celestial sphere and a map of the stars. It is ordinarily
mounted in a framework which represents the horizon and the
meridian, in the manner shown in Fig. 83.

The "horizon," HH' in the figure, is usually a wooden
ring three or four inches wide and perhaps three-quarters of
an inch thick, directly supported by the pedestal. It carries
upon its upper surface at the inner edge a circle marked with
degrees for measuring the azimuth of any heavenly body, and
outside this the so-called zodiacal circles, which give the sun's
longitude and the equation of time for every day of the year.

The meridian ring, MM', is a circular ring of metal which
carries the bearings upon which the globe revolves. Things
are so arranged, or ought to be, that the mathematical axis
of the globe is exactly in the same plane as the graduated face

of the ring, which is divided into degrees.   The meridian ring
is held underneath the globe by a support, with a clamp which
enables us to fix it securely in any desired position.

The surface of the globe is marked first with the celestial
equator, next with the ecliptic, crossing the equator at an

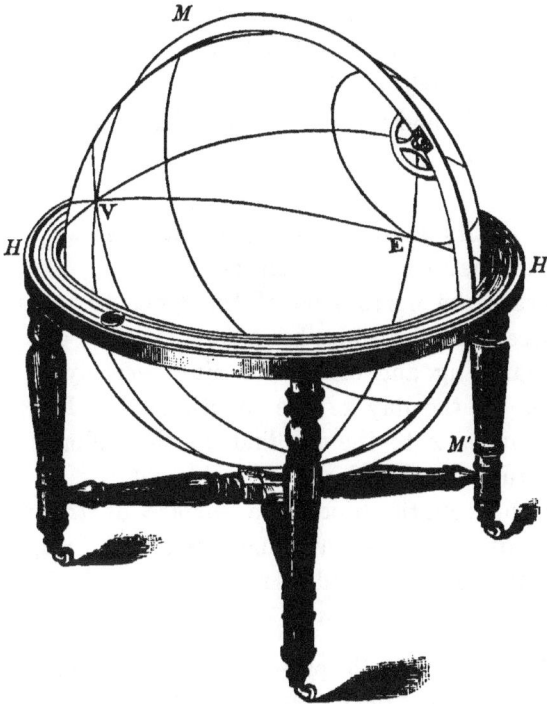

FIG. 83. — The Celestial Globe.

angle of 23½° at V (as the figure is drawn, V happens to be the
autumnal equinox, not the vernal), and each of these circles is
divided into degrees.   The equinoctial and solstitial colures
are also always represented.   As to the other circles, usage
differs.   The ordinary way at present is to mark the globe
with twenty-four hour-circles 15° apart (the colures, Art. 117,
being four of them), and with parallels of declination 10°

apart. On the surface of the globe are plotted the positions of the stars and the outl:nes of the constellations.

It is perhaps worth noting that ι˙any of the spirited figures of the constellations upon our present globes are copied from designs drawn by Albert Dürer for a star-map published in his time.

The *Hour-index* is a small circle of thin metal, about four inches in diameter, which is fitted to the northern pole of the globe with a stiffish friction, so that it can be set like the hands of a clock, and when once set will turn with the globe without shifting.

401. To rectify a Globe, — *i.e.,* to set it so as to show the aspect of the heavens at any time : —

(1) Elevate the north pole of the globe to an angle equal to the observer's latitude by means of the graduation on the meridian ring, and clamp the ring securely.

(2) Look up the day of the month on the horizon of the globe, and opposite to the day find on the zodiacal circle the sun's longitude for that day.

(3) On the ecliptic (upon the surface of the globe) find the degree of longitude thus indicated, and bring it to the graduated face of the meridian ring. The globe is thus set to correspond to *apparent noon* of the day in question.

It may be well to mark the place of the sun temporarily with a bit of paper gummed on at the proper place in the ecliptic. It can easily be wiped off after using.

(4) Hold the globe fast, so as to keep the place of the sun exactly on the meridian, and turn the *hour-index* until it shows at the edge of the meridian ring the *mean time* of apparent noon (*i.e.,* 12^h ± the equation of time given on the wooden horizon for the day in question).

If *standard* time is used, the hour-index must be set to the *standard* time for apparent noon instead of the local mean time.

(5) Finally, turn the globe upon its axis until the hour-index shows at the meridian the hour for which it is to be set. The globe will then represent the true aspect of the heavens at that time.

The positions of the moon and planets are not given by this operation, since they have no fixed places in the sky, and therefore cannot be put in by the globe-maker. If one wants them represented, he must look up their right ascensions and declinations in some almanac, and mark the proper places on the globe with bits of wax or paper.

TELESCOPES.

402. Telescopes are of two kinds, — refracting and reflecting. The refractor was first invented, early in the seventeenth century, and is much more used, but the largest instruments ever made are reflectors.   In both, the fundamental principle is the same.   The large *lens* of the instrument (or else its concave mirror) forms *a real image* of the object looked at, and this image is then examined and magnified by the eye-piece, which in principle is only a magnifying-glass.

In the form of instrument, however, which was originally devised by Galileo and is still used as the "opera-glass," the rays from the object-glass are intercepted, and brought to parallelism, by the concave lens which serves as an eye-glass, *before* they form the image. Telescopes of this construction are never made of much power, being inconvenient on account of the smallness of the field of view.

403. The Simple Refracting Telescope. — This consists essentially, as shown in Fig. 84, of two convex lenses : one, the object-glass *A*, of large size and long focus ; the other, the eye-glass *B*, of short focus, — the two being set at a distance nearly equal to the sum of their focal lengths.   Recalling the optical principles relating to the formation of images by lenses, we see that if the instrument is pointed towards the moon, for instance, all the rays that strike the object glass from the *top* of the crescent will be collected to a focus at a, while those from the *bottom* will come to a focus at b; and similarly with rays from the other points on the surface of the moon. We shall, therefore, get in the "focal plane" of the object-glass a small inverted "image" of the moon. The image is a *real*

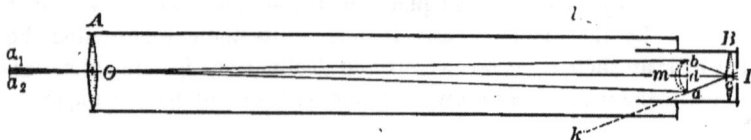

FIG. 84.— The Simple Refracting Telescope.

one; i.e., the rays really meet at the focal points, so that if we insert a photographic plate in the focal plane at ab and properly expose it, we shall get a picture of the object. The *size* of the picture will depend upon the apparent angular diameter of the object and the distance from the object-glass to the image ab.

If the focal length of the lens A is ten feet, then the image of the moon will be a little more than one inch in diameter.

404. Magnifying Power. — If we use the naked eye, we cannot see the image distinctly from a distance much less than a foot; but if we use a magnifying lens of, say, one inch focus, we can view it from a distance of only an inch, and it will look correspondingly larger. Without stopping to prove the principle, we may say that the magnifying power is simply equal to the *quotient obtained by dividing the focal length of the object-glass by that of the eye-lens.*

It is to be noted, however, that a magnifying power of *unity* is sometimes spoken of as no magnifying power at all, since the image appears of the same size as the object.

The magnifying power of a telescope is changed at pleasure by simply interchanging the eye-pieces, of which every telescope of any pretensions always has a considerable stock, giving various powers.

405. Brightness of the Image. — This depends not upon the focal length of the object-glass, but upon its diameter; or, more strictly, its *area*. If we estimate the diameter of the pupil of the eye at one-fifth of an inch, as it is usually reckoned, then (neglecting the loss from want of perfect transparency in the lenses) a telescope one inch in diameter collects into the image of a star 25 times as much light as the naked eye receives; and the great Lick telescope of 36 inches in diameter, 32,400 times as much, or about 30,000 after allowing for the losses. The amount of light is proportional to the *square* of the diameter of the object-glass.

The *apparent brightness* of an object which, like the moon or a planet, *shows a disc*, is not, however, increased in any such ratio, because the light gathered by the object-glass is spread out by the magnifying power of the eye-piece. But the total quantity of light in the image of the object greatly exceeds that which is available for vision with the naked eye, and objects which, like the stars, are mere luminous *points*, have their brightness immensely increased, so that with the telescope millions otherwise invisible are brought to light. With the telescope, also, *the brighter stars are easily seen in the daytime.*

406. The Achromatic Telescope. — A single lens cannot bring the rays which emanate from a single point in the object to any exact focus, since the rays of each different color are differently refracted, — the blue more than the green, and this more than the red. In consequence of this so-called "chromatic aberration," the simple refracting telescope is a very poor[1] instrument.

---

[1] By making it extremely long in proportion to its diameter, the indistinctness of the image is considerably diminished, and in the middle of the seventeenth century instruments more than 100 feet in length were used by Huyghens and others. Saturn's rings and several of his satellites were discovered with instruments of this kind.

About 1760, it was discovered in England that by making the object-glass of two or more lenses of different kinds of glass, the chromatic aberration can be nearly corrected. Object-glasses so made — none others are now in common use — are called *achromatic*. In practice, only two lenses are ordinarily used in the construction of an astronomical glass, — a convex of *crown* glass, and a concave of *flint* glass, the curves of the two lenses and the distances between them being so chosen as to give the most perfect possible correction of the "spherical" aberration ("Physics," p. 363) as well as of the chromatic.

407. Achromatism not Perfect. — It is not possible with the kinds of glass hitherto available to obtain a *perfect* correction of color. Even the best achromatic telescopes show a purple halo around the image of a bright star, which, though usually regarded as "very beautiful" by tyros, seriously injures the definition, and is especially obnoxious in large instruments.

This imperfection of achromatism makes it impossible to get satisfactory photographs with an ordinary object-glass, corrected for *vision*. An instrument for photography must have an object-glass specially corrected for the purpose, since the rays most efficient in impressing the image upon the photographic plate are the blue and violet rays, which in the ordinary object-glass are left to wander very wildly.

Much is hoped from the new kinds of glass now being made for optical purposes at Jena, Germany, as the results of the experiments conducted by Professor Abbé at the expense of the German government. Though the new glass is especially intended for use in the construction of microscopes, a few telescope lenses from three to six inches in diameter have been already made with it, which appear to be nearly perfect in their color correction.

408. Diffraction and Spurious Discs. — Even if a lens were absolutely perfect as regards the correction of aberrations, both spherical and chromatic, it would still be unable to give vision absolutely distinct. Since light consists of waves of finite length, the image of a luminous point can never be also a *point*, but must of mathematical necessity be a *disc* of finite diameter surrounded by a series of 'diffraction' rings. The diameter of the "spurious disc" of a star, as it is called, varies inversely with the diameter of the object-glass: the larger the telescope, the *smaller* the image of a star with a given magnifying power.

With a good telescope and a power of about 30 to the inch of aperture (120 for a 4-inch telescope) the image of a star, when the air is steady (a condition unfortunately seldom fulfilled), should be a clean, round disc, with a bright ring around it, separated from the disc by a clear black space. According to Dawes, the disc of a star with a 4½-inch telescope should be about 1″ in diameter; with a 9-inch instrument 0″.5, and ⅛″ for a 36-inch glass.

409. Eye-pieces. — For some purposes the simple convex lens is the best "eye-piece" possible; but it performs well only for a small object, like a close double star, placed exactly

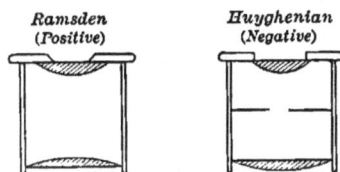

Ramsden
(Positive)

Huyghenian
(Negative)

in the centre of the field of view. Generally, therefore, we employ "eye-pieces" composed of two or more lenses, which give a larger field of view than a single lens, and define satisfactorily over the whole extent

FIG. 85. — Telescope Eye-pieces.

of the field. They fall into two general classes, the *positive* and the *negative*.

The *positive* eye-pieces are much more generally useful. They act as simple magnifying-glasses, and can be taken out of the telescope and used as hand-magnifiers if desired. The image of the object formed by the object-glass lies *outside of* this kind of eye-piece, between it and the object-glass.

In the *negative* eye-piece, on the other hand, the rays from the object-glass are intercepted by the so-called "field-lens" before reaching the focus, and the image is formed between the two lenses of the eye-piece. It cannot therefore be used as a hand-magnifier.

Fig. 85 shows the two most usual forms of eye-piece.

These eye-pieces show the object in an inverted position; but this is of no importance as regards astronomical observations.

410. Reticle. — When the telescope is used for pointing upon an object, as it is in most astronomical instruments, it must be provided with a 'reticle' of some sort. The simplest form is a metallic frame with *spider lines* stretched across it, the intersection of the spider lines being the point of reference. This reticle is placed not at or near the object-glass, as is often supposed, but *in its focal plane*, as *ab* in Fig. 84. Sometimes a glass plate with fine lines ruled upon it is used instead of spider lines. Some provision must be made for illuminating the lines, or "wires," as they are usually called, by reflecting into the instrument a faint light from a lamp suitably placed.

411. The Reflecting Telescope. — About 1670, when the chromatic aberration of refractors first came to be understood (in consequence of Newton's discovery of the "decomposition of light"), the reflecting telescope was invented. For nearly 150 years it held its place as the chief instrument for star-gazing, until about 1820, when large achromatics began to be made. There are several varieties of reflecting telescope, differing in the way in which the image formed by the mirror is brought within reach of the magnifying eye-piece.

Until about 1870, the large mirror (technically "speculum") was always made of speculum metal, a composition of copper and tin. It is now usually made of glass, silvered on the front by a chemical process. When new, these silvered films reflect much more light than the old speculum metal : they tarnish rather easily, but fortunately can be easily renewed.

412. Large Telescopes. — The largest telescopes ever made have been reflectors. At the head stands the enormous instrument of Lord Rosse of Birr Castle, Ireland, six feet in diameter and sixty feet long,

made in 1842, and still used. Next in size, but probably superior in
power, comes the five-foot silver-on-glass reflector of Mr. Common, at
Ealing, England, completed in 1889; and then follow a number (four
or five) of four-foot telescopes, — that of Herschel (erected in 1789, but
long ago dismantled) being the first, while the great instrument at
Melbourne is the only instrument of this size now in active use.

Of the refractors, the largest is that of the Yerkes Observatory at
Lake Geneva, Wisconsin, with an object-glass 40 inches in diameter,
and a tube nearly 70 feet long. The next in size is the telescope of
the Lick Observatory (see frontispiece), which has an aperture of 36
inches. Next to this come the great telescopes at Pulkowa, Meudon
and Nice, with apertures of about 30 inches; the Vienna telescope,
27 inches; the two telescopes at Washington and the University of
Virginia, $26\frac{1}{4}$ inches; and four or five others with apertures of from
26 to 23 inches, at Cambridge (England), Greenwich, Paris and
Princeton. Most of these large object-glasses were made by the
Clarks of Cambridge (U.S.).

413. Relative Advantages of Reflectors and Refractors. —
There is no little discussion on this point, each form of instrument
having its earnest partisans.

In favor of the reflector we have *first*, its cheapness and com-
parative ease of construction, since there is but one surface to grind and
polish, as against four in an achromatic object-glass; *second*, the fact
that reflectors can be made *larger* than refractors; *third*, the reflector
is absolutely achromatic.

On the other hand, a refractor gives a much brighter image than a
reflector of the same size; it also generally defines much better,
because, for optical reasons into which we cannot enter here, any
slight distortion or malformation of the speculum of a reflector dam-
ages the image many times more than the same amount of distortion
of an object-glass. Then a lens hardly deteriorates at all with age,
while a speculum soon tarnishes, and must be re-silvered or re-polished
every few years.

As a rule, also, refractors are lighter and more convenient than
reflectors of equal power.

414. Mounting of a Telescope, — the Equatorial. — A telescope, however excellent optically, is not good for much unless firmly and conveniently mounted.[1]

At present some form of *equatorial* mounting is practically universal. Fig. 86 represents schematically the ordinary arrangement of the instrument. Its essential feature is that its "principal axis" (*i.e.*, the one which turns in fixed bearings attached to the pier, and is called the *polar axis*) is placed parallel to the earth's axis, pointing to the celestial pole, so that the circle *H*, attached to it, is parallel to the celestial equator. This circle is sometimes called the *hour-circle*, sometimes the *right-ascension circle*. At the extremity of the polar axis a "sleeve" is fastened, which carries within it the declination axis *D*, and to this declination axis is attached the telescope tube *T*, and also the declination circle *C*.

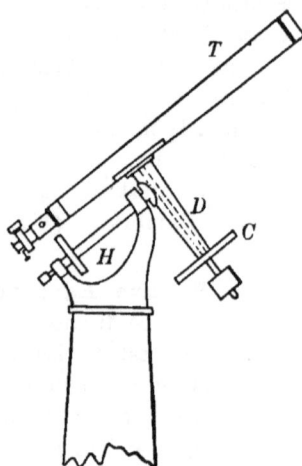

The advantages of this mounting are very great. In the first place, when the telescope is once pointed upon an object, it is not

FIG. 86. — The Equatorial.

necessary to move the declination axis at all in order to keep the object in the field, but only to turn the *polar* axis with a perfectly uniform motion, which motion can be, and usually is, given by *clock-work* (not shown in the figure).

In the next place, it is very easy to *find* an object even if

---

[1] We may add that it must, of course, be mounted where it can be pointed directly at the stars, without any intervening window-glass between it and the object. We have known purchasers of telescopes to complain bitterly because they could not see Saturn well through a closed window.

APPENDIX. [§ 414

invisible to the eye (like a faint comet, or a star in the day-time), provided we know its right ascension and declination, and have the sidereal time, — a sidereal clock or chronometer being an indispensable accessory of the instrument.

The frontispiece shows the actual mounting of the Lick telescope. Fig. 71, Art. 337, represents another form of equatorial mounting, which has been adopted for the instruments of the photographic campaign.

415. The Micrometer. — This is an instrument for measuring small angles, usually not exceeding 15' or 20'. Various kinds are employed, all of them small pieces of apparatus, which, when used, are secured to the eye-end of a telescope. The most common is the parallel-wire micrometer, which is a pair of parallel spider threads, one or both of which can be moved with a fine screw with a graduated head, so that the distance between the two 'wires' can be varied at pleasure, and then "read off" by looking at the micrometer head. Fig. 87 represents such an instrument attached to a telescope: the spider threads are in the box *BB*, and are viewed through the eye-piece.

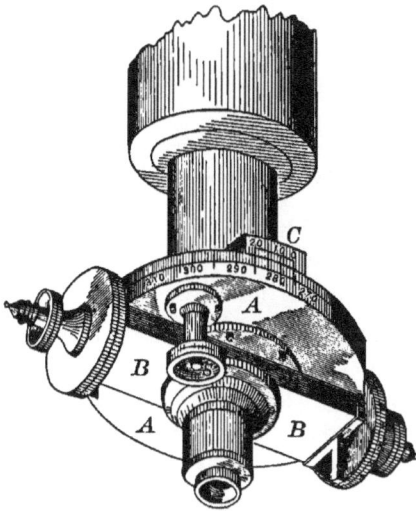

FIG. 87. — The Filar Position Micrometer.

416. The Transit Instrument (Fig. 88). — This consists of a telescope carrying at the eye-end a reticle, and mounted on a stiff axis with pivots that are perfectly true. They turn in

Y's, which are firmly set upon some sort of framework or on the top of solid piers, and so placed that the axis will be exactly east and west and precisely level. When the telescope is turned on its axis, the middle wire of the reticle, if everything is correctly adjusted, will follow the celestial meridian, and whenever a star crosses the wire, we know that it is exactly on the meridian. Instead of a single wire, the reticle generally contains a number of wires equally spaced, as shown in Fig. 89. The object is then observed upon each of the wires, and the mean of the observations is taken as giving the moment when the star crossed the middle wire.

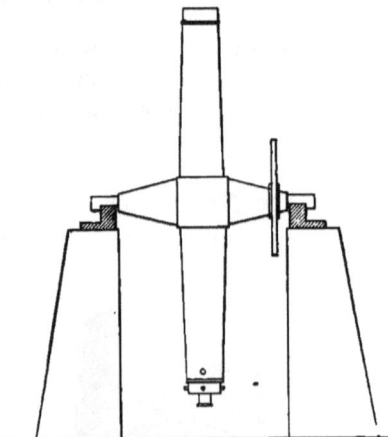

FIG. 88. — The Transit Instrument.

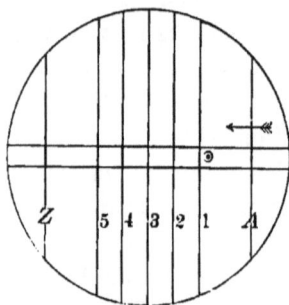

FIG. 89. — Reticle of the Transit Instrument.

A delicate spirit-level, to be placed on the pivots and test the horizontality of the axis, is an indispensable accessory.

So far as the theory of the instrument is concerned, a graduated circle is not essential; but practically it is necessary to have one attached to the axis in order to enable the observer to set for a star in preparing for the observation.

417. The Astronomical Clock, Chronometer, and Chronograph. — A good timepiece is an essential adjunct of the transit instrument, and equally so of most other astronomical instruments. The invention of the pendulum clock by Huyghens was almost as important an event in the history of practical astronomy as that of the telescope itself.

The astronomical clock differs in no essential respect from any other, except that it is made with extreme care, and has a "compensated" pendulum so constructed that the rate of the clock will not be affected by changes of temperature. It is almost invariably made to beat seconds, and usually has its face divided into twenty-four hours instead of twelve.

Excellence in a clock consists essentially in the *constancy of its 'rate'*; *i.e.*, it should gain or lose precisely the same amount each day, and as a matter of convenience the daily rate should be small, not to exceed a second or two. The rate is adjusted by slightly raising or lowering the pendulum bob, or putting little weights upon a small shelf attached to the rod;— the 'error,' when necessary, by simply setting the hands.

The error of a timepiece is the difference between the time shown by the clock-face and the true time at the moment; the rate is the amount it gains or loses in twenty-four hours.

The chronometer is simply a carefully made watch, and has the advantage of portability, though in accuracy it cannot quite compete with a well-made clock.

Formerly transit-instrument observations were made by simply noting with eye and ear the time indicated by the clock at the moment when the star observed was crossing the wire or reticle. A skilful observer can do this within about a tenth of a second. At present the observer usually presses a telegraph-key at the moment of the transit, and so telegraphs the instant to an instrument called a "*chronograph*," which makes a permanent record of the observation upon a sheet of paper, — thus making the observation much more accurate as well as easier. (For the description of the chronograph, see General Astronomy, Art. 56.)

418. The Meridian Circle. — In many respects this is the fundamental instrument of a working observatory. It is

simply the transit instrument *plus* a finely graduated circle
or circles attached to the axis, and provided with microscopes
for reading the graduation with precision.   In the accurate
construction of the pivots
of the instrument and of
the circles, with their
graduation, the utmost re-
sources of the mechanical
art are taxed.   Fig. 90
shows the instrument in
principle.   Fig. 91 is a
small meridian circle, as
actually constructed, with
a four-inch telescope and
twenty-four-inch circles.

Its main purpose is to
determine the right ascension and declination of
objects as they cross the

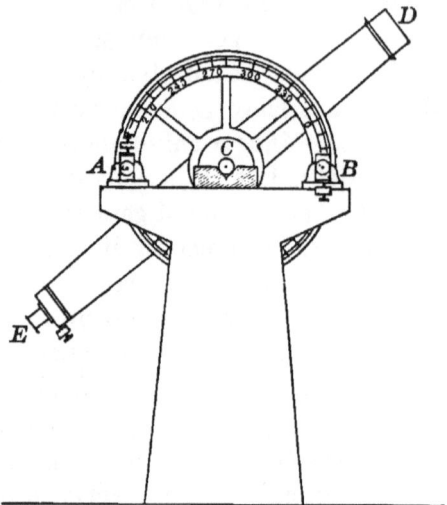

FIG. 90. — The Meridian Circle (Schematic).

meridian.   The declination is determined by measuring how
many degrees the object is north or south of the celestial equator at the moment of transit.   The "circle-reading" for the
equator must first be determined as a *zero point;* and this is
done by observing a star near the pole and getting the circle-
reading as it crosses the meridian above the pole, and twelve
hours later, when it crosses again below it.   The mean of these
two readings, corrected for refraction, will be the circle-reading
for the pole, or the *polar point*, which is, of course, just 90°
from the equatorial zero point.

419. The Nadir Point. — To get the latitude of the observer
with this instrument (Art. 81), it is necessary also to have the
nadir point as a zero; *i.e.*, the circle-reading which corresponds
to the vertical position of the telescope.   This point is found
by pointing the telescope down towards a basin of mercury

beneath it, and setting it so that the image of the east and west wire in the reticle coincides with itself. Then the telescope will be exactly vertical. The horizontal point is just 90° from the nadir point, and the difference between the

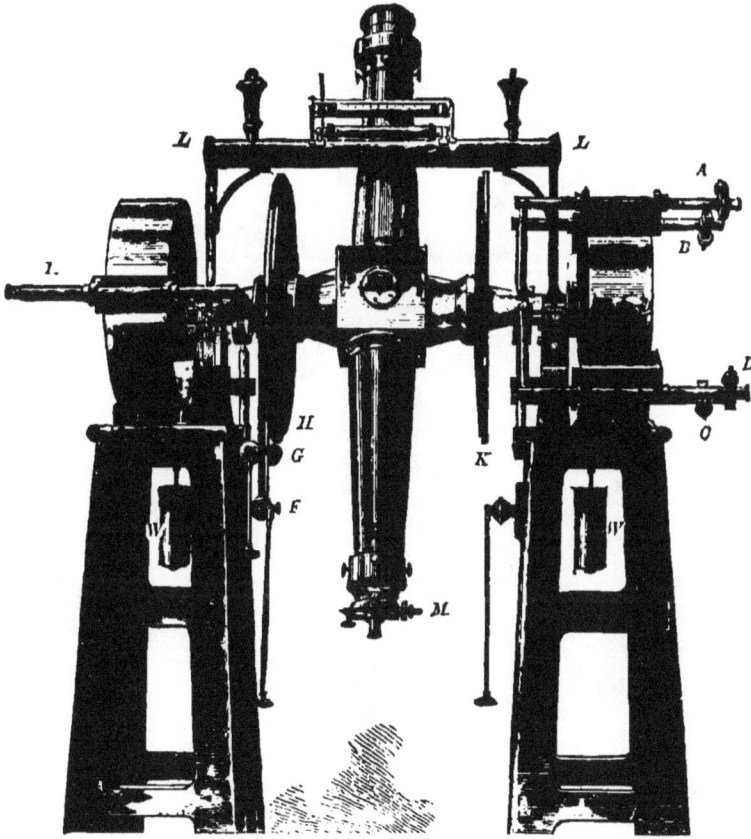

Fig. 91. — A Meridian Circle.

(north) horizontal point and the polar point is the latitude of the observatory.

Obviously the instrument can also be used as a simple transit instrument in connection with a clock, so that (Art. 99) the observer can determine both the right ascension and declination of any object which is visible when it crosses the meridian.

420. The Sextant. — All the instruments so far mentioned, except the chronometer, require firmly fixed supports, and are, therefore, useless at sea. The sextant is the only instrument for measurement upon which the mariner can rely. By means of it he can measure the angular distance between any two points (as, for instance, the sun and the visible horizon), not

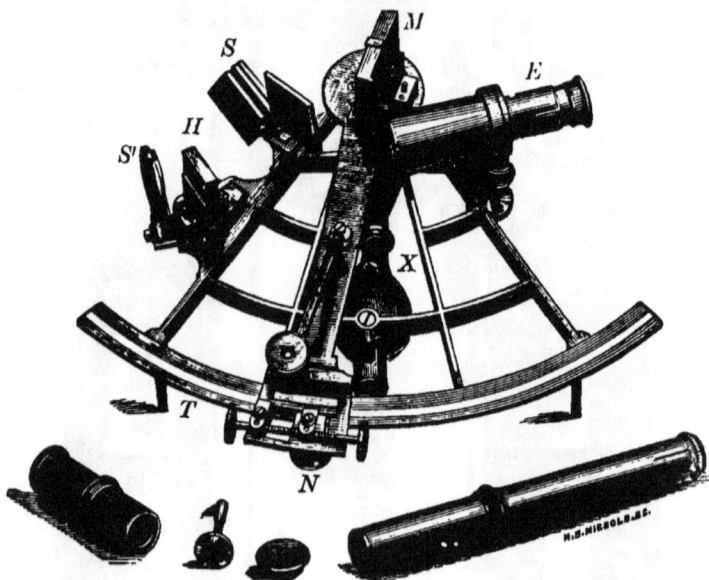

FIG. 92. — The Sextant.

by pointing first on one and afterwards on the other, but by *sighting them both simultaneously* and in apparent coincidence. This observation can be accurately made even if he has no stable footing, but is swinging about on the deck of a vessel. Fig. 92 represents the instrument. For a detailed description and explanation, see General Astronomy, Arts. 76–80.

421. Use of the Instrument. — The principal use of the instrument is in measuring the altitude of the sun. At sea, an

observer holding the instrument in his right hand, and keeping the plane of the arc vertical, looks directly towards the
visible horizon through the horizon-glass, *H*, at the point under
the sun. Then by moving the index, *N*, with his left hand,
he inclines the index mirror upward, until he sees the reflected image of the sun, and the lower edge of this image is
brought to touch the horizon-line. The reading of the graduation, after due correction for refraction, etc., gives the sun's
true altitude at the moment. If the observation is made near
noon, for the purpose of determining the latitude, it will not
be necessary to read the chronometer at the same time. If,
however, the observation is made for the purpose of determining the longitude (Art. 497), the instant of observation, as
shown by the chronometer, must be carefully noted.

The skilful use of the sextant requires considerable dexterity,
and from the small size of the telescope, the angles measured
are less precisely measured than with large fixed instruments;
but the portability of the instrument and its applicability at
sea render it absolutely invaluable. It was invented by
Gregory, of Philadelphia, in 1730, but an earlier design of an
instrument on the same principle has been found (unpublished)
among the papers of Newton.

CHAPTER XIV.

MISCELLANEOUS.

HOUR-ANGLE AND TIME. — TWILIGHT. — DETERMINATION
OF LATITUDE. — SHIP'S PLACE AT SEA. — FINDING THE
FORM OF THE EARTH'S ORBIT. — THE ELLIPSE. — ILLUS-
TRATIONS OF KEPLER'S THIRD LAW. — THE EQUATION
OF LIGHT AND THE SUN'S DISTANCE. — ABERRATION OF
LIGHT. — DE L'ISLE'S METHOD OF GETTING THE SOLAR
PARALLAX FROM THE TRANSIT OF VENUS. — THE
CONIC SECTIONS. — STELLAR PARALLAX. —

422. Hour-angle and Time (supplementary to Arts. 89–91).
— There is another way of looking at the matter of time,
which has great advantages. If we face towards the north
pole and consider the star m (Fig. 93) as carried at the end
of the arc mP of the hour-circle, which connects it to the
pole, we may regard this arc as a sort of clock-hand; and if
we produce it to the celestial equator and mark off the equator into 15° spaces, or ' hours,' the angle QPm, or the arc QY,
will measure the time which has elapsed since m was on the
meridian PQ. The angle mPQ is called the *hour-angle* of the
star m. It is the *angle at the pole between the meridian and
the hour-circle which passes through the body.*

Having now this definition of the hour-angle, we may define
sidereal time (Art. 91) at any moment as the *hour-angle of the
vernal equinox* at that moment. In the same way, the *apparent
solar time* (Art. 88) is the hour-angle of the sun's centre; the

mean solar time (Art. 89) is the hour-angle of a *fictitious sun* which moves around the heavens uniformly, once a year, in the equator, keeping its right ascension equal to the mean longitude of the real sun. For some purposes, as in dealing with the tides, it is convenient to use *lunar* time, which is simply the hour-angle of the *moon* at any moment.

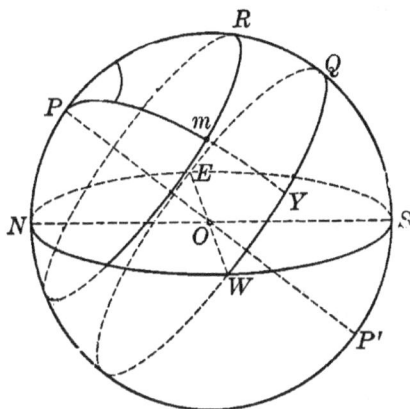

FIG. 93. — Hour-Angle.

423. Twilight is caused by the *reflection* of sunlight from the upper portions of the earth's atmosphere. After the sun has set, its rays still continue to shine through the air above the observer's head, and twilight continues as long as any portion of this illuminated air can be seen from where he stands. It is considered to end when stars of the sixth magnitude become visible near the zenith, which does not occur until the sun is about 18° below the horizon; but this is not strictly the same for all places.

The duration of twilight varies with the season and with the observer's latitude. In latitude 40° it is about 90 minutes on March 1st and Oct. 12th; but more than two hours at the summer solstice. In latitudes above 50°, when the days are longest, twilight never quite disappears, even at midnight. On the mountains of Peru, on the other hand, it is said never to last more than half an hour.

424. Methods of determining Latitude by Other Observations than those of Circumpolar Stars (supplementary to Art. 81). — To determine the latitude by observations of a circumpolar star, the observer must remain at the same station at least twelve hours. The latitude can be determined, however, with a good instrument, with almost equal precision, by observing the *meridian altitude, or zenith distance, of a body whose* *declination is accurately known.* In Fig. 94 the circle $AQPB$ is the meridian, Q and P being respectively the equator and the pole, and Z the zenith. QZ is evidently the *declination of the zenith* (*i.e.*, the distance of the zenith from the celestial equator) and is equal to PB, the *latitude of the observer*, or height of the pole. Suppose now that we observe Zs, *i.e.*, the zenith distance of the star s, south of the zenith, as it crosses the me-

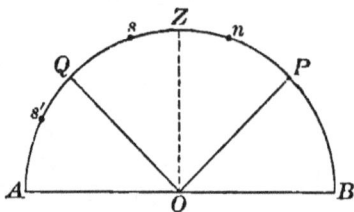

FIG. 94. — Determination of Latitude.

ridian, and that we know Qs, the declination of the star. Evidently $QZ = Qs + sZ$; *i.e.*, the *latitude equals the declination of the star plus its zenith distance.* If the star were at s', south of the equator, the same equation would hold good algebraically, because the declination, Qs', is a minus quantity. If the star were at n, between the zenith and the pole, we should have : Latitude equals the declination of the star *minus* the zenith distance. This is the method actually used at sea (Art. 426), the sun being the object observed.

There are many other methods in use, as, for instance, that by the *zenith telescope* and that by the *prime-vertical instrument,* which are practically more convenient and more accurate than either of the two described, but they are more complicated, and their explanation would take us too far. The reader is referred to General Astronomy, Arts. 104–107.

FINDING THE PLACE OF A SHIP.

425. The determination of the place of a ship at sea is, from the economic point of view, the most important problem of Astronomy. National observatories and nautical almanacs were established, and are maintained, principally to supply the mariner with the data needed to make this determination accurately and promptly. The methods employed are neces-

sarily such that the required observations can be made with
the sextant and chronometer, since fixed instruments, like the
transit instrument and meridian circle, are obviously out of
the question on board a vessel.

426. Latitude at Sea. — This is obtained by observing with
the sextant the sun's maximum altitude, which is reached
when the sun is crossing the meridian.

Since at sea the sailor seldom knows beforehand the precise
time which will be shown by his chronometer at noon, he
takes care not to be too late, and begins to measure the sun's
altitude a little before noon, repeating his observations every
minute or two. At first the altitude will keep increasing, but
when noon comes the sun will cease rising, and then begin to
descend. The observer uses, therefore, the *maximum* altitude
obtained, which, with due allowance for refraction and some
other corrections (for details, see larger works) gives him the
true altitude of the sun's centre. Taking this from 90°, we
get its zenith distance.

Referring now to Fig. 94, in which the circle $AQZPB$ is
the meridian, P the pole, Z the zenith, and OQ the celestial
equator seen edgewise, we see that PB, the altitude of the
pole, is necessarily equal to ZQ, the distance from the zenith
to the equator. Now from the almanac we find the declination of the sun, Qs, for the day on which the observations are
made.[1] We have only to add to this, Zs, the measured distance of the sun from the zenith, to obtain QZ, which is the
observer's latitude.

It is easy in this way, with a good sextant, to get the latitude within about half a minute of arc, or, roughly, about
half a mile, which is quite sufficiently accurate for nautical
purposes.

---

[1] If the sun happened to be south of the equator (in the winter), as at
s', we should have ZQ equals $Zs - s'Q$.

427. Determination of Local Time and Longitude at Sea.
— The usual method now employed for the longitude depends
upon the chronometer.   This is carefully 'rated' in port;
i.e., its error and its daily gain or loss are determined by comparisons with an accurate clock for a week or two, the clock
itself being kept correct to Greenwich time by transit observations.   By merely allowing for the gain or loss since leaving
port, and adding this gain or loss to the 'error' (Art. 417),
which the chronometer had when brought on board, the sea-
man at once obtains the error of the chronometer on Greenwich time at any moment; and allowing for this error, he has
the *Greenwich time* itself, with an accuracy which depends
only on the *constancy* of the chronometer's rate : it makes no
difference whether it is gaining much or little, provided its
daily rate is steady.

He must also determine his own *local time;* and this must
be done with the sextant, since, as was said before, an instrument like the transit cannot be used at sea.   He does it by
measuring the altitude of the sun, *not at or near noon*, as often
supposed, but when the sun is as near due east or west as circumstances permit.   From such an observation the sun's hour-
angle, *i.e.*, the apparent solar time (Art. 422), is easily found,
by a trigonometrical calculation, provided the ship's latitude
is known.   (For the method of calculation, see General Astronomy, Art. 116.)

The longitude follows at once, being simply the difference
between the Greenwich time and the local time.

In certain cases where the chronometers have been for
some reason disturbed, the mariner is obliged to get his Greenwich time by observing with a sextant the distance of the
moon from some neighboring fixed star, but the results thus
obtained are comparatively inaccurate and unsatisfactory.

428. To find the Form of the Earth's Orbit (supplementary
to Art. 119). — Take the point *S* (Fig. 95) for the sun, and

draw from it a line, SO, directed toward the vernal equinox, from which longitudes are measured. Lay off from S lines indefinite in length, making angles with SO equal to the earth's longitude as seen from the sun on each of the days when the observations are made (earth's longitude equals sun's longitude $+ 180°$). We shall thus get a sort of "spider," showing the *direction* of the earth as seen from the sun on each of those days.

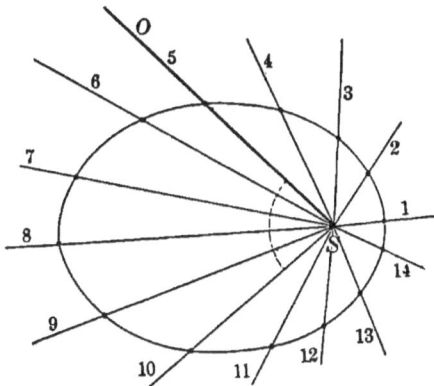

FIG. 95.

Determination of the Form of the Earth's Orbit.

Next, as to the distances. While the apparent diameter of the sun does not tell us its *absolute* distance from the earth, unless we know his diameter in miles, yet the *changes* in the apparent diameter do inform us as to the *relative* distance at different times, since the nearer we are to the sun, the larger it looks. If, then, on the legs of the "spider" we lay off distances *inversely proportional*[1] to the number of seconds of arc in the sun's measured diameter at each date, these distances will be proportional to the true distance of the earth from the sun, and the curve joining the points thus obtained will be a true map of the earth's orbit, though without any scale of miles. When the operation is performed, we find that the orbit is an ellipse of small eccentricity, with the sun in one of the two foci.

429. The Ellipse, and Definitions relating to it (supplementary to Arts. 119, 120). — If we drive two pins into a board, as at F and S in Fig. 96, and put a looped thread around the

---

[1] *I.e.*, lay off S_1, S_2, etc., each equal to $\dfrac{10000''}{\text{diameter}}$.

pins, attached to the point of a pencil, *P*, then on carrying the pencil around it will mark out an ellipse. The pins, *F* and *S*, are the "foci" of the ellipse, and *C* is its centre. From the manner in which the ellipse is constructed, it is clear that at any point, *P*, on its outline, the sum of the two lines, *PS* and *PF*, will always be the same, and equal to the line *AA'*. The length of the ellipse, *AA'*, is called

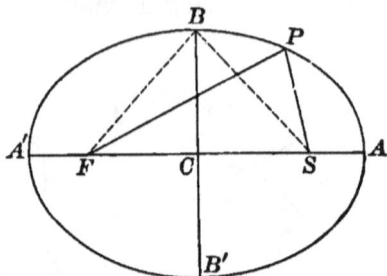

FIG. 96.—The Ellipse.

its *major axis*, and *AC* its *semi*-major axis, which is usually designated by *a*, while the semi-minor axis, *BC*, is lettered *b*. The fraction, $\dfrac{CS}{AC}$, is called the eccentricity of the ellipse, and determines the shape of the oval. Its usual symbol is *e*. If *e* is nearly unity, — *i.e.*, if *CS* is nearly equal to *CA*, — the oval will be very narrow compared with its length; but if *CS* is very small compared with *CA*, the ellipse will be almost round. Taken together, *a* and *e* determine the size and form of the oval. The ellipse is called a 'conic,' because when a cone is cut across obliquely the section is elliptical (see Art. 440).

430. Problems illustrating the 'Harmonic Law' (supplementary to Art. 220). — To aid the student in apprehending the meaning and scope of Kepler's third law, we give a few simple examples of its application.

1. What would be the period of a planet having a mean distance from the sun of one hundred astronomical units; *i.e.*, a distance a hundred times that of the earth?

$$1^8 : 100^8 = 1^2(\text{year}) : X^2;$$

whence, X (in years) $= \sqrt{100^8} = 1000$ years.

2. What would be the distance from the sun of a planet having a period of 125 years?

$$1^2 \,(\text{year}) : 125^2 = 1^8 : X^8; \text{ whence } X = \sqrt[3]{125^2} = 25 \text{ astron. units.}$$

3. What would be the period of a satellite revolving close to the earth's surface ?

$(\text{Moon's Dist.})^3 : (\text{Dist. of Satellite})^3 = (27.3 \text{ days})^2 : X^2,$

$\text{or, } 60^3 : 1^3 = 27.3^2 : X^2 ;$

$\text{whence, } X = \dfrac{27.3}{\sqrt{60^3}} \text{ days} = 0^d.587 = 1^h \, 24.5^m.$

4. How much would an increase of 10 per cent in the earth's distance from the sun lengthen the year?

$100^3 : 110^3 = (365\tfrac{1}{4})^2 : X^2, \text{ whence } X = \sqrt{\dfrac{110^3 \times 365\tfrac{1}{4}^2}{100^3}},$

X being the new length of the year. X is found by logarithmic computation to be 421.38 days. The increase is 56.13 days.

5. What is the distance from the sun of an asteroid with a period of $3\tfrac{1}{2}$ years ?

$1^2 (\text{year}) : 3.5^2 = 1^3 : \text{Dist.}^3$

$\therefore \text{Dist.} = \sqrt[3]{(3.5)^2} = \sqrt[3]{12.25} = 2.305 \text{ astron. units.}$

431. The Equation of Light. — When we observe a celestial body, we see it not as it *is* at the moment of observation, but as it *was* at the moment when the light which we see left it. If we know its distance in astronomical units, and know how long light takes to traverse that unit, we can at once correct our observation by simply *dating it back* to the time when the light started from the object. The necessary correction is called the "*equation of light*," and *the time required by light to traverse the astronomical unit of distance is called the "Constant of the Light-equation"* (not quite 500 seconds, as we shall see).

It was in 1675 that Roemer, the Danish astronomer (the inventor of the transit instrument, meridian-circle, and prime-vertical instrument, — a man almost a century in advance of his day), found that the eclipses of Jupiter's satellites show a peculiar variation in their times of occurrence, which he explained as due to the *time taken by light to pass through space* His bold and original suggestion was neglected for more than fifty years, until long after his death, when Bradley's discovery of aberration (Art. 435) proved the correctness of his views.

432. Determination of the Constant of the Equation of Light. — Eclipses of the satellites of Jupiter recur at intervals which are really almost exactly equal (the perturbations being very slight), and the interval can easily be determined and the times tabulated. But if we thus predict the times of the eclipses during a whole synodic period of the planet, then, beginning at the time of opposition, it is found that as the planet recedes from the earth, the eclipses, *as observed*, fall constantly more and more behindhand, and by precisely the same amount for all four satellites. The difference between the predicted and observed time continues to increase until the planet is near conjunction, when the eclipses are about 16ᵐ 38ˢ later than the prediction. After the conjunction they quicken their pace, and make up the loss, so that when opposition is reached once more they are again on time.

It is easy to see from Fig. 97 that at opposition

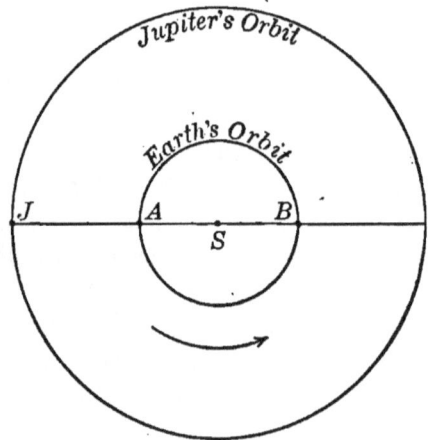

FIG. 97. — The Equation of Light.

the planet is nearer the earth than at conjunction by just two astronomical units. At opposition the distance between Jupiter and the earth is *JA*, while six and a half months later, at the time of Jupiter's superior conjunction, it is *JB*. The difference between *JA* and *JB* is just twice the distance from *S* to *A*.

The whole apparent retardation of eclipses between opposition and conjunction must therefore be exactly *twice the time*[1] *required for light to come from the sun to the earth.* In this way the "light-equation constant" is found to be very nearly 499 seconds, or 8 minutes 19 seconds, with a probable error of perhaps two seconds.

433. Since these eclipses are *gradual* phenomena, the determination of the exact moment of a satellite's disappearance or reappearance is very difficult, and this renders the result somewhat uncertain. Prof. E. C. Pickering of Cambridge has proposed to utilize *photometric* observations for the purpose of making the determination more precise, and two series of observations of this sort, and for this purpose, are now in progress, — one in Cambridge, United States, and the other at Paris under the direction of Cornu, who has devised a similar plan. Pickering has also applied *photography* to the observation of these eclipses with encouraging success.

434. The Distance of the Sun determined by the "Light-equation." — Until 1849 our only knowledge of the *velocity of light* was obtained from such observations of Jupiter's satellites. By assuming as known *the earth's distance from the sun,* the velocity of light can be obtained when we know the *time* occupied by light in coming from the sun.

At present, however, the case is reversed. We can determine the velocity of light by two independent *experimental* methods, and with a surprising degree of accuracy. Then, knowing this velocity and the "light-equation constant," we can deduce the distance of the sun. According to the latest determinations the velocity of light is 186,330 miles per second. Multiplying this by 499 we get 92,979,000 miles for the sun's distance (compare Art. 436).

---

[1] The student's attention is specially directed to the point that the observations of the eclipses of Jupiter's satellites give *directly* neither the velocity of light nor the distance of the sun : they give only the *time* required by light to make the journey from the sun. Many elementary text books, especially the older ones, state the case carelessly.

435. Aberration of Light. — The fact that light is not transmitted instantaneously causes the apparent displacement of
an object viewed from any moving station, unless the motion
is directly towards or from that object. If the motion of
the observer is not rapid, this displacement, or "aberration,"
is insensible; but the earth moves so swiftly (18½ miles
per second) that it is easily observable in the case of the
stars. Astronomical aberration may be defined, therefore, as
*the apparent displacement of a heavenly body due to the combina-
tion of the orbital motion of the earth with that of light* — the
direction in which we have to point our telescope in observing
a star is not the same as if the earth were at rest.

We may illustrate this by considering what would happen in the
case of falling rain-drops. Suppose the observer standing with a tube
in his hand while the drops
are falling straight down: if
he wishes to have the drops
descend through the middle of
the tube without touching the
sides, he must keep it vertical
so long as he stands still; but
if he advances in any direction
the drops will strike the side of
the tube, and he must thrust
forward its upper end (Fig. 98)
by an amount which equals
the advance he makes while a
drop is falling through it; *i.e.*,

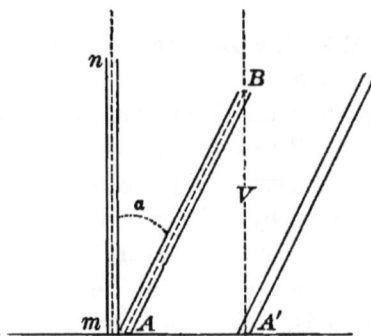

FIG. 98. — Aberration.

he must incline the tube *forward* at an angle, depending both upon
the velocity of the rain-drop and the swiftness of his own motion,
so that when the drop, which entered the tube at B, reaches A', the
bottom of the tube will be there also.

It is true that this illustration is not a *demonstration*, because light
does not consist of *particles* coming towards us, but of *waves* transmitted through the ether of space. But it has been shown (though
the proof is by no means elementary) that within very narrow limits,
the apparent direction of a *wave* is affected in precisely the same way
as that of a moving projectile.

The best observations show that a star situated on a line at right angles to the direction of the earth's motion, is thus apparently displaced by an angle of about 20″.5. The Pulkowa observations give 20″.493, while, according to Newcomb, the mean of all other determinations is 20″.463. This is the so-called "CONSTANT OF ABERRATION."

If the star is in a different part of the sky, its displacement will be less, the amount being easily calculated when the star's position is given.

436. Determination of the Sun's Distance by Means of the Aberration of Light. — The constant of aberration, a, and the two velocities, that of the earth in its orbit, u, and the velocity of light, V, are connected by the very simple equation

$$a = 206265 \times \frac{u}{V}; \text{ whence } u = \frac{a}{206265} \times V.$$

When, therefore, we have ascertained the value of a (20″.492) from observations of the stars, and of V (186,330 miles, according to the most recent determinations by Michelson and Newcomb) by physical experiments, we can immediately find u, the velocity of the earth in her orbit. The *circumference* of the earth's orbit is then found by multiplying this velocity, u, by the number of seconds in a *sidereal* year (Art. 127); and from this we get the *radius* of the orbit, or *the earth's mean distance from the sun*, by dividing the circumference by 2π ($\pi =$ 3.14159). Taking $a = 20″.478$, the mean distance of the sun comes out **92,913000** miles.

But the uncertainty of a is probably as much as 0″.03, and this affects the distance proportionally, say one part in 600, or 150,000 miles. Still, the method is one of the very best of all that we possess for determining in miles the value of "the Astronomical Unit."

437. De l'Isle's Method of determining the Sun's Parallax by a Transit of Venus. — We have thus (Arts. 434 and 436) two methods by which the mean distance of the sun from the earth can be determined. They both depend upon a knowledge of the velocity of light, and of course were unavailable before 1849, when Fizeau first succeeded in actually measuring it. Before that time it was necessary to rely entirely upon observations of either Mars or Venus, made at times when they come specially near us.

Most of the methods of getting the sun's parallax and distance from such observations depend upon our having a previous knowledge of the relative distances of the planets from the sun. These *relative* distances were ascertained centuries ago. Copernicus knew them nearly as accurately as we have them now; but since we have not explained in this book how

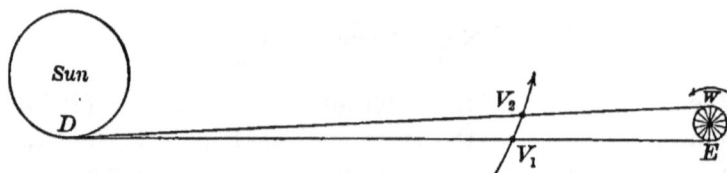

FIG. 99. — Transit of Venus.

they are found (the explanation involves a little Trigonometry), we limit ourselves to giving here a single very simple method, which requires a previous knowledge not of the relative distances of Venus and the earth from the sun, but only of the *synodic period* of the planet (Art. 228); *i.e.*, the time in which she gains one entire revolution upon the earth. This is 584 days, as has been known from remote antiquity.

Fig. 99 represents things at a transit of Venus, as they would be seen by one looking down from an infinitely distant point above the earth's north pole. As seen from the earth itself, Venus would appear to cross the sun, striking the disc on the east side and moving straight across to the west, making four 'contacts' with the edge of the sun as shown in Fig. 100.

438. Suppose, now, that two observers, E and W (Fig. 99), are stationed opposite each other, and near the earth's equator.

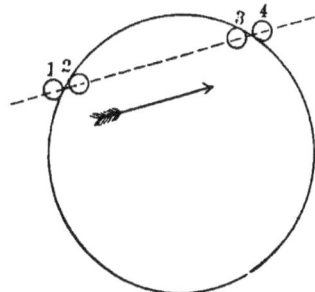

E will see Venus strike the sun's disc before W does, and if they both observe the moment of contact, in *Greenwich time*, the difference between their records will be the time it takes Venus to move over the arc from V_1 to V_2. From the figure it is clear that the angle, V_1DV_2, is the same as EDW, *the earth's apparent diameter seen from the sun*, and this is at once known when we have the time from V_1 to V_2.

FIG. 100.

Contacts in a Transit of Venus.

Since Venus gains one revolution in 584 days, in one day she will gain $\frac{1}{584}$ of a revolution, or 37' (very nearly), and this will make her gain $1''.54$ in one minute. Now it is found that the difference between the moments of contact at two stations situated like E and W is about $11^m 25^s$, and hence that the diameter of the earth as seen from the sun is $17''.6$, *or the sun's horizontal parallax* (Art. 139) is $8''.8$; from which its distance is easily found (Art. 140).

The reader will see that the two observers must know their *longitudes* accurately, in order to be sure of the correct Greenwich time. Moreover, the two stations can never be quite exactly opposite each other, but stations a little nearer together must be taken and proper allowances made. Finally, we are very sorry to add that the necessary observations of the moment when Venus reaches the edge of the sun's disc cannot be made with the accuracy which is desirable, owing to the effect of the planet's atmosphere (see Art. 248); so that practically the method is less accurate than might be hoped. For further details, see General Astronomy, Chapter XVI.

439. The Parabola (supplementary to Arts. 292-298). — This differs from the ellipse in never coming around into itself.

In Fig. 101, the curves PA_1, PA_2 and PA_3, are ellipses of different length, all having S at one of their foci. The first and smallest of the ellipses is nearly circular, and shaped about like the orbit of Mercury; the next, more eccentric than the orbit of any asteroid; and the third still more so. Now if we

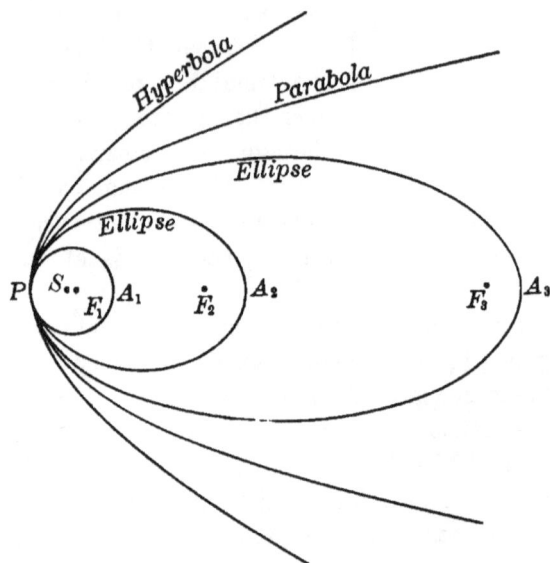

FIG. 101. — Ellipse, Parabola, and Hyperbola.

imagine the point F carried farther and farther to the right, the ellipse will grow larger and longer, until when F is infinitely far away the curve will become a *parabola*.

Of course if the point F is *very* distant, even if not *infinitely* so, the part of the curve near S will agree with the parabola so closely that no one could distinguish between them.

All *ellipses* that have S for the focus and P for the perihelion lie inside of the parabola, while another set of conic curves called *hyperbolas*, with the same focus and perihelion, lie entirely outside of it, which is, so to speak, a sort of boundary or division line between the ellipses and hyperbolas which have this focus and perihelion.

440. The Conic Sections. — The way in which these curves,
— the ellipse, parabola, and hyperbola — are formed by sections of the cone is shown by Fig. 102.

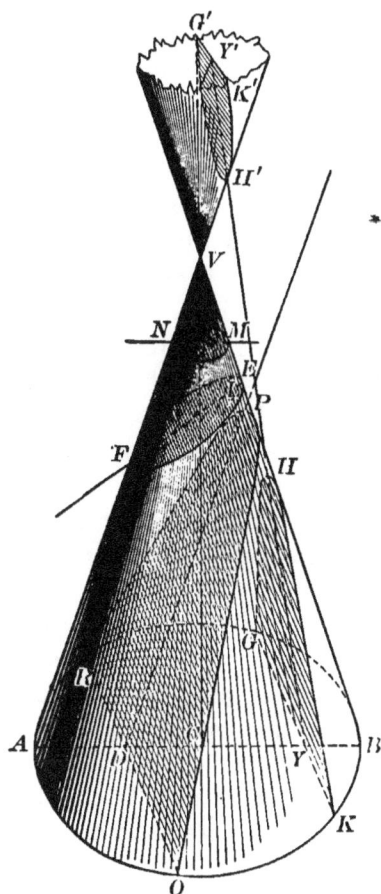

(*a*) If the cone be cut by a
plane which makes with its
axis, *VC*, an angle *greater* than
BVC, the plane of the section
will cut completely across the
cone, and the section *EF* will
be an *ellipse*, which will vary
in shape and size according to
the position of the plane. The
circle is simply a special case
when the cutting plane is perpendicular to the axis, as *NM*.

(*b*) When the cutting plane
makes with the axis an angle
less than *BVC* (the semi-angle
of the cone), it plunges continually deeper and deeper into
the cone and never comes out
on the other side (the cone is
supposed to be indefinitely
prolonged). The section in
this case is an *hyperbola, GHK*.
If the plane of the section be
produced *upward*, however, it
encounters the "cone pro-
duced," cutting out from it
a second hyperbola, *G'H'K'*,
precisely like the original one,

Fig. 102. — The Conics.

but turned in the opposite direction.

The axis of the hyperbola is always reckoned as negative,
lying outside of the curve itself: in the figure, it is the line
HH'. The centre of the hyperbola is the middle point of this
axis, a point also outside of the curve.

(c) When the angle made by the cutting plane with the axis is exactly equal to the cone's semi-angle, the plane will be parallel to the side of the cone, and we then get the special case of the *parabola*, *RPO*, which forms a partition, so to speak, between the infinite variety of ellipses and hyperbolas which can be cut from a given cone. *All parabolas are of the same shape*, just as all circles are, differing only in size. The fact is by no means self-evident, and we cannot stop to prove it, but it is true.

441. Determination of the Parallax of a Star (supplementary to Art. 343). — The determination of the parallax of stars had been attempted over and over again from the time of Tycho Brahe down, but without success until, in 1838, Bessel at last demonstrated and measured the parallax of 61 Cygni; and the next year Henderson, of the Cape of Good Hope, determined that of Alpha Centauri. The operation of measuring the parallax of a star is on the whole the most delicate in the whole range of practical Astronomy. Two methods have been used so far, known as *the absolute and the differential*.

442. The Absolute Method consists in making the most scrupulously precise observations of the star's right ascension and declination with the meridian circle at different times through the course of an entire year, applying rigidly all known corrections (for precession, aberration, proper motion, etc.), and then examining the deduced positions. If the star is without parallax, these positions will all agree. If the star has a sensible parallax, they will show, on the other hand, when plotted on a chart, *an apparent annual orbital motion of the star in a little ellipse*, the major axis of which is twice the star's annual parallax, as can easily be shown.

Theoretically, the method is perfect; practically, it seldom gives satisfactory results, because the annual changes of temperature and moisture disturb the instrument in such a way that the instrumental errors intertwine themselves with the parallax of a star in a manner that defies disentanglement. No process of multiplying observations and taking averages helps the matter very much, because the instrumental errors are themselves periodic annually, just as is the parallax; still, in a few cases the method has proved successful, as in the case of Alpha Centauri, above cited.

443. The Differential Method. — This, the method which has principally proved successful thus far, consists in measuring the annual displacement of the star whose parallax we are seeking, with respect to other small stars near it in apparent position (*i.e.*, within a few minutes of arc), but presumably so far beyond as to have no sensible parallax of their own.

If, for instance, the observer notes the apparent place of an object at no great distance from him with reference to the trees on a distant hill-side, and then moves a few feet one way or the other, he will see that the nearer object shifts its position with reference to the trees. In the same way, on account of the earth's orbital motion, those stars which are very near the earth appear every year to shift slightly backwards and forwards with respect to those that are far beyond them; and by measuring the amount of this shift it is possible to deduce approximately the parallax and distance of the nearer stars.

We say *approximately*, because the shift thus measured is not really the whole parallax of the nearer star, but only the *difference* between that parallax and the parallax of the remote objects with which it is compared; so that observations, if accurately made, will always give us for the nearer star a parallax *too small*, if anything, — never too large; and, as a consequence, the distance of the nearer star determined in this way will come out a little too large, and never too small.

444. The necessary measurements, if the comparison stars are within a minute or two of arc, may be made with the wire

micrometer (Art. 415); but if the distance exceeds a few minutes, we must resort to the "heliometer" (see General Astronomy, Art. 677), with which Bessel first succeeded; or we may
employ photography, which Professor Pritchard at Oxford
has recently been doing with remarkable success.

On the whole, the differential method, notwithstanding the
fundamental objection to it, that it never gives us the entire
parallax of the star, is at present more trustworthy than the
other.

It is obviously necessary to choose for observation by either
method those stars that are presumably near us. The most
important indication of nearness in a star is a large proper
motion; brightness, also, is of course confirmatory. Still,
neither of these indications is certain. A star which happens
to be moving directly towards or from us shows no proper
motion at all, however near it may be; and the faint stars are
so very much more numerous than the brighter ones that
among their millions it is quite likely that we shall ultimately
find individuals which are even nearer than Alpha Centauri.

445. Spectroscopic Method. — In time it will be possible to
determine the distance of certain *binary* stars by the help of
the spectroscope. The velocity of one or both of the two
stars "in the line of sight" can be measured by the spectroscope at different parts of the star's orbit, and this will enable
us to compute the size of the orbit in miles; at the same time
the micrometer measures will give its angular dimensions,
and from these data the distance can be found. It will
probably be many years, however, before any results can be
obtained in this way, because the periods of most of the
binaries are very long.

SUGGESTIVE QUESTIONS

FOR USE IN REVIEWS.

To many of these questions *direct* answers will not be found in the book; but the principles upon which the answers depend have been given, and the student will have to use his own thinking in order to make the proper application.

1. What point in the celestial sphere has both its right ascension and declination zero?

2. What angle does the (celestial) equator make with the horizon at this place?

3. Name the (fourteen) principal points in the celestial sphere (zenith, etc.).

4. What important circles in the heavens have no correlatives on the surface of the earth?

5. What constellation of the zodiac rises at sunset to-day, and which one is then on the meridian? (Use the star-maps.)

6. If Vega comes to the meridian at 8 o'clock to-night, at what time (approximately) will it transit eight days hence?

7. What bright star can I observe on the meridian between 3 and 4 P.M., in the middle of August? (See star-maps.)

8. At what time of the year will Sirius be on the meridian at midnight?

9. The declination of Vega is 38° 41'; does it pass the meridian north of your zenith, or south of it?

10. What are the right ascension and declination of the north pole of the ecliptic?

11. What are the longitude and latitude (celestial) of the north celestial pole (the one near the Pole-star)?

12. Can the sun ever be directly overhead where you live? If not, why not?

13. What is the zenith distance of the sun at noon on June 22d in New York City (lat. 40° 42')?

14. What are the greatest and least angles made by the ecliptic with the horizon at New York? Why does the angle vary?

15. If the obliquity of the ecliptic were 30°, how wide would the temperate zone be? How wide if the obliquity were 50°? What must the obliquity be to make the two temperate zones each as wide as the torrid zone?

16. Does the equinox always occur on the same days of March and September? If not, why not; and how much can the date vary?

17. Was the sun's declination at noon on March 10th, 1887, precisely the same as on the same date in 1889?

18. In what season of the year is New Year's Day in Chili?

19. When the sun is in the constellation Taurus, in what *sign* of the zodiac is he?

20. In what constellation is the sun when he is vertically over the tropic of Cancer? Near what star? .(See star-map.)

21. When are day and night most unequal?

22. In what part of the earth are the days longest on March 20th? On June 20th? On Dec. 20th?

23. Why is it warmest in the United States when the earth is farthest from the sun?

24. What will be the Russian date corresponding to Feb. 28th, 1900, of our calendar? To May 28th?

25. Why are the intervals from sunrise to noon and from noon to sunset usually unequal as given in the almanac? (For example, see Feb. 20th and Nov. 20th.)

26. If the earth were to shrink to half its present diameter, what would be its mean density?

27. Is it absolutely necessary, as often stated, to know the diameter of the earth in order to find the distance of the sun from the earth?

28. How will a projectile fired horizontally on the earth deviate from the line it would follow if the earth did not rotate on its axis?

29. If the earth were to contract in diameter, how would the weight of bodies on its surface be affected?

30. What keeps up the speed of the earth in its motion around the sun?

31. Why is the sidereal month shorter than the synodic?

32. Does the moon rise every day of the month?

33. If the moon rises at 11.45 Tuesday night, when will it rise next?

34. How many times does the moon turn on its axis in a year?

35. What determines the direction of the horns of the moon?

36. Does the earth rise and set for an observer on the moon? If so, at what intervals?

37. How do we know that the moon is not self-luminous?

38. How do we know that there is no water on the moon?

39. How much information does the spectroscope give us about the moon?

40. What conditions must concur to produce a lunar eclipse?

41. Can an eclipse of the moon occur in the daytime?

42. Why can there not be an annular eclipse of the moon?

43. Which are most frequent at New York, solar eclipses or lunar?

44. Can an occultation of Venus by the moon occur during a lunar eclipse? Would an occultation of Jupiter be possible under the same circumstances?

45. Which of the heavenly bodies are not self-luminous?

46. When is a planet an evening star?

47. What planets have synodic periods longer than their sidereal periods?

48. When a planet is at its least distance from the earth, what is its apparent motion in right ascension?

49. A planet is seen 120° distant from the sun; is it an inferior or a superior planet?

50. Can there be a transit of Mars across the sun's disc?

51. When Jupiter is visible in the evening, do the shadows of the satellites precede or follow the satellites themselves as they cross the planet's disc?

52. What would be the length of the month if the moon were four times as far away as now? (Apply Kepler's third law.)

53. What is the distance from the sun of an asteroid which has a period of eight years? (Kepler's third law.)

54. Upon what circumstances does the apparent length of a comet's tail depend?

55. How can the distance of a meteor from the observer, and its height above the earth, be determined?

56. What heavenly bodies are not included in the solar system?

57. How do we know that stars are suns?   How much is meant by the assertion that they are?

58. Suppose that in attempting to measure the parallax of a bright star by the differential method (Art. 443) it should turn out that the small star taken as the point to measure from, and supposed to be far beyond the bright one, should really prove to be nearer. How would the measures show the fact?

59. If Alpha Centauri were to travel straight towards the sun with a uniform velocity equal to that of the earth in its orbit, how long would the journey take, on the assumption that the star's parallax is $0''.75$?

60. If Altair were ten times as distant from us, what would be its apparent "magnitude"?   What, if it were a thousand times as remote? (See Arts. 346, 347 ; and remember that the apparent brightness varies inversely with the square of the distance.)

TABLES OF ASTRONOMICAL DATA.

TABLE I. — ASTRONOMICAL CONSTANTS.

TIME CONSTANTS.

The sidereal day $= 23^h\ 56^m\ 4^s.090$ of mean solar time.
The mean solar day $= 24^h\ 3^m\ 56^s.556$ of sidereal time.

To reduce a time interval expressed in units of *mean solar time* to *units of sidereal time*, multiply by 1.00273791; Log. of 0.00273791 = [7.4374191].

To reduce a time interval expressed in units of *sidereal time* to units of *mean solar time*, multiply by 0.99726957 = $(1 - 0.00273043)$; Log. 0.00273043 = [7.4362316].

Tropical year (Leverrier, reduced to 1900), $365^d\ 5^h\ 48^m\ 45^s.51.$
Sidereal year   "   "   "   365   6   9   8.97.
Anomalistic year   "   "   "   365   6 13   48.09.

Mean synodical month (new moon to new), $29^d\ 12^h\ 44^m\ 2^s.684.$
Sidereal month, ... 27   7   43   11.545.
Tropical month (equinox to equinox),   . 27   7   43   4.68.
Anomalistic month (perigee to perigee),. 27 13 18   37.44.
Nodical month (node to node),   .   . 27   5   5   35.81.

Obliquity of the ecliptic (Leverrier),
        $23°\ 27'\ 08''.0 - 0''.4757\ (t - 1900).$
Constant of precession (Struve),
        $50''.264 + 0''.000227\ (t - 1900).$
Constant of nutation (Peters), 9''.223.
Constant of aberration (Nyrén), 20''.492.

Equatorial semi-diameter of the earth (Clarke's spheroid of 1880), — 20 926 202 ^{feet} = 6 378 190 ^{metres} = 3963.296 ^{miles}.

Polar semi-diameter, —
        20 854 895 ^{feet} = 6 356 456 ^{metres} = 3949.790 ^{miles}.

Ellipticity, or Polar Compression, $\frac{1}{293.46}$

TABLE II.—PRINCIPAL ELEMENTS OF THE SOLAR SYSTEM (1850).

	Symbol	Name	Semi-Major Axis of Orbit.	Mean Dist. Millions of Miles.	Sidereal Period (mean solar days).	Period in Years.	Eccentricity.	Inclination to Ecliptic.	Longitude of Ascending Node.	Longitude of Perihelion.
Terrestrial Planets.	☿	Mercury	0.387099	36.0	87.96926	0.24	.20560	7° 00' 8"	46° 33' 9"	75° 7' 14"
	♀	Venus	0.723332	67.2	224.7008	0.62	.00684	3 23 35	75 19 52	129 27 15
	⊕	The Earth	1.000000	92.9	365.2564	1.00	.01677	0 00 00	0 00 00	100 21 22
	♂	Mars	1.523691	141.5	686.9505	1.88	.09326	1 51 2	48 23 53	333 17 54
	①	Ceres	2.767265	257.1	1681.414	4.60	.07631	10 37 10	80 46 39	149 37 49
Major Planets.	♃	Jupiter	5.202800	483.3	4332.580	11.86	.04825	1 18 41	98 56 17	11 54 58
	♄	Saturn	9.538861	886.0	10759.22	29.46	.05607	2 29 40	112 20 53	90 6 38
	♅	Uranus	19.18329	1781.9	30686.82	84.02	.04634	0 46 20	73 13 54	170 50 7
	♆	Neptune	30.05508	2791.6	60181.11	164.78	.00896	1 47 2	130 6 25	45 59 43

	Symbol	Name	Apparent Angular Diameter.	Mean Diameter in Miles.	Mass. ☉=1.	Volume. ⊕=1.	Density. ⊕=1.	Time of Axial Rotation.	Inclination Equator to Orbit.	Oblateness, or Ellipticity.	Gravity at Surface.
	☉	Sun	32' 04" mean	866,400	1.00000	1 310.000	0.25	25ᵈ 7ʰ 48ᵐ	7° 15'	?	27.65
	☾	Moon	31' 07" "	2,163	$\frac{1}{88181782}$	0.020	0.61	27 7 43	6 33 ±	?	☾
Terrestrial Planets.	☿	Mercury	5" to 13"	3,000	$\frac{1}{4865600}$?	0.055	0.67 ?	88ᵈ	?	?	0.25 ?
	♀	Venus	11" to 67"	7,700	$\frac{1}{411008}$	0.920	0.86	?	?	?	0.83
	⊕	The Earth		7,918	$\frac{1}{354936}$	1.000	1.00	23 56 4.09	23° 27' 12"	$\frac{1}{303}$	1.00
	♂	Mars	3".6 to 25"	4300	$\frac{1}{3501806}$	0.152	0.72	24 37 22.67	24 50	$\frac{1}{257}$	0.38

TABLE III.—THE SATELLITES OF THE SOLAR SYSTEM.

Primary	Name	Discovery	Distance in Miles	Sidereal Period				Inclination to Ecliptic			Diameter in Miles
				27^d	7^h	43^m	11^s.5	5°	08'	40"	
⊕	The Moon	—	238,840	27	7	43	11.5	5°	08'	40"	2163
♂	1 Phobos	Hall, 1877	5,850		7	39	15.1	26	17'.2		7 ?
	2 Deimos	"	14,650	1	6	17	54.0	25	47'.2		5 ?
♃	1 Io	Galileo, 1610	261,000	1	18	27	33.5	2	08'	3"	2500
	2 Europa	"	415,000	3	13	13	42.1	1	38	57	2100
	3 Ganymede	"	664,000	7	3	42	33.4	1	59	53	3550
	4 Callisto	"	1167,000	16	16	32	11.2	1	57	00	2960
	5	Barnard, 1892	112,500		11	57	22.6		?		100 ?
♄	1 Mimas	W. Herschel, 1789	117,000		22	37	5.7	28	10		600 ?
	2 Enceladus	"	150,000	1	8	53	6.9	28	10		800 ?
	3 Tethys	J. D. Cassini, 1684	186,000	1	21	18	25.6	28	10		1290 ?
	4 Dione	"	238,000	2	17	41	9.3	28	10		1100 ?
	5 Rhea	" 1672	332,000	4	12	25	11.6	28	10		1500 ?
	6 Titan	Huyghens, 1655	771,000	15	22	41	23.2	27	38	49	3500 ?
	7 Hyperion	Bond, 1848	934,000	21	6	39	27.0	27	4	48	500 ?
	8 Iapetus	J. D. Cassini, 1671	2225,000	79	7	54	17.1	78	31	30	2000 ?
⛢	1 Ariel	Lassell, 1851	120,000	2	12	29	21.1	97	51 =		500 ?
	2 Umbriel	"	167,000	4	3	27	37.2	− 82	09		400 ?
	3 Titania	W. Herschel, 1787	273,000	8	16	56	29.5	—	—		1000 ?
	4 Oberon	"	365,000	13	11	7	6.4	—	—		800 ?
♆	Nameless	Lassell, 1846	225,000	5	21	2	44.2	145	12		2000 ?

TABLE IV.—THE PRINCIPAL VARIABLE STARS.

A selection from S. C. Chandler's catalogue of variable stars, containing such as, at the maximum, are easily visible to the naked eye, have a range of variation exceeding half a magnitude, and can be seen in the United States.

No.	Name.	Place, 1900. α	Place, 1900. δ	Range of Variation.		Period (days).	Remarks.
		h m					
1	R Andromedæ	0 18.8	+38° 1'	5.6 to 13		411.2	Mira. Varia-
2	o Ceti. . . .	2 14.3	− 3 26	1.7	9.5	331.3363	tions in length
3	ρ Persei . . .	2 58.7	+38 27	3.4	4.2	33	of period.
4	β Persei . . .	3 1.6	+40 34	2.3	3.5	2ᵈ 20ʰ 48ᵐ 55ˢ.43	Algol. Period now shortening.
5	λ Tauri . . .	3 55.1	+12 12	3.4	4.2	3ᵈ 22ʰ 52ᵐ 12ˢ	Algol type, but
6	ε Aurigæ . .	4 54.8	+43 41	3	4.5	Irregular	irregular.
7	α Orionis . .	5 49.7	+ 7 23	1	1.6	196 ?	Irregular.
8	η Geminorum .	6 8.8	+22 32	3.2	4.2	229.1	
9	ζ Geminorum .	6 58.2	+20 43	3.7	4.5	10ᵈ 3ʰ 41ᵐ 30ˢ	
10	R Canis Maj. .	7 14.9	−16 12	5.9	6.7	1ᵈ 3ʰ 15ᵐ 55ˢ	Algol type.
11	R Leonis . .	9 42.2	+11 54	5.2	10	312.87	
12	U Hydræ . .	10 32.6	−12 52	4.5	6.3	194.65	
13	R Hydræ . .	13 24.2	−22 46	3.5	9.7	496.91	Period short'ing
14	δ Libræ . . .	14 55.6	− 8 7	5.0	6.2	2ᵈ 7ʰ 51ᵐ 22ˢ.8	Algol type.
15	R Coronæ . .	15 44.4	+28 28	5.8	13	Irregular	
16	R Serpentis .	15 46.1	+15 26	5.6	13	357.6	
17	α Herculis . .	17 10.1	+14 30	3.1	3.9	Two or three months, but very irreg.	
18	U Ophiuchi .	17 11.5	+ 1 19	6.0	6.7	20ʰ 7ᵐ 41ˢ.6	
19	X Sagittarii .	17 41.3	−27 48	4	6	7.01185	
20	W Sagittarii .	17 58.6	−29 35	5	6.5	7.59445	
21	R Scuti . . .	18 42.1	− 5 49	4.7	9	71.10	Secondary mini-
22	β Lyræ . . .	18 46.4	+33 15	3.4	4.5	12ᵈ 21ʰ 46ᵐ 58ˢ.3	mum about mid-
23	χ Cygni . . .	19 46.7	+32 40	4.0	13.5	406.045	way. Period length'ng
24	η Aquilæ . .	19 47.4	+ 0 45	3.5	4.7	7ᵈ 4ʰ 14ᵐ 0ˢ.0	
25	S Sagittæ . .	19 51.4	+16 22	5.6	6.4	8ᵈ 9ʰ 11ᵐ	
26	T Vulpeculæ .	20 47.2	+27 52	5.5	6.5	4ᵈ 10ʰ 20ᵐ	
27	T Cephei . .	21 8.2	+68 5	5.6	9.9	383.20	
28	μ Cephei . .	21 40.4	+58 19	4	5	432 ?	
29	δ Cephei . .	22 25.4	+57 54	3.7	4.9	5ᵈ 8ʰ 47ᵐ 39ˢ.97	
30	β Pegasi . . .	22 58.9	+27 32	2.2	2.7	Irregular	
31	R Cassiopeiæ .	23 53.3	+50 50	4.8	12	429.00	

TABLE V.—STELLAR PARALLAXES AND PROPER MOTIONS.

(From Oudeman's Table, Ast. Nach., Aug., 1889.)

No.	Name.	Mag.	Proper Motion.	Annual Parallax.	Distance Light Years.
1	α Centauri .	0.7	3″.67	0″.75	4
2	Ll. 21185 . .	6.9	4.75	0.50	6.5
3	61 Cygni . .	5.1	5.16	0.40	8
4	Sirius . . .	−1.4	1.31	0.39	8.3
5	Σ 2398 . . .	8.2	2.40	0.35	9.3
6	Ll. 9352 . .	7.5	6.96	0.28	12
7	Procyon . .	0.5	1.25	0.27	12.3
8	Ll. 21258 . .	8.5	4.40	0.26	12.5
9	Altair . . .	1.0	0.65	0.20	16.3
10	ε Indi . . .	5.2	4.60	0.20	16.3
11	o² Eridani .	4.5	4.05	0.19	17
12	Vega . . .	0.2	0.36	0.16	20
13	β Cassiopeiæ,	2.4	0.55	0.16	20
14	70 Ophiuchi .	4.1	1.13	0.15	21
15	e Eridani . .	4.4	3.03	0.14	23
16	Aldebaran .	1.0	0.19	0.12	27
17	Capella . .	0.2	0.43	0.11	29
18	Regulus . .	1.4	0.27	0.10	32
19	Polaris .	2.1	0.05	0.07	47

These are not all the stars upon Oudeman's list which are given as having parallaxes exceeding 0″.1; but they are probably the best determined ones. A parallax of 0″.45 is assigned to Eta Cassiopeiæ by a recent determination by Davis.

THE GREEK ALPHABET.

Letters.	Name.	Letters.	Name.	Letters.	Name.
A, α,	Alpha.	I, ι,	Iota.	P, ρ, ϱ,	Rho.
B, β,	Beta.	K, κ,	Kappa.	Σ, σ, ς,	Sigma.
Γ, γ,	Gamma.	Λ, λ,	Lambda.	T, τ,	Tau.
Δ, δ,	Delta.	M, μ,	Mu.	Y, υ,	Upsilon.
E, ε,	Epsilon.	N, ν,	Nu.	Φ, φ,	Phi.
Z, ζ,	Zeta.	Ξ, ξ,	Xi.	X, χ,	Chi.
H, η,	Eta.	O, ο,	Omicron.	Ψ, ψ,	Psi.
Θ, θ, ϑ,	Theta.	Π, π, ϖ,	Pi.	Ω, ω,	Omega.

MISCELLANEOUS SYMBOLS.

☌, Conjunction.

□, Quadrature.

☍, Opposition.

☊, Ascending Node.

☋, Descending Node.

A.R., or α, Right Ascension.

Decl., or δ, Declination.

λ, Longitude (Celestial).

β, Latitude (Celestial).

φ, Latitude (Terrestrial).

ω, Angle between line of nodes and line of apsides; also the obliquity of the ecliptic.

INDEX.

INDEX.

[All references, unless expressly stated to the contrary, are to *articles*, not to *pages*.]

A.

Aberration, of light, 435; determining distance of sun, 436.

Absolute scale of star magnitudes, 346.

Acceleration of rotation at the sun's equator, 163.

Achromatic telescope, 406, 407.

ADAMS, J. C. (and LEVERRIER), discovery of Neptune, 283; orbit of the Leonids, 327.

Aerolite. See **Meteorite.**

Age of the sun and planetary system, 193, 397-399.

Albedo defined, 149, 235; of the moon (Zöllner), 149; of the planets (Zöllner), 242, 247, 253, 268, 276, 281, 285.

Algol, or β Persei, 40, 351, 358, 360.

Alphabet, the Greek, page 344.

Altitude defined, 11; parallels of, 11; of the pole equals latitude, 80.

Andromeda, constellation of, 35; nebula of, 377, 378, 392, note; nebula of, temporary star in, 355.

Andromedes, or Bielids, 312, 326.

Angular measurements, units of, 8.

Annual or heliocentric parallax defined, 343; methods of determining it for the stars by observation, 441-444.

Annular eclipses, 201.

Anomalistic year, 127.

Anomalous phenomena in comets, 308.

Apex of the sun's way, 342.

Aphelion defined, 120.

Apogee defined, 137.

Apparent motion of a planet, 225-229; motion of the sun, 115-117; solar time, 88.

Apsides, line of, defined, 20, 137; of the moon's orbit, 137.

Aquarius, 78, 118.

Aquila, 71.

Arcs of meridian, measurement of, 105, 110.

Areas, equal, law of, 121, 137, 220.

Argo Navis, 51.

Ariel, a satellite of Uranus, 282.

Aries, first of, defined, 17; constellation of, 38, 118.

Asteroids, or minor planets, 260-263.

Astronomical constants, table of, page 339; day, beginning of, 90; symbols, page 344; unit, — see **Distance** of the sun.

Astronomy, utility of, 1.

Atmosphere of the moon, 148; of Mars, 253; of Mercury, 242; of Venus, 248.

Attraction of gravitation, its law, 220, 221.

Auriga, 41.

Axis of the earth, 13, 109; its permanence, 109.

Azimuth defined, 11.

B.

BAYER, his system of lettering the stars, 24.

Beginning of the century (Ceres discovered), 260; of the day, 90, 98.

BESSEL, dark stars, 350, 360; first measures stellar parallax, 441, 444.

355

For Supplementary Index see page 366.

[All references, unless expressly stated to the contrary, are to *articles*, not to *pages*.]

Bethlehem, the star of, 355.
BIELA's comet, 311, 312.
Bielids, or Andromedes, 312, 324, 328.
Binary stars, 368–371.
Bissextile year, 129.
BODE's law, 219.
BOND, W. C., discovery of the "gauze ring" of Saturn, 277; discovery of Hyperion, 280.
Boötes, 59.
BREDICHIN, his theory of comets' tails, 307.
Brightness of comets, 291; of meteors, 318; of stars, and causes of difference, 345–350.
BROOKS, his comets, 290, 299.

C.

CÆSAR, JULIUS, reformation of the calendar, 129.
Calendar, the, 128–130.
Calory, the, defined, 187.
Camelopardus, 31.
Canals of Mars, 256.
Cancer, 52, 118.
Canes Venatici, 58.
Canis Major, 49.
Canis Minor, 48.
Capricornus, 73, 118.
Capture theory of comets, 298.
Cardinal points defined, 16.
CARRINGTON, discovery of the peculiar law of the sun's rotation, 163.
CASSINI, J. D., discovers division in Saturn's ring, 277.
Cassiopeia, 28.
Catalogues of stars, 335.
Celestial globe described, 400, 401; sphere, infinite, 6.
Centaurus, 62.
Centrifugal force due to earth's rotation, 111.
Cepheus, 29.
Ceres, the first of the asteroids, 260.
Cetus, 39.
CHANDLER, S. C., identification of Lexell's comet, 299; his catalogue of variable stars, 361.

Changes, gradual, in the brightness of stars, 353; on the surface of the moon, 155.
Chemical constitution of the sun, 175, 176.
Chromosphere of the sun, 180, 194; and prominences made visible by the spectroscope, 182.
Chronograph, the, 417.
Chronometer, the, 417; longitude by, 96, 427.
Circle, meridian, the, 81, 99, 418.
Circles, hour, defined, 15.
Circumpolar stars, latitude by, 81.
Civil day and astronomical day, 90.
Classification of the planets, Humboldt, 217; of stellar spectra, Secchi, 363; of variable stars, 352.
Clock, the astronomical, 417; its rate and error, 92, 93, 417.
Clusters of stars, 376.
Columba, 45.
Colures defined, 117.
Coma Berenices, 57.
Comet, Biela's, 311, 312; Donati's, 289; Encke's, 293, 311; Lexell-Brooks, 299; Halley's, 293; of 1882, 313, 314.
Comets, anomalous phenomena shown by, 308; attendant companions, 314; brightness and visibility, 291; capture theory of their origin, 298; central stripe in tail, 308; connection with meteors, 327–329; constitution of, 300; danger from, 310; density of, 303; designation and nomenclature, 290; dimensions of, 301; elliptic, 293, 297; envelopes in head, 305; families of, 297; formation of the tail, 306; their light and spectra, 304; mass of, 302; nature of, 309; number of, 289; orbits of, 292, 293; periodic, their origin, 297, 298; sheath of comet of 1882, 314; tails or trains, 300, 306–308; visitors to the solar system, 296.
Comet-groups, 294.
Conic sections, the, 440.
Conjunction defined, 132, 227.

[All references, unless expressly stated to the contrary, are to *articles*, not to *pages*.]

Constant, solar, defined and discussed, 187.

Constellations, the, 4, 333. (For detailed description, see Chap. II.)

Constitution of comets, 300; of the sun, 194.

Contraction of a comet nearing the sun, 301; of the sun, Helmholtz's theory, 192, 396, 397.

COPERNICUS, rotation of the earth, 106; his system, 230.

Corona Borealis, 60.

Corona, the solar, 183-185.

Coronium, hypothetical element of the corona, 184.

Correction or error of a timepiece, 92, 427.

Corvus, 55.

Cosmogony, 389-396.

Crater, 55.

Cygnus, 68.

D.

Dark stars, 350, 360.

DARWIN, G. H., demonstrates that a meteoric swarm behaves like a gaseous nebula, 394.

Day, beginning of, 98; civil and astronomical, 90.

Declination defined, 14; determination of, 99, 100; parallels of, 14.

Degrees of latitude, length of, 110.

Deimos, a satellite of Mars, 258.

DE L'ISLE, his method of observing a transit of Venus, 437, 438.

Delphinus, 74.

Density of comets, 303; of the earth, 113; of the moon, 143; of the sun, 161.

Designation and nomenclature of comets, 290; and nomenclature of the stars, 24, 334; and nomenclature of variable stars, 361.

Diameter of a planet, how determined, 232.

Difference of brightness in stars, its causes, 350.

Diffraction, telescopic, 408.

Diffraction grating, the, 171, note.

Dione, a satellite of Saturn, 280.

Disc, spurious, of a star, 408.

Displacement of spectrum lines by motion in line of sight, 179, 341, 373.

Distance of a body as depending on its parallax, 140; of the moon, 141; of the nebulæ, 382; of the planets from the sun, Table II., page 340; of the stars, 343, 441-444; of the sun, by aberration of light, 436; of the sun, by the equation of light, 434; of the sun, by its parallax, 437.

Distribution of the nebulæ, 382; of the stars in the heavens, 384; of sun spots, 169.

Diurnal or geocentric parallax defined, 139; rotation of the heavens, 12.

DOPPLER's principle, 179.

Double stars, 366, 367; optical and physical, distinguished, 367.

Draco, 30.

DRAPER, H., photograph of the nebula of Orion, 378; photographs of star spectra, 364.

Duration of solar eclipses, 203; probable, of the solar system, 193, 397-399.

E.

Earth, the, astronomical facts relating to it, 102; its density, 113; dimensions of, 105, 110, Table I.; ellipticity or oblateness determined, 110; its interior constitution, 114; mass, 113; orbital motion of, 115-122, 428; its orbit, changes in, 122; its rotation, invariability of, 108; its rotation, proofs of, 107; shadow of, its dimensions, 196; surface area and volume, 112; velocity in its orbit, 158.

Earth-shine on the moon, 147.

Ebb defined, 210.

Eccentricity of the earth's orbit, 119; of an ellipse defined, 119, 429.

Eclipses, frequency of, 206; of Jupiter's satellites, 273; lunar, 197-199; Oppolzer's canon of, 205; number in a year, 206; recurrence of, 207; so-

358

[All references, unless expressly stated to the contrary, are to *articles*, not to *pages*.]

lar, duration of, 203; solar, phenomena of, 204; solar, varieties of, total, annular, and partial, 201, 202.
Ecliptic, the, defined, 116; obliquity of, 116; poles of, 117.
Elements, chemical, recognized in the stars, 362; chemical, recognized in the sun, 176; of the planets' orbits, Table II., page 340.
Ellipse, the, defined and described, 429, 439, 440.
Elliptic comets, 292, 293.
Ellipticity, or oblateness of the earth, 110.
Elongation defined, 132, 227.
Enceladus, a satellite of Saturn, 280.
Encke's comet, 293, 311.
Energy of the solar radiation, 188, 189.
Envelopes in the head of a comet, 305, 314.
Equation of light, 431–433; of time, 89.
Equator, celestial or equinoctial, defined, 14.
Equatorial acceleration of the sun's surface rotation, 163; instrument, the, 414; use in determining the place of a heavenly body, 100.
Equinoctial, the, or celestial equator, defined, 14.
Equinox, vernal, defined, 17, 116.
Equinoxes, precession of, 125, 126.
Equuleus, 75.
Eridanus, 44.
Error or correction of a timepiece, 92, 93, 417.
Eruptive prominences on the sun, 182.
Establishment of a port, 210.
Eye-pieces, telescopic, various forms, 409.

F.

Faculæ, solar, 165.
Families of comets, 297.
Faye, depth of sun spots, 168; modification of the nebular hypothesis, 393.
Filar micrometer, the, 415.
Flood tide, 210.

Form of the earth's orbit determined, 428.
Foucault, his pendulum experiment, 107.
Fraunhofer lines in the solar spectrum, 175, note.
Frequency of eclipses, 206.

G.

Galaxy, the, 383.
Galileo, his discovery of Jupiter's satellites, 272; discovery of phases of Venus, 247; discovery of Saturn's ring, 277; discovery of sun spots, 169; his telescope, 402.
Gemination of the canals of Mars, 256.
Gemini, 47, 118.
Genesis of the planetary system, 390, 391.
Geocentric parallax, 139.
Gibbous phase defined, 146.
Globe, the celestial, described, 400, 401.
Grating, diffraction, 171, note.
Gravitation, 221, 222.
Gravity, at the moon's surface, 143; at the pole and equator of the earth, 111; at the sun's surface, 161; superficial, of a planet, how determined, 233.
Greek alphabet, the, page 344.
Gregorian calendar, the, 130.
Groups, cometary, 294.
Grus, 79.
Gyroscope illustrating the cause of the seasons, 123.

H.

Habitability of Mars, 259.
Hall, A., discovery of the satellites of Mars, 258.
Halley discovers the proper motion of stars, 339; his periodic comet, 293.
Harmonic law, Kepler's, 220, 430.
Harvest and hunter's moons, the, 136.
Heat of meteors, its explanation, 318; from the moon, 150; from the stars, 348, note; of the sun, its constancy, 191; of the sun, its intensity, 190; of the sun, its maintenance, 192; of the sun, its quantity, 187, 189.

[All references, unless expressly stated to the contrary, are to *articles*, not to *pages*.]

Heavenly bodies defined and enumerated, 2; apparent place of, 7.
Heliocentric, or annual parallax, defined, 139, 343.
Helium, hypothetical element in the sun, 181.
HELMHOLTZ, his theory of the sun's heat, 192.
Hercules, 66.
HERSCHEL, SIR J., illustration of the solar system, 238; his names for the satellites of Saturn and Uranus, 280, 282.
HERSCHEL, SIR W., discovery of Uranus, 281; his great telescope, 412; relation between nebulæ and stars, 395.
HERSCHELS, the, their star-gauges, 384.
HIPPARCHUS, 120, 125, 335, 345.
Horizon defined, rational and visible, 10.
Horizontal parallax, 139.
Hour-angle defined, 422.
Hour-Circles defined, 15.
Hourly number of meteors, 321.
HUGGINS, W., observes spectrum of Mars, 253; observes spectrum of Mercury, 242; observes spectrum of nebulæ, 380; observes spectrum of stars, 362; observes spectrum of temporary star of 1866, 355; spectroscopic measures of star motions, 341.
HUMBOLDT, his classification of the planets, 217.
Hunter's moon, the, 136.
HUYGHENS, his discovery of Saturn's ring, 277; discovery of Titan, 280; invention of the pendulum clock, 417.
Hydra, 55.
Hyperbola, the, 439, 440.
Hyperion, a satellite of Saturn, 280.

I.

Iapetus, the remotest satellite of Saturn, 280.
Identification of the orbits of certain comets and meteors, 328.

Illuminating power of a telescope, 405.
Illumination of the moon's disc during a lunar eclipse, 198.
Illustration of the proportions of the solar system, 238.
Influence of the moon on the earth, 151; of sun spots on the earth, 170.
Intensity of the sun's heat, 189–190; of the sun's light, 186.
Intra-Mercurian planets, 264.
Invariability of the earth's rotation, 108; of the length of the year and distance from the sun, 122.
Iron in comets, 314; in meteorites, 316; in stars, 362; in the sun, 175.

J.

Julian calendar, the, 129.
Juno, the third asteroid, 260.
Jupiter (the planet), 266–271; his belts, red spot, and other markings, 268, 271; his rotation, 270; his satellites, and their eclipses, 272, 273.
Jupiter's family of comets, 297.

K.

KANT, a proposer of the nebular hypothesis, 391.
KEPLER, his laws of planetary motion, 121, 220, 430.
KIRCHHOFF, fundamental principles of spectrum analysis, 173.

L.

Lacerta, 76.
LANGLEY, S. P., his value of the solar constant, 188.
LAPLACE, his capture theory of comets, 298; his nebular hypothesis, 392, 393; stability of the solar system, 288*.
LASSELL, his discovery of Ariel and Umbriel, 282; his discovery of the satellite of Neptune, 286.
Latitude (celestial) defined, 20; (terrestrial) defined, 80; length of degrees, 110; methods of determining, 81, 424, 426; variations of, 109.
Law, Bode's, 219; of the earth's orbital motion, 121; of gravitation, 221, 222.

360 INDEX.

[All references, unless expressly stated to the contrary, are to *articles*, not to *pages*.]

Laws, Kepler's, 121, 220, 430.
Leap-year, 129, 130.
Leo, 53, 118.
Leo minor, 54.
Leonids, the, 324, 325, 326, 329.
Lepus, 45.
LEVERRIER (and ADAMS), discovery of Neptune, 283; on the origin of the Leonids, 329.
Libra, 61, 118.
Librations of the moon, 145.
LICK telescope, the, 412.
Light, aberration of, 435, 436; of comets, 291; equation of, the, 432, 433; of the moon, 149; of the sun, its intensity, 186; of the stars, 348-350; velocity of, used to determine the distance of the sun, 434, 436; the zodiacal, 265.
Light-ratio of the scale of stellar magnitude, 346.
Light-year, the, 344.
Local time, 97; time from altitude of the sun, 427; time by transit instrument, 93, 416.
LOCKYER, J. N., his meteoritic hypothesis, 330, 394; on spectra of nebulæ, 380.
Longitude and latitude (celestial) 20; (terrestrial), defined, 94; (terrestrial), methods of determining it, 95, 96, 427.
Lunar. See Moon.
Lupus, 62.
Lynx, 46.
Lyra, 67.

M.

Magnesium in nebulæ (Lockyer), 380; in the stars, 362; in the sun, 176.
Magnifying power of a telescope, 404.
Magnitudes, star, 345-347; star, absolute scale of, 346; star, and telescopic power, 347.
Mars (the planet), 251-257; habitability of, 259; map of the planet, 257; satellites, 258; Schiaparelli's observations, etc., 256; telescopic aspect, rotation, etc., 253, 254.

Mass, definition, 113; of comets, 302; of earth, 113; of the moon, 143; of a planet, how determined, 233; of shooting stars, how estimated, 323; of the sun, 161.
Masses of binary stars, 371.
Mazapil, meteorite of, 326.
Mean and apparent places of stars, 336; and apparent solar time, 88-89.
Melbourne reflector, 412.
Mercury (the planet), 239-244; rotation of, 243; transits of, 244.
Meridian (celestial) defined, 11, 15, 16; (terrestrial), arcs of, measured, 105, 110; circle, the, 81, 99, 418.
Meteoritic hypothesis (Lockyer), 330, 394; showers, 324-326.
Meteorite of Mazapil, 326.
Meteorites, 315; their constituents, 316; their fall, 315.
Meteors, ashes of, 323; connection with comets, 327-329; heat and light, 318; observation of, 317; origin of, 319; path and velocity, 317.
Micrometer, the, 415.
Midnight sun, the, 86.
Milky Way, the, 383.
Mimas, the inner satellite of Saturn, 280.
Mira Ceti, 356.
Missing and new stars, 353.
Monoceros, 50.
Month, sidereal and synodic, 133.
Moon, its albedo, 149; its atmosphere discussed, 148; changes on its surface, 155; character of its surface, 153; density, 143; diameter, surface area and bulk, 142; distance and parallax, 141; eclipses of, 195-199; heat, 150; influence on the earth, 151; librations, 145; light and albedo, 149; map, 154, 156; mass, density, and gravity, 143; motion (in general), 132-135; nomenclature of objects on surface, 156; perturbations of, 134; phases, 146; rotation, 144; shadow of, 200; surface structure, 153; telescopic appearance, 152; temperature, 150; water not present, 148.

[All references, unless expressly stated to the contrary, are to *articles*, not to *pages*.]

Motion, apparent diurnal, of the heavens, 12, 13; of the moon, 132-134; of a planet, 225, 226, 229; of the sun, 115-117; in line of vision, effect on spectrum, 179, 341; of the sun in space, 342.
Motions of stars, 338-341.
Mountains, lunar, 153, 156.
Mounting of a telescope, 414.
Multiple stars, 375.

N.

Nadir defined, 10.
Nadir-point of meridian circle, 419.
Names of planets, 218; of satellites of Saturn, 280; of satellites of Uranus, 282.
Neap tide, 210.
Nebulæ, the, 377-382; changes in, 379; distance and distribution, 382; spectra of, 380, 381.
Nebular hypothesis, the, 392, 393.
Negative eye-pieces, 409.
Neptune (the planet), 283-287.
NEWCOMB, S., on the age and duration of the system, 193; and **MICHELSON,** the velocity of light, 436.
NEWTON, H. A., estimate of the daily number of meteors, 321; investigation of the orbit of the Leonids, 327; nature of comets, 309.
NEWTON, SIR ISAAC, law of gravitation, 221, 222.
Nodes of the moon's orbit and their regression, 134; of the planetary orbits, 224.
NORDENSKIOLD, ashes of meteors, 323.
Norma, 64.
Number of comets, 289; of eclipses in a saros, 207; of eclipses in a year, 206; of the stars, 332.
NYRÉN, his value of the aberration constant, 435.

O.

Oberon, a satellite of Uranus, 282.
Oblateness or ellipticity of the earth defined, 110.
Oblique sphere, 85.
Obliquity of the ecliptic, 116.

OLBERS, discovers Pallas and Vesta, 260.
Ophiuchus, 65.
OPPOLZER, his canon of eclipses, 205.
Opposition defined, 132, 227.
Orbit of the earth, its form, etc., 115, 122, 428; of the moon, 137; parallactic, of a star, 442.
Orbital motion of the earth, proof of it, 115.
Orbits of binary stars, 370; of comets, 292; of the planets, 223.
Origin of the asteroids, 263; of meteors, 319; of periodic comets, 297.
Orion, 43.

P.

PALISA, discovery of asteroids, 260.
Pallas, the second asteroid, 260.
Parabola, the, 439, 440.
Parallax, annual or heliocentric, of the stars, 139, 343, 441-444; diurnal or geocentric, 139; solar, by transit of Venus, de l'Isle's method, 437; stellar, how determined, 441-444.
Parallaxes, stellar, table of, Table V., page 343.
Parallel sphere, 84.
Pegasus, 77.
Pendulum used to determine earth's form, 111; Foucault, 107.
Perigee defined, 137.
Perihelion defined, 120.
Periodicity of sun spots, 169.
Periods of the Planets, 218; sidereal and synodic, 133, 162, 228.
Perseids, the, 324-326, 328, 329.
Perseus, 40.
Perturbations, lunar, 134; planetary, 122, 288*.
PETERS, asteroid discoveries, 260.
Phase of Mars, 253.
Phases of Mercury and Venus, 242, 247; of the moon, 146; of Saturn's rings, 278.
Phobos, a satellite of Mars, 258.
Phœnix, 39.
Photographic power of eclipsed moon, 198; star-charts, 337; telescopes, 337.

[All references, unless expressly stated to the contrary, are to *articles*, not to *pages*.]

Photographs of nebulæ, 378; of star-spectra, 341, 364.
Photography, solar, 164.
Photometry, stellar, 348, 349.
Photosphere, the, 165, 194.
PIAZZI discovers Ceres, 260.
PICKERING, E. C., photographs of star-spectra, 364, 373; photometric observations of eclipses of Jupiter's satellites, 433; photometric measures of stellar magnitudes, 346.
Pisces, 36, 118.
Piscis Australis, 79.
Place of a heavenly body defined, 7; of a heavenly body, how determined by observation, 99, 100; of a ship, how determined, 426, 427.
Planet, albedo of, defined, 231, 235; apparent motion of, 225-229; diameter and volume, how measured, 232; mass and density, how determined, 233; rotation on axis determined, 234; satellite system, how investigated, 236; superficial gravity determined, 233.
Planetary data, their relative accuracy, 237; system, its genesis, age, and duration, 390-398; its stability, 288*.
Planetoids. See Asteroids.
Planets, Humboldt's classification, 217; the list of, 218; intra-Mercurian, 264; minor, 260-263; possibly attending stars, 372; table of elements, Appendix, Table II., page 340; table of names, symbols, etc., 218.
Pleiades, the, 42, 376.
Pointers, the, 12, 26.
Pole (celestial), altitude of, equals latitude, 80; defined, 13; effect of precession, 126; (terrestrial), diurnal phenomena near it, 83.
Pole-star, former, α Draconis, 126; how recognized, 12.
Positive eye-pieces, 409.
Precession of the equinoxes, 125, 126.
Prime vertical, the, 11.
PROCTOR, sun spots, 168; theory of comets, 208.

Prominences, the solar, 181, 182, 194
Proper motion of stars, 339.
Ptolemaic system, the, 230.
PTOLEMY, 4, 230.

Q.

Quadrature defined, 132, 227.
Quiescent prominences, 182.

R.

Radiant, the, of a meteoric shower 324.
Radius vector defined, 120.
Rate of a timepiece defined, 417.
Rectification of a globe, 401.
Recurrence of eclipses, 207.
Red spot of Jupiter, 271.
Reflecting telescope, the, 411, 413.
Refracting telescope, the, 403-407, 413.
Refraction, astronomical, 82.
Reticle, the, 410, 416.
Retrograde and retrogression defined, 226.
Reversing layer, 177.
Rhea, a satellite of Saturn, 280.
Right ascension defined, 18, 93; how determined by observation, 99, 100.
Right sphere, the, 83.
Rings of Saturn, the, 277-279.
ROBERTS, photographs of nebulæ, 378.
ROSSE, LORD, his great reflector, 412.
Rotation, apparent diurnal, of the heavens, 12; definition of, 144; distinguished from revolution, 106, note; of earth, its effect on gravity, 111; of earth, proofs of, 107; of earth, variability of, 108; of the moon, 144; of the sun, 162, 163.
Rotation-period of Jupiter, 270; of Mars, 254; of Mercury, 243; of a planet, how ascertained, 234; of Saturn, 275; of Venus, 249.

S.

Sagitta, 70.
Sagittarius, 72, 118.
Saros, the, 207.

[All references, unless expressly stated to the contrary, are to *articles*, not to *pages*.]

Satellite system, how investigated, 236; systems, table of, Table III., page 341.
Satellites of Jupiter, 272; of Mars, 258; of Neptune, 286; of Saturn, 280; of Uranus, 282.
Saturn (the planet), 274-280.
Scale of stellar magnitudes, 346.
SCHIAPARELLI, identification of cometary and meteoric orbits, 328; observations of Mars, 256; rotation of Mercury and Venus, 243, 249.
SCHWABE, discovers periodicity of sun spots, 169.
Scintillation of the stars, 365.
Scorpio, 63, 118.
Sea, position at, how found, 426, 427.
Seasons, explanation of, 123-124.
SECCHI, on stellar spectra, 363; on sun spots, 168.
Secondary spectrum of achromatic object-glass, 407.
Serpens, 65.
Serpentarius, 65.
Sextant, the, 420, 421.
Shadow of the earth, its dimensions, 196; of the moon, its dimensions, 200; of the moon, its velocity, 203.
Ship at sea, determination of its position, 426, 427.
Shooting stars (see also Meteors) 320-324; ashes of, 323; brightness of, 323; elevation and path, 322; mass of, 323; materials of, 323; nature of, 320; number, daily and hourly, 321; radiant, 324; showers of, 324-326; spectrum of, 323; velocity of, 322.
Showers, meteoric, 324-326.
Sidereal and synodic months, 133; and synodic periods of planets, 228; time defined, 91; year, 127.
Signs of the zodiac, 118; effect of precession on them, 126.
Sirius, its companion, 369; light compared with that of the sun, 349; its mass compared with that of the sun, 370.
Solar constant, the, 187; parallax, 158; time, mean and apparent, 88, 89.

Solstice defined, 117.
SOSIGENES and the calendar, 129.
Spectroscope, its principle and construction, 171, 172; slitless, 364, 445; used to observe the solar prominences, 182; used to measure motions in line of sight, 178, 179, 341, 373, 374.
Spectrum of the chromosphere and prominences, 181; of comets in general, 304; of the comet of 1882, 314; of meteors, 323; of nebulæ, 380, 381; of a shooting star, 323; of stars, 362-364; the solar, 172-175; of the solar corona, 184; of a sun spot, 178.
Spectrum analysis, fundamental principles, 173.
Speculum of a reflecting telescope, 411.
Sphere, celestial, the, 6; doctrine of the, 9-20.
Spots, solar. See Sun spots.
Spring tide defined, 210.
Stability of the planetary system, 288*.
Standard time, 97.
Stars, binary, 368-371; catalogues of, 335; charts of, 337; clusters of, 376; dark, 350, 360; designation and nomenclature, 24, 334; dimensions of, 351, 360; distance of, 343, 344; distribution of, 384; double, 366, 367; gravitation among them, 368, 371, 386; heat from them, 348, note; light of certain stars compared with sunlight, 348, 349; magnitudes and brightness, 345-350; mean and apparent places of, 336; missing and new, 353; motions of, 338-342; multiple, 375; new, 353; number of, 332; parallax of, 343, 441-444; Table V., page 343; shooting (see Shooting stars, also Meteors); spectra of, 362-364; system of the, 386; temporary, 355; total amount of light from the, 348; twinkling of, 365; variable, 352-361; Table IV., page 342.
Star-gauges of the Herschels, 384.

[All references, unless expressly stated to the contrary, are to *articles*, not to *pages*.]

Starlight, its total amount, 348.
Stellar parallaxes, table of, Table V., page 343; photometry, 348, 349.
Structure of the stellar universe, 385.
Sun, the age and duration of, 193, 397, 398; apparent motion in the heavens, 115–117; its chromosphere, 180; its constitution, 194; its corona, 183–185; its density, 161; dimensions of, 160; distance of, 158, 159, 434–438; elements recognized in it, 176; faculæ, 165; gravity on its surface, 161; heat of, quantity, intensity and maintenance, 187–192; light of, its intensity, 186; mass of, 161; motion in space, 342; parallax of, 159; prominences, 181, 182, 194; reversing layer, the, 177, 194; rotation of, 162, 163; spectrum of, 172, 175; temperature of, 190; temperature diminishing, Lockyer, 396, note.
Sun spots, appearance and nature, 166, 170; cause of, 168; distribution of, 169; influence on the earth, 170; periodicity of, 169; spectrum of, 178.
Superficial gravity of a planet, how determined, 233.
Surface structure of the moon, 153, 154.
Swarms, meteoric, 324–329.
Synodic and sidereal months, 133; and sidereal periods of planets, 228.
System, planetary, its age and duration, 397–399; its genesis and evolution, 390–393; its stability, 288*; stellar, its probable nature, 386–388.
Syzygy defined, 132.

T

Tables, astronomical constants, Table I., page 339; astronomical symbols, page 344; binary stars, orbits and masses, 370; Bode's law, 219; constellations, showing place in heavens, page 54; Greek alphabet, page 344; moon, names of principal objects, 155; planet's elements, Table II., page 340; planets' names, distances,

etc., approximate, 218; satellite systems, Table III., page 341; stellar parallaxes and proper motions, Table V., page 343; variable stars, Table IV., page 342.
Tails of comets, 300, 301, 305–308.
Taurus, 42.
Telegraph, longitude by, 95.
Telescope, achromatic, 406, 407; eyepieces of, 409; general principles of, 402; illuminating power, 405; magnifying power, 404; magnitude of stars visible with a given aperture, 347; mounting of, 414; reflecting, 411; simple refracting, 403.
Telescopes, great, 412.
Temperature of the moon, 150; of the sun, 190.
Temporary stars, 355.
Terminator, the, defined and described, 146.
Tethys, a satellite of Saturn, 280.
Thomson, Sir W., the internal heat of the earth, 396; the heat of meteors, 318; the rigidity of the earth, 114.
Tidal-wave, course of, 213.
Tides, the definitions relating to, 210; due mainly to moon's action, 209; explanation of, 208, 209, 211, 212; height of, 214; motion of, 211, 213; in rivers, 215.
Time, equation of, 89; local, from sun's altitude, 427; methods of determining, 92, 93, 427; relation to hour-angle, 422; sidereal, defined, 91; solar — mean and apparent, 88, 89; standard, defined, 97.
Titan, satellite of Saturn, 280.
Titania, satellite of Uranus, 282.
Total and annular eclipses, 201.
Trains of meteors, 315.
Transit or meridian circle, 81, 99, 418.
Transit instrument, the, 92, 416.
Transits of Mercury, 244; of Venus, 250.
Triangulum, 37.
Tropical year, the, 127.
Twinkling of the stars, 365.
Tycho Brahe, his temporary star, 355.

[All references, unless expressly stated to the contrary, are to *articles*, not to *pages*.]

U.

Ultra-Neptunian planet, 288.
Umbriel, a satellite of Uranus, 282.
Universe, stellar, its structure, 385.
Uranography defined, 5.
Uranolith, or Uranolite. See Meteorite.
Uranus (the planet), 281, 282.
Ursa Major, 26.
Ursa Minor, 27.
Utility of astronomy, 1.

V.

Vanishing point, 6, note.
Variable stars, 352–361; table of, Table IV., page 342.
Velocity of earth in its orbit, 102, 158; of light, 436; of moon's shadow, 203; of meteors and shooting stars, 317, 322; of star motions, 340, 341.
Venus (the planet),.245–250; phases of, 247; transits of, 250.
Vernal equinox, the, 17, 36, 116.
Vertical circles, 11.
Vesta, the fourth asteroid, 260.
Virgo, 56, 118.
Visible horizon defined, 10.
VOGEL, H. C., spectroscopic determination of star motions in the line of sight, 341; spectroscopic observations of Algol and Spica, 360, 374.

Volcanoes on the moon, 153.
Vulcan, the hypothetical intra-Mercurian planet, 264.
Vulpecula, 69.

W.

Water absent from the moon, 148.
Wave-length of a light-ray affected by motion in the line of sight, Doppler's principle, 179, 341.
Wave, tidal, its course, 213.
Way, the sun's, 342.
Weather, the moon's influence on, 151.
Weight, loss of, between pole and equator, 111.

Y.

Year, the sidereal, tropical, and anomalistic, 127, and Table I., page 339.

Z.

Zenith, the, defined, 10.
Zenith distance defined, 11.
Zero-points of the meridian circle, 418, 419.
Zodiac, the, and its signs, 118; its signs as affected by precession, 126.
Zodiacal light, the, 265.
ZÖLLNER, determination of planets' albedoes, 242, 247, 253, 268, 276, 281, 285; measurement of moonlight, 149; measures of light of stars, 348.

SUPPLEMENTARY INDEX.

———•◦•———

[All references are to *articles*.]

BARNARD, E. E., measures of diameters of planets, 262, 267, 275, 285; discovery of the fifth satellite of Jupiter, 272; of comets, 314.*

BOYS, V., determination of density of the earth, 113.

Calcium, in the sun, 176; in faculæ, 165; in chromosphere and prominences, 182, 182.*

CAMPBELL, W. C., spectrum of Mars, 253; of nebulæ, 380.

CHANDLER, S. C., variation of latitude, 109.

CHARLOIS, discoverer of asteroids by photography, 260.

Chromosphere, photography of, 182.*

DESLANDRES, photography of solar prominences, 182.*

Faculæ, bright lines of calcium in spectrum, 165.

H and K lines of calcium, 165, 176, 182, 182.*

HALE, G. E., photography of solar prominences, 182.*

Helium, identification of, in uraninite, 181; in variable stars, 355, 356; in nebulæ, 380.

KEELER, J. E., spectroscopic observation of the rings of Saturn, 279; of nebulæ, 380.

KELVIN, LORD (formerly Sir Wm. Thomson), 114, 318, 396.

LICK Observatory, telescope, 412; various observations, 156, 253, 256, 262, 267, 272, 275, 299, 314*, 380.

LOWELL, P., observations on Mars, 256.

Oases, on Mars, 256.

Parallax, stellar, determination of, by spectroscopic observations on binary stars, 445.

Photography, of solar prominences, 182*; applied to discovery of asteroids, 260; of comets, 314.*

Pole (terrestrial), motion and displacement of, 109.

RAMSAY, PROF., identifies Helium, 181.

SEE, T. J. J., evolution of binary stars, 370.

SPOERER, peculiar law of sun-spot latitude, 169.

STRUVE, H., mass of Saturn's rings, 277.

WILSON and GRAY, temperature of the sun, 190.

WOLF, photographic discovery of asteroids, 260.

YERKES telescope, 412.

366

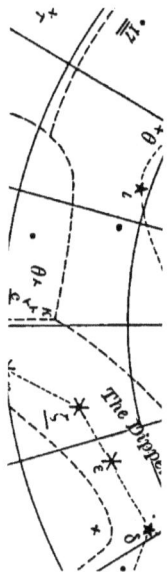

MAP I.

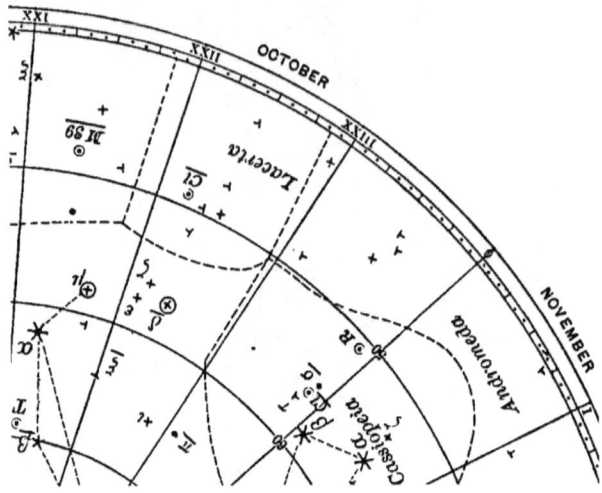

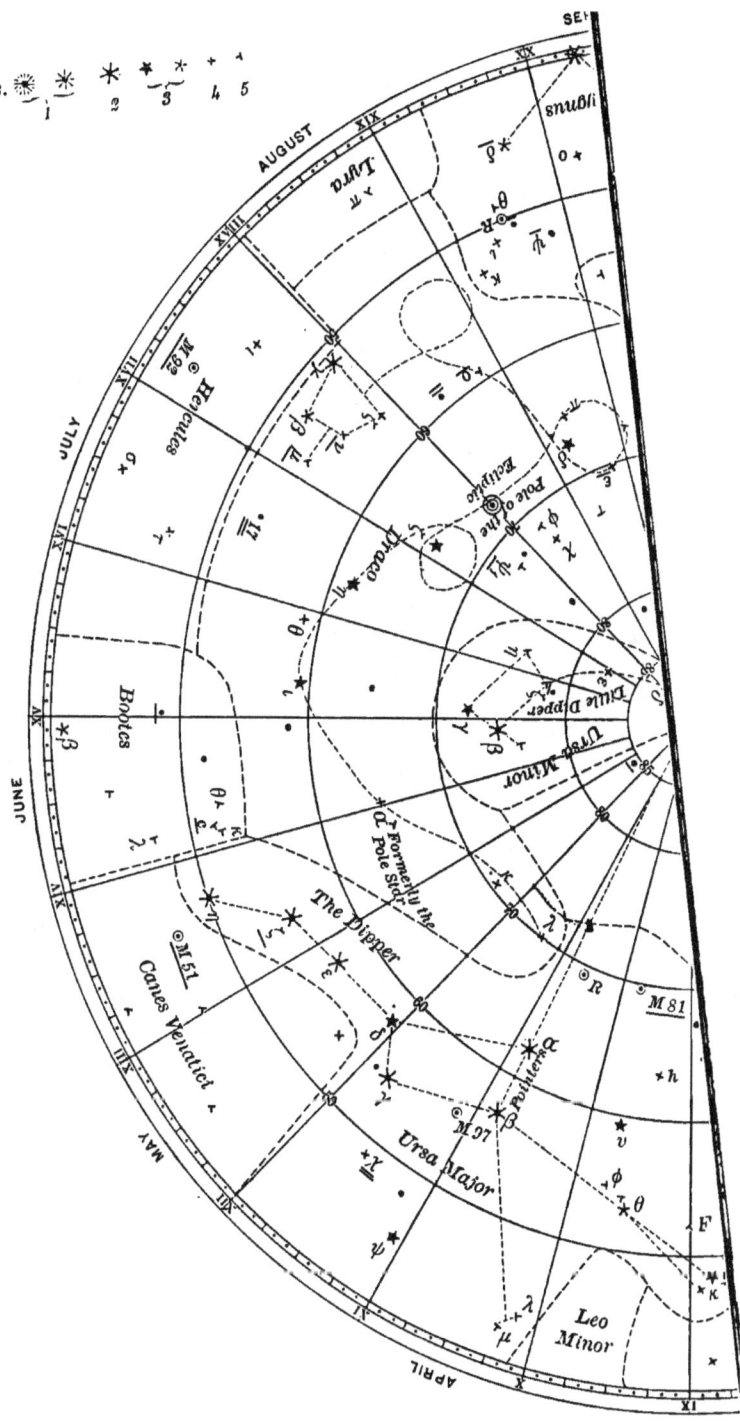

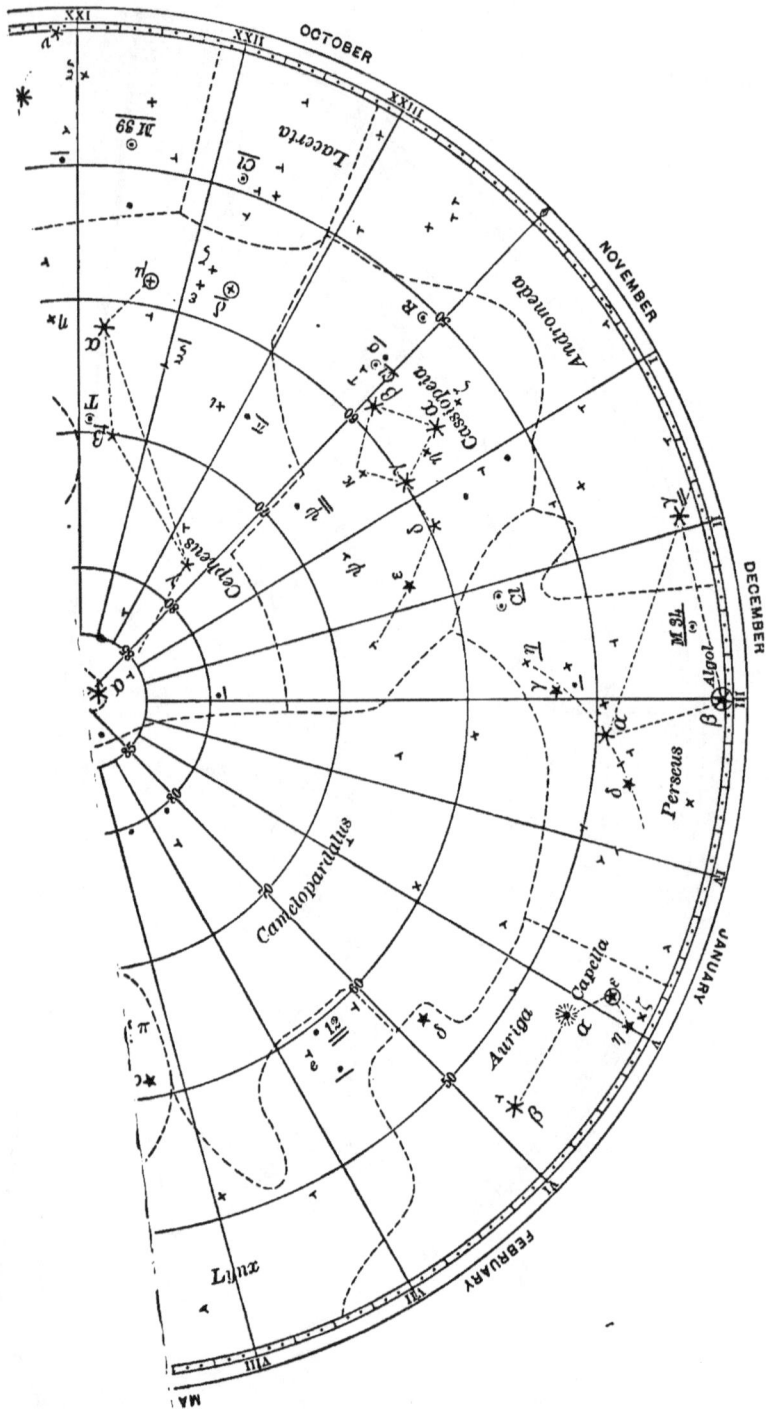

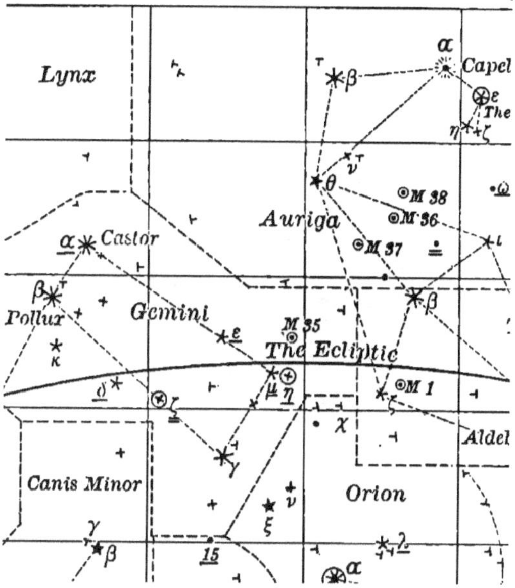
Lynx
Capel
The
Auriga
M 38
M 36
M 37
Castor
Pollux
Gemini
M 35
The Ecliptic
M 1
Aldel
Canis Minor
Orion
15

MAP II

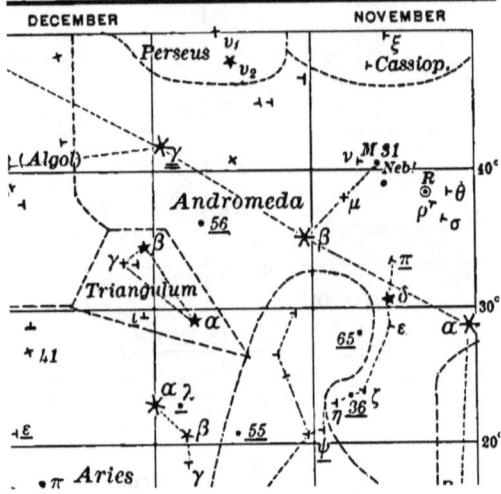

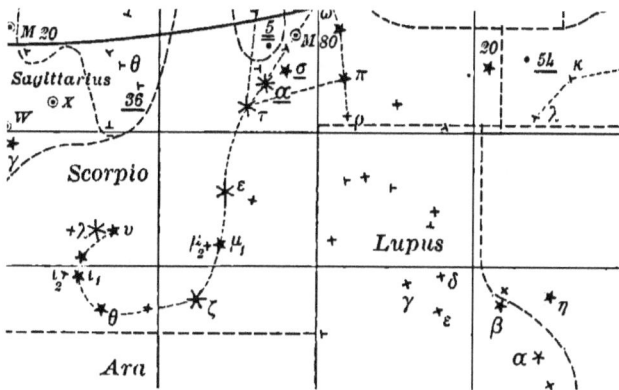

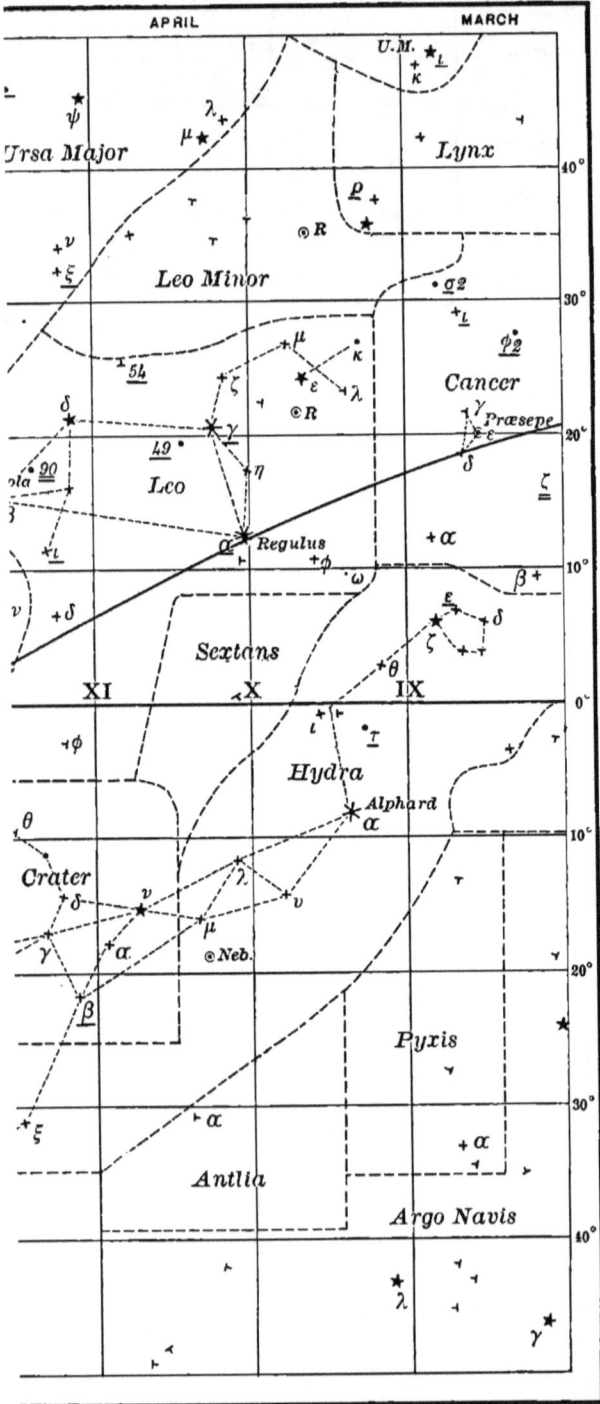

MAP III.

APRIL

MARCH

U.M. ★ *ι*
+
κ

ψ ★

λ
+
μ ★

Jrsa Major

Lynx

40°

ρ +

⊚ *R*
★

+ *ν*

+ *ξ*

Leo Minor

• *σ2*

30°

+*ι*

φ2

+ *μ*

• κ

Cancer

54

+ ζ

★ ε
λ

γ *Præsepe*
ε

δ ★

δ

R

δ

20°

γ

ζ

49 •

η

90

Leo

+*ι*

α *Regulus*

+ *α*

+ *φ*

• ω

β +

10°

ν

+ δ

Sextans

ε
★
ζ
θ

δ
+ τ

XI

X

IX

0°

+ φ

ι

• *τ*

Hydra

★ *Alphard*
α

θ

10°

Crater

+ δ

ν ★

λ

ν

+ *α*

μ

⊚ *Neb.*

γ

★

20°

+ β

Pyxis

30°

+ ξ

α

α

+ α

Antlia

Argo Navis

40°

★

λ

γ ★

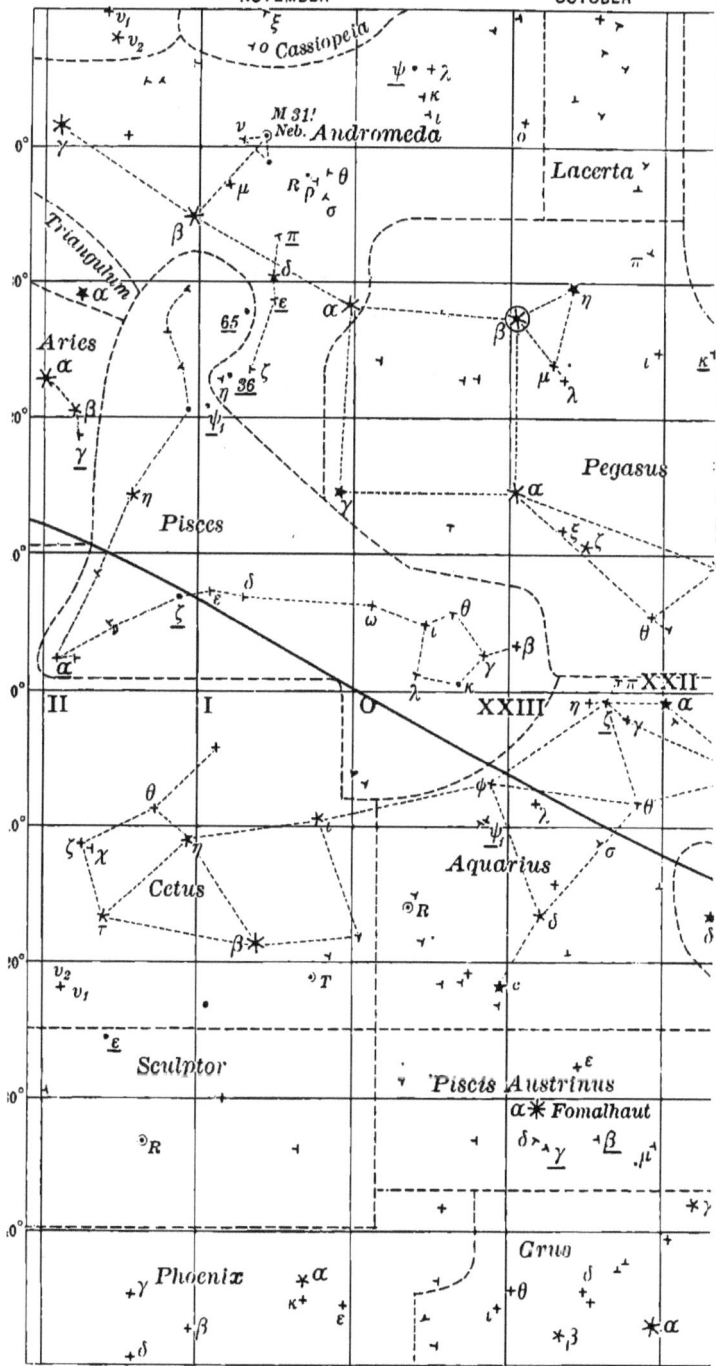

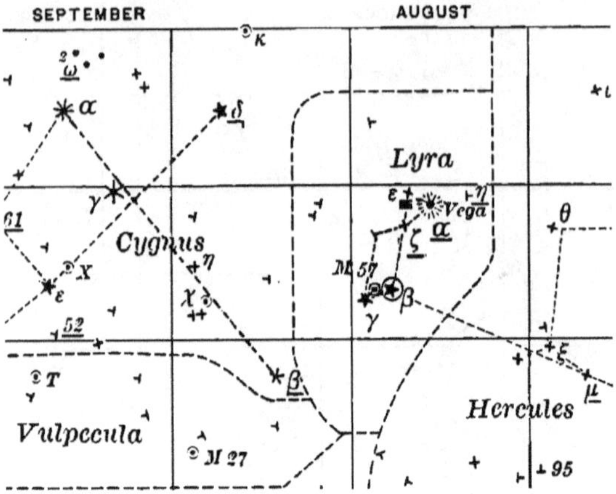

SEPTEMBER
AUGUST
Lyra
Vega
Cygnus
M 57
Hercules
Vulpecula
M 27

NATURAL SCIENCE TEXT-BOOKS.

Principles of Physics. A Text-book for High Schools and Academies. By ALFRED P. GAGE, *Instructor of Physics in the English High School, Boston.* $1.30.

Elements of Physics. A Text-book for High Schools and Academies. By ALFRED P. GAGE. $1.12.

Introduction to Physical Science. By ALFRED P. GAGE. $1.00.

Physical Laboratory Manual and Note-Book. By ALFRED P. GAGE. 35 cents.

Introduction to Chemical Science. By R. P. WILLIAMS, *Instructor in Chemistry in the English High School, Boston.* 80 cents.

Laboratory Manual of General Chemistry. By R. P. WILLIAMS. 25 cents.

Chemical Experiments. General and Analytical. By R. P. WILLIAMS. For the use of students in the laboratory. 50 cents.

Elementary Chemistry. By GEORGE R. WHITE, *Instructor of Chemistry, Phillips Exeter Academy.* $1.00.

General Astronomy. A Text-book for Colleges and Technical Schools. By CHARLES A. YOUNG, *Professor of Astronomy in the College of New Jersey,* and author of "The Sun," etc. $2.25.

Elements of Astronomy. A Text-book for High Schools and Academies, with a Uranography. By Professor CHARLES A. YOUNG. $1.40. **Uranography.** 30 cents.

Lessons in Astronomy. Including Uranography. By Professor CHARLES A. YOUNG. Prepared for schools that desire a brief course free from mathematics. $1.20.

An Introduction to Spherical and Practical Astronomy. By DASCOM GREENE, *Professor of Mathematics and Astronomy in the Rensselaer Polytechnic Institute, Troy, N.Y.* $1.50.

Elements of Structural and Systematic Botany. For High Schools and Elementary College Courses. By DOUGLAS HOUGHTON CAMPBELL, *Professor of Botany in the Leland Stanford Junior University.* $1.12.

Elements of Botany. By J. Y. BERGEN, Jr., *Instructor in Biology in the English High School, Boston.* $1.10.

Laboratory Course in Physical Measurements. By W. C. SABINE, *Instructor in Harvard University.* $1.25.

Elementary Meteorology. By WILLIAM M. DAVIS, *Professor of Physical Geography in Harvard University.* With maps, charts, and exercises. $2.50.

Blaisdell's Physiologies: Our Bodies and How We Live, 65 cents; How to Keep Well, 45 cents; Child's Book of Health, 30 cents.

A Hygienic Physiology. For the Use of Schools. By D. F. LINCOLN, M.D., author of "School and Industrial Hygiene," etc. 80 cents.

Copies will be sent, postpaid, to teachers for examination on receipt of the introduction prices given above.

GINN & COMPANY, Publishers, Boston, New York, Chicago, Atlanta.

THE BEST HISTORIES.

Myers's History of Greece. — Introduction price, $1.25.

Myers's Eastern Nations and Greece. — Introduction price, $1.00.

Allen's Short History of the Roman People. — Introduction price, $1.00.

Myers and Allen's Ancient History. — Introduction price, $1.50. This book consists of Myers's Eastern Nations and Greece and Allen's History of Rome bound together.

Myers's History of Rome. — Introduction price, $1.00.

Myers's Ancient History. — Introduction price, $1.50. This book consists of Myers's Eastern Nations and Greece and Myers's History of Rome bound together.

Myers's Mediæval and Modern History. — Introduction price, $1.50.

Myers's General History. — Introduction price, $1.50.

Emerton's Introduction to the Study of the Middle Ages. — Introduction price, $1.12.

Emerton's Mediæval Europe (814-1300). — Introduction price, $1.50.

Fielden's Short Constitutional History of England. — Introduction price, $1.25.

Montgomery's Leading Facts of English History. — Introduction price, $1.12.

Montgomery's Leading Facts of French History. — Introduction price, $1.12.

Montgomery's Beginner's American History. — Introduction price, 60 cents.

Montgomery's Leading Facts of American History. — Introduction price, $1.00.

Cooper, Estill and Lemmon's History of Our Country. — Introduction price, $1.00.

For the most part, these books are furnished with colored and sketch maps, illustrations, tables, summaries, analyses and other helps for teachers and students.

GINN & COMPANY, Publishers.

BOSTON.   NEW YORK.   CHICAGO.   ATLANTA.

THE CLASSIC MYTHS

IN

ENGLISH LITERATURE.

By CHARLES MILLS GAYLEY,

Professor of the English Language and Literature in the University of California
and formerly Assistant-Professor of Latin in the University of Michigan.

12mo. Half leather. 540 pages. For introduction, $1.50.
New Edition with 16 full-page illustrations.

This work, based chiefly on Bulfinch's "Age of Fable"
(1855), has here been adapted to school use and in large
part rewritten. It is recommended both as the best manual
of mythology and as indispensable to the student of our
literature.

Special features of this edition are :

1. An introduction on the indebtedness of English poetry to the
literature of fable; and on methods of teaching mythology.

2. An elementary account of myth-making and of the principal
poets of mythology, and of the beginnings of the world, of gods and
of men among the Greeks.

3. A thorough revision and systematization of Bulfinch's Stories of
Gods and Heroes : with additional stories, and with selections from
English poems based upon the myths.

4. Illustrative cuts from Baumeister, Roscher, and other standard
authorities on mythology.

5. The requisite maps.

6. Certain necessary modifications in Bulfinch's treatment of the
mythology of nations other than the Greeks and Romans.

7. Notes, following the text (as in the school editions of Latin and
Greek authors), containing an historical and interpretative commentary
upon certain myths, supplementary poetical citations, a list of the better
known allusions to mythological fiction, references to works of art
and hints to teachers and students.

GINN & COMPANY, Publishers,

Boston, New York, and Chicago.

ENGLISH LITERATURE.

Arnold's English Literature. 558 pages. Price, $1.50.

Baker's Plot-Book of Some Elizabethan Plays. *In press.*

Baldwin's Inflection and Syntax of Malory's Morte d'Arthur.

Browne's Shakspere's Versification. 34 pages. Price, 25 cts.

Corson's Primer of English Verse. 232 pages. Price, $1.00.

Emery's Notes on English Literature. 152 pages. Price, $1.00.

Garnett's Selections in English Prose from Elizabeth to Victoria. 701 pages. Price, $1.50.

Gayley's Classic Myths in English Literature. 540 pages. Price, $1.50

Gayley's Introduction to Study of Literary Criticism. *In press.*

Gummere's Handbook of Poetics. 250 pages. Price, $1.00.

Hudson's Life, Art, and Characters of Shakespeare. 2 vols. 1003 pages. Price, $4.00.

Hudson's Classical English Reader. 467 pages. Price, $1.00.

Hudson's Text-Book of Prose. 648 pages. Price, $1.25.

Hudson's Text-Book of Poetry. 704 pages. Price, $1.25.

Hudson's Essays on English Studies in Shakespeare, etc. 118 pages. Price, 25 cts.

Lee's Graphic Chart of English Literature. 25 cts.

Minto's Manual of English Prose Literature. 566 pages. Price, $1.50.

Minto's Characteristics of the English Poets. (From Chaucer to Shirley.) 382 pages. Price, $1.50.

Montgomery's Heroic Ballads. Poems of War and Patriotism. Edited with Notes by D. H. Montgomery. 319 pages. Boards, 40 cts.; Cloth, 50 cts.

Phelps's Beginnings of the English Romantic Movement. 192 pages. Price, $1.00.

Sherman's Analytics of Literature. 468 pages. Price, $1.25.

Smith's Synopsis of English and American Literature. 125 pages. Price, 80 cts.

Thayer's Best Elizabethan Plays. 611 pages. Price, $1.25.

Thom's Shakespeare and Chaucer Examinations. 346 pages. Price, $1.00.

White's Philosophy of American Literature. 66 pages. Price, 30 cts.

Winchester's Five Short Courses of Reading in English Literature. 99 pages. Price, 40 cents.

Wylie's Studies in the Evolution of English Criticism. 212 pages. Price, $1.00.

Descriptive Circulars of these books sent postpaid to any address.

GINN & COMPANY, Publishers, Boston, New York, and Chicago.

The
Athenæum Press Series.

ISSUED UNDER THE GENERAL EDITORSHIP OF

PROFESSOR GEORGE LYMAN KITTREDGE,
Of Harvard University,

AND

PROFESSOR C. T. WINCHESTER,
Of Wesleyan University.

THIS series is intended primarily for use in colleges and higher schools; but it will furnish also to the general reader a library of the best things in English letters in editions at once popular and scholarly. The works selected will represent, with some degree of completeness the course of English Literature from Chaucer to our own times.

The volumes will be moderate in price, yet attractive in appearance, and as nearly as possible uniform in size and style. Each volume will contain, in addition to an unabridged and critically accurate text, an Introduction and a body of Notes. The amount and nature of the annotation will, of course, vary with the age and character of the work edited. The notes will be full enough to explain every difficulty of language, allusion, or interpretation. Full glossaries will be furnished when necessary.

The introductions are meant to be a distinctive feature of the series. Each introduction will give a brief biographical sketch of the author edited, and a somewhat extended study of his genius, his relation to his age, and his position in English literary history. The introductory matter will usually include a bibliography of the author or the work in hand, as well as a select list of critical and biographical books and articles.

This Series is intended to furnish a library of the best English literature, from Chaucer to the present time, in a form adapted to the needs of both the student and the general reader.

GINN & COMPANY, PUBLISHERS,
BOSTON, NEW YORK, AND CHICAGO.

THE ATHENÆUM PRESS SERIES

The following volumes are now ready:

Sidney's Defense of Poesy. Edited, with Introduction and Notes, by ALBERT S. COOK, Professor of English in Yale University. 103 pages. For introduction, 80 cents.

Ben Jonson's Timber; or Discoveries. Edited, with Introduction and Notes, by FELIX E. SCHELLING, Professor in the University of Pennsylvania. 166 pages. For introduction, 80 cents.

Selections from the Essays of Francis Jeffrey. Edited, with Introduction and Notes, by LEWIS E. GATES, Instructor in English in Harvard University. 213 pages. For introduction, 90 cents.

Old English Ballads. Selected and edited, with Introduction and Notes, by Professor F. B. GUMMERE of Haverford College. 380 pages. For introduction, $1.25.

Selections from the Poetry and Prose of Thomas Gray. Edited, with Introduction and Notes, by WM. LYON PHELPS, Instructor in English Literature in Yale College. 179 pages. For introduction, 90 cents.

A Book of Elizabethan Lyrics. Selected and edited, with Introduction and Notes, by F. E. SCHELLING, Professor in the University of Pennsylvania. 327 pages. For introduction, $1.12.

Herrick: Selections from the Hesperides and the Noble Numbers. Edited, with Introduction, Notes, and Glossary, by Professor EDWARD E. HALE, Jr., of Union College. 200 pages. For introduction, 90 cents.

Selections from the Poems of Keats. Edited, with Introduction and Notes, by ARLO BATES, Professor of English Literature, Massachusetts Institute of Technology. 302 pages. For introduction, $100.

Carlyle's Sartor Resartus. Edited, with Introduction and Commentary, by ARCHIBALD MACMECHAN, Professor of English in Dalhousie College, Halifax, N.S.   pages. For introduction,

Selections from Wordsworth's Poems. Edited, with Introduction and Notes, by Professor EDWARD DOWDEN, of the University of Dublin,   pages. For introduction,   .

GINN & COMPANY, Publishers.

www.ingramcontent.com/pod-product-compliance
Lightning Source LLC
Chambersburg PA
CBHW021351210326
41599CB00011B/835